Encyclopaedia of
Mathematical Sciences

Volume 24

T0181640

Editor-in-Chief: R.V. Gamkrelidze

Springer
Berlin
Heidelberg
New York
Hong Kong
London
Milan
Paris
Tokyo

S. P. Novikov   V. A. Rokhlin (Eds.)

Topology II

Homotopy and Homology.
Classical Manifolds

Springer

Title of the Russian original edition:
Itogi nauki i tekhniki, Sovremennye problemy matematiki,
Fundamental'nye napravleniya, Vol. 24, Topologiya-2,
and Part II of Vol. 12, Topologiya-1
Publisher VINITI, Moscow

Mathematics Subject Classification (2000): 55Pxx, 55Nxx, 57Nxx

ISBN 978-3-642-08084-5

Springer-Verlag Berlin Heidelberg New York
a member of BertelsmannSpringer Science + Business Media GmbH
http://www.springer.de
© Springer-Verlag Berlin Heidelberg 2010
Printed in Germany

List of Editors, Authors and Translator

Editor-in-Chief

R. V. Gamkrelidze, Russian Academy of Sciences, Steklov Mathematical Institute,
ul. Gubkina 8, 117966 Moscow; Institute for Scientific Information (VINITI),
ul. Usievicha 20a, 125219 Moscow, Russia, e-mail: gam@ipsun.ras.ru

Consulting Editors

S. P. Novikov, Department of Mathematics, Institute for Physical Sciences
and Technology, University of Maryland at College Park, College Park,
MD 20742-2431, USA, e-mail: novikov@ipst.umd.edu
V. A. Rokhlin†

Authors

D. B. Fuchs, Department of Mathematics, University of California, Davis,
CA 95616-8633, USA, e-mail: fuchs@math.ucdavis.edu
O. Ya. Viro, Department of Mathematics, Uppsala University, P.O. Box 480,
75106 Uppsala, Sweden, e-mail: oleg.viro@math.uu.se

Translator

C. J. Shaddock, 39 Drummond Place, Edinburgh EH3 6NR, United Kingdom

Contents

Contents

Preface*

Algebraic topology, which went through a period of intense development from the forties to the sixties of the last century, has now reached a comparatively stable state. A body of concepts and facts of general mathematical interest has been clearly demarcated, and at the same time the area of applications of topology has been significantly widened to include theoretical physics and a number of applied disciplines, as well as geometry and analysis.

The subject matter of the two parts of this volume can be characterized as "elementary topology". This term has a quite precise meaning and denotes those parts of topology in which only comparatively simple algebra is used. The most important topics in this volume are: homotopy groups, bundles, cellular spaces, homology, Poincare duality, characteristic classes, and Steenrod squares. In most cases proofs are omitted, but they are not difficult as a rule, and the reader can reconstruct them if desired, obtaining all the necessary ideas from the text. Thus the book may be regarded as the synopsis of a textbook on topology.

The textbook itself has been written only in part: we have in mind *Beginner's course in topology: geometric chapters* by D.B. Fuks and V.A. Rokhlin. In writing the present work we have used not only this book, but also the numerous drafts of its second part, on homology, work on which was broken off on the death of V.A. Rokhlin in December 1984.

It was originally intended that V.A. Rokhlin would be one of the authors of both parts of this volume (as well as of other volumes in the *Encyclopaedia of Mathematical Sciences*). He played an active part in preparing the detailed plan of this volume and in discussions of some of its key sections. While writing this book the authors have continually referred to his texts, both published and unpublished. Unfortunately for purely formal reasons V.A. Rokhlin cannot be considered to be our coauthor; indeed, we very much doubt that our text would meet with his approval. In spite of this, we dedicate this volume with gratitude to the memory of Vladimir Abramovich Rokhlin.

---

* Publisher's note: This is the Preface to the Russian edition of Enc. Math. Sc. 24, thus referring only to Parts I and II of the present volume. For organisational reasons the second part of the Russian edition of Enc. Math. Sc. 12 was added to this volume as Part III.

I. Introduction
to Homotopy Theory

O.Ya. Viro, D.B. Fuchs

Translated from the Russian
by C.J. Shaddock

Contents

Chapter 1
Basic Concepts

§1. Terminology and Notations

1.1. Set Theory.

In addition to the standard set-theoretical terminology and notations, whose use is unambiguous, we shall use the following.

If A is a subset of a set X, the inclusion of A in X may be regarded as the map defined by $x \mapsto x$. Notation: in : $A \to X$. If there is no ambiguity about A and X, we simply write in.

If A is a subset of X and B a subset of Y, then to each mapping $f : X \to Y$ such that $f(A) \subset B$, there corresponds the map $f|_{A,B} : A \to B$ defined by $x \mapsto f(x)$, called a *submap* of f. If there is no ambiguity about A and B, we may just write $f|$ instead of $f|_{A,B}$. If $B = Y$, then $f|_{A,B}$ is also called the *restriction* of f to A and denoted by $f|_A$.

The quotient (or factor) set of X under a partition S is denoted by X/S. The map $X \to X/S$ that takes each point to the element of the partition containing it is called the *projection*, denoted by pr.

If S and T are partitions of sets X and Y, and $f : X \to Y$ is a map that maps the elements of S to the elements of T, then there is a corresponding map $X/S \to Y/T$, taking an element A of S to the element of T that contains $f(A)$. This map is denoted by $f/S, T$, and is called the *quotient map* of f. In particular, it is defined when T is the partition into single points, and f is constant on the elements of S. Thus, to each map $f : X \to Y$ constant on the elements of a partition S of X, there corresponds a map $X/S \to Y$; it is denoted by f/S. If there is no ambiguity about S and T, we simply write $f/$ instead of $f/S, T$.

The *sum of a family of sets* $\{X_\mu\}_{\mu \in M}$ is the union of disjoint copies of the sets X_μ, that is, the set of pairs (x_μ, μ) such that x_μ is an element of the set X_μ. Notation: $\coprod_{\mu \in M} X_\mu$. The map of X_ν ($\nu \in M$) into $\coprod_{\mu \in M} X_\mu$ defined by $x \mapsto (x, \nu)$ is denoted by in$_\nu$. Each family of maps $\{f_\mu : X_\mu \to Y_\mu\}_{\mu \in M}$ determines a map $\coprod_{\mu \in M} X_\mu \to \coprod_{\mu \in M} Y_\mu$ in a natural way; it is called the *sum of the maps* f_μ and denoted by $\coprod_{\mu \in M} f_\mu$. If M consists of the numbers $1, \ldots, n$, then we write $X_1 \coprod \ldots \coprod X_n$, $f_1 \coprod \ldots \coprod f_n$ as well as $\coprod X_\mu$ and $\coprod f_\mu$.

The map $X_1 \times \ldots \times X_n \to X_i : (x_1, \ldots, x_n) \to x_i$ is called the *i*th *projection*, denoted by pr$_i$. If we have maps $f_1 : X_1 \to Y_1, \ldots, f_n : X_n \to Y_n$, then there is a map $X_1 \times \ldots \times X_n \to Y_1 \times \ldots \times Y_n : (x_1, \ldots, x_n) \mapsto (f_1(x_1), \ldots, f_n(x_n))$, called the *product of the maps* $f_1, \ldots, f_n$ and denoted by $f_1 \times \cdots \times f_n$

1.2. Logical Equivalence.

We shall use the expression "iff" to mean "if and only if".

1.3. Topological Spaces.

A. If A is a subset of a topological space X, then its interior will be denoted by Int A, or more precisely $\text{Int}_X A$, its closure by Cl A, or $\text{Cl}_X A$, and finally, its frontier, that is, Cl $A \setminus$ Int A by Fr A, or $\text{Fr}_X A$.

B. Our notations for the standard topological spaces will follows those of D.B. Fuchs in Part III of the present volume. In particular, the fields of real and complex numbers are denoted by $\mathbb{R}$ and $\mathbb{C}$, the skew field of quaternions by $\mathbb{H}$, and the algebra of Cayley numbers by Ca. The corresponding n-dimensional spaces, that is, the n-fold products $\mathbb{R} \times \cdots \times \mathbb{R}$, $\mathbb{C} \times \cdots \times \mathbb{C}$, $\mathbb{H} \times \cdots \mathbb{H}$ and Ca $\times \cdots \times$ Ca are denoted by $\mathbb{R}^n$, $\mathbb{C}^n$, $\mathbb{H}^n$ and Ca^n. We regard $\mathbb{R}^n$ as a metric space with the distance between $(x_1, \ldots, x_n)$ and $(y_1, \ldots, y_n)$ defined as $[\sum_{i=1}^{n}(x_i - y_i)^2]^{1/2}$ The spaces $\mathbb{C}^n$, $\mathbb{H}^n$, and Ca^n can be naturally identified with $\mathbb{R}^{2n}$, $\mathbb{R}^{4n}$, and $\mathbb{R}^{8n}$, and in particular have natural metrics and topologies. The closed ball and sphere in $\mathbb{R}^n$ with centre $(0, 0, \ldots, 0)$ and radius 1 are called simply the n-*ball* and $(n - 1)$-*sphere*, and denoted by D^n and S^{n-1}. In particular, d^0 is a point, S^0 a pair of points, and $S^{-1} = \emptyset$. The unit interval $[0, 1] \subset \mathbb{R}$ is denoted by I. I^n denotes the *unit n-cube* $\{(x_1, \ldots, x_n) \in \mathbb{R}^n | 0 \leq x_i \leq 1, i = 1, \ldots, n\}$; its frontier (in $\mathbb{R}^n$) is denoted by ∂I_n. Real projective n-dimensional space is denoted by $\mathbb{R}P^n$, complex by $\mathbb{C}P^n$, quaternionic projective space by $\mathbb{H}P^n$, and the Cayley projective line and plane by $\text{Ca}P^1$ and $\text{Ca}P^2$. Recall that $\mathbb{R}P^1$, $\mathbb{C}P^1$, $\mathbb{H}P^1$, and $\text{Ca}P^1$ are canonically homeomorphic to the spheres S^1, S^2, S^4, and S^8. The real Grassmann manifolds are denoted by $\mathbb{R}G(m, n)$ or $G(m, n)$. By definition, $G(m, n)$ is the set of n-dimensional (vector) subspaces of the space $\mathbb{R}^{m+n}$. The manifold of oriented n-dimensional subspaces of $\mathbb{R}^{m+n}$ is denoted by $G_+(m, n)$. The complex Grassmann manifold of n-dimensional (complex vector) subspaces of $\mathbb{C}^{m+n}$ is denoted by $\mathbb{C}G(m, n)$. The corresponding quaternionic Grassmann manifold is $\mathbb{H}G(m, n)$.

C. The appearance of the symbol ∞ as a dimensional parameter denotes passage to the inductive limit. Thus $\mathbb{R}^\infty$ is the inductive limit of the sequence of spaces $\mathbb{R}^k$ with the natural inclusion $\mathbb{R}^k \to \mathbb{R}^{k+1} : (x_1, \ldots, x_k) \mapsto (x_1, \ldots, x_k, 0)$. The points of $\mathbb{R}^\infty$ may be interpreted as infinite sequences $(x_1, x_2, \ldots)$ of real numbers, in which only finitely many terms are non-zero. A topology is introduced into $\mathbb{R}^\infty$ by the rule: a set $U \subset \mathbb{R}^\infty$ is open if all the intersections $U \cap \mathbb{R}^n$ are open in the spaces $\mathbb{R}^n$. The symbols $\mathbb{C}^\infty$, $\mathbb{H}^\infty$, D^∞, S^∞, $\mathbb{R}P^\infty$, $\mathbb{C}P^\infty$, $\mathbb{H}P^\infty$, $G(\infty, n)$, $G(\infty, \infty)$ etc. are interpreted similarly. None of these spaces are metrizable.

1.4. Operations on Topological Spaces.

A. The sum $\coprod_{\mu \in M} X_\mu$ of a family of topological spaces is canonically provided with a topology: a subset of the sum is declared to be open if its inverse images under all the maps $\text{in}_\nu : X_\nu \to \coprod_{\mu \in M} X_\mu$ are open. It is clear that each of the maps in_ν is an embedding and that the images $\text{in}_\nu(X_\nu)$ are simultaneously open and closed in $\coprod X_\mu$. It is also clear that if $f_\mu : X_\mu \to Y_\mu$, $\mu \in M$, are continuous maps, then their sum $\coprod_{\mu \in M} f_\mu : \coprod_{\mu \in M} X_\mu \to \coprod_{\mu \in M} Y_\mu$ is continuous.

B. The product $X_1 \times \cdots \times X_n$ of topological spaces $X_1, \ldots, X_n$ is canonically provided with a topology: a basis for the open sets in $X_1 \times \cdots \times X_n$ consists of the sets $U_1 \times \cdots \times U_n \subset X_1 \times \cdots \times X_n$, where $U_1, \ldots, U_n$ are open sets in $X_1, \ldots, X_n$. It is clear that the projections $\mathrm{pr}_i : X_1 \times \cdots \times X_n \to X_i$ are continuous open maps for any spaces $X_1, \ldots, X_n$. If $Y, X_1, \ldots, X_n$ are any sets whatever, then to each map $f : Y \to X_1 \times \cdots \times X_n$ there correspond the maps $\mathrm{pr}_i \circ f : Y \to X_i$, and for any given maps $f_i : Y \to X_i$, there exists a unique map $f : Y \to X_1 \times \cdots \times X_n$ with $\mathrm{pr}_i \circ f = f_i$. It is clear that if $Y, X_1, \ldots, X_n$ are topological spaces, then f is continuous iff all the $\mathrm{pr}_i \circ f$ are continuous. It is also evident that the product $f_1 \times \cdots \times f : X_1 \times \cdots \times X_n \to Y_1 \times .. \times Y_n$ of continuous maps $f_1 : X_1 \to Y_1, \ldots, f_n : X_n \to Y_n$ is continuous.

C. The quotient space X/S of a topological space X with respect to any partition S has a natural topology: a set is open if its inverse image under the map $\mathrm{pr} : X \to X/S$ is open. This natural topology is called the *quotient topology*, and the set X/S endowed with this topology is called the *quotient space of X with respect to the partition S*. The map $\mathrm{pr} : X \to X/S$ is clearly continuous.

In the special case when all the elements of S are points except for a single set A, the space X/S is called the *quotient of X by A* and denoted by X/A.

It follows from the definition of the quotient topology that if X and Y are any topological spaces with partitions S and T, and $f : X \to Y$ is a continuous map taking the elements of S into elements of T, then the map $f/S, T : X/S \to Y/T$ is continuous.

D. Let X and Y be topological spaces, A a subset of Y, and $\phi : A \to X$ be a continuous map. The quotient space of the sum $X \coprod Y$ with respect to the partition into one-point subsets of $\phi \coprod (Y \setminus A)$ and sets of the form $x \coprod \phi^{-1}(x)$ with $x \in X$ is denoted by $X \cup_\phi Y$; we say that it is obtained by *attaching Y to X by ϕ*. It is clear that the natural injection $X \to X \cup_\phi Y$ is a topological embedding. In the case when X is a point, attaching Y to X by $\phi : A \to X$ is clearly equivalent to forming the quotient space Y/A.

E. The product of the interval $I = [0, 1]$ with the space X is called the *cylinder over X*. The subsets $X \times 0$ and $X \times 1$ of $X \times I$ are called its (lower and upper) *bases* (they are copies of X), and a subset of the form $x \times I$, $x \in X$, is called a *generator* (it is a copy of I). If all the points of the base $X \times 0$ are identified to each other, we obtain the *cone $CX = X \times I / X \times 0$ over X*. The cone CX has a *base*, usually identified with X – the image of the upper base of $X \times I$ – and a *vertex*, the point obtained from the lower base $X \times 0$. The images of the generators of the cylinder under the map $\mathrm{pr} : X \times I \to CX$ are called *generators* of the cone. If we take the quotient of the cone with respect to its base, we obtain the *suspension ΣX over X*; thus $\Sigma X = CX/X$. Alternatively we may say that ΣX is obtained as the quotient space of the cylinder $X \times I$ with respect to the partition whose elements are the bases $X \times 0$ and $X \times 1$ and the one-point subsets of $X \times (0, 1)$. The images of the bases are called the *vertices of the suspension*. The sets $\mathrm{pr}(x \times I)$ are the *generators of the suspension*. The set $\mathrm{pr}(X \times \frac{1}{2})$ is the *base of the suspension*, and is a copy of X. The suspension X may be regarded as two cones over X joined together by their bases. The joined bases form the base of the suspension. It is clear that CS^n and S^n are homeomorphic to D^{n+1} and S^{n+1}.

To each map $f : X \to Y$ there corresponds the map $f \times \mathrm{id} : X \times I \to Y \times I$ and its quotient maps $CX \to CY$ and $\Sigma X \to \Sigma Y$ are continuous if f is continuous. The map $(f \times \mathrm{id})/ : \Sigma X \to \Sigma Y$ is denoted by Σf and called the *suspension of the map* f.

F. It is convenient to regard the *join* $X * Y$ of the spaces X and Y as the union of the line segments joining each point of X to each point of Y. For example, the join of two segments lying on skew lines in $\mathbb{R}^3$ is a tetrahedron. A formal definition of the join is the following: it is obtained as the quotient of $X \times Y \times I$ with respect to the partition whose elements are the sets $x \times Y \times 0$ $(x \in X)$ and $X \times y \times 1$ $(y \in Y)$ and the points of $X \times Y \times (0, 1)$. The set $\mathrm{pr}(x \times y \times I) \subset X * Y$, $x \in X$, $y \in Y$, is called a *generator of the join*; it is just the segment joining $x \in X$ and $y \in Y$. X and Y themselves are embedded in $X * Y$ as follows: $X \to X * Y : x \mapsto \mathrm{pr}(x \times Y \times 0)$ and $Y \to X * Y : y \mapsto \mathrm{pr}(X \times y \times 1)$. Their images under these embeddings are called the *bases of the join*. The generators cover the join. Each of them is determined by the points of the base that they join. Two distinct generators can intersect only in a single point and this point can only lie in one of the bases. Equivalently the join $X * Y$ can be defined as $(X \coprod Y) \cup_\phi (X \times Y \times I)$, where ϕ is the map $X \times Y \times (0 \cup 1) \to X \coprod Y$ with $\phi(x, y, 0) = x$, $\phi(x, y, 1) = y$. The quotient space of $X * Y$ with respect to the partition consisting of the bases X, Y and the points of the complement $(X * Y) \setminus (X \cup Y)$ is the same as the suspension $\Sigma(X \times Y)$.

The operation $*$ (like $\times$) is commutative: there is an obvious canonical homeomorphism $Y * X \to X * Y$. It can be shown that for Hausdorff locally compact spaces the operation $*$ is associative, but this is not true in general. In fact, if X and Y are Hausdorff and locally compact, the maps $X * Y \to CX \times CY : \mathrm{pr}(x, y, t) \mapsto (\mathrm{pr}(x, 1 - t), \mathrm{pr}(y, t))$ is a toplogical embedding with image $\{(\mathrm{pr}(x, s), \mathrm{pr}(y, t)) \in CX \times CY | s + t = 1\}$. By repeating this construction in the case of locally compact Hausdorff spaces $X_1, \ldots, X_n$, we can obtain a homeomorphism of $((X_1 * X_2) * \ldots) * X_n$ onto the space

$$\{(\mathrm{pr}(x_1, t_1), \ldots, \mathrm{pr}(x_n, t_n)) \in CX_1 \times \cdots \times CX_n | t_1 + \ldots + t_n = 1\}$$

and then use the associativity of the operation $\times$.

The join $X * D^0$ is canonically homeomorphic to the cone CX, and $X * S^0$ to the suspension ΣX. The join $X * S^k$ is canonically homeomorphic to the multiple suspension $\Sigma^{k+1} X$; in particular, $S^p * S^q$ is canonically homeomorphic to S^{p+q+1}.

To each pair of maps $f_1 : X_1 \to Y_1$, $f_2 : X_2 \to Y_2$ there corresponds the map $(f_1 \times f_2 \times \mathrm{id}_I)/ : X_1 * X_2 \to Y_1 * Y_2$, which is continuous if f_1 and f_2 are continuous. It is denoted by $f_1 * f_2$.

G. Let $f : X \to Y$ be a continuous map. The result of attaching the product $X \times I$ to Y by the map $X \times 1 \to Y : (x, 1) \to f(x)$ is called the *mapping cylinder* of f, and denoted by $\mathrm{Cyl} f$. The subsets of $\mathrm{Cyl} f$ obtained from $X \times 0$ and Y are called the *lower and upper bases* of $\mathrm{Cyl} f$. The bases are related to X and Y by obvious canonical homeomorphisms and are usually identified with X and Y. The subsets of $\mathrm{Cyl} f$ obtained from $x \times I$ with $x \in X$ are called the *generators* of $\mathrm{Cyl} f$; they are canonically homeomorphic to I. There is a canonical map $\text{Cyl} f \to Y$, taking each generator to its point of intersection with Y. It is clear that the composition of the inclusion $X \to \text{Cyl} f$ and this map $\text{Cyl} f \to Y$ is equal to f.

The result of attaching the cone CX to Y by the map f of its base is called the *mapping cone* of f, and denoted by $\text{Con} f$. Clearly $\text{Con} f = \text{Cyl} f / X$. The subset of $\text{Con} f$ obtained from Y is called the *base* of $\text{Con} f$; it is obviously canonically homeomorphic to Y and is usually identified with Y.

If $Y = X$ and $f = \text{id}_X$, then $\text{Cyl} f$ is canonically homeomorphic to $X \times I$, and $\text{Con} f$ to CX.

H. If X and Y are topological spaces, let $C(X, Y)$ denote the set of all continuous maps $X \to Y$. If $A_1, \ldots, A_n$ are subsets of X, and $B_1, \ldots, B_n$ subsets of Y, then $C(X, A_1, \ldots, A_n; Y, B_1, \ldots, B_n)$ denotes the subset of $C(X, Y)$ consisting of the maps ϕ for which $\phi(A_1) \subset B_1, \ldots, \phi(A_n) \subset B_n$. The notation $\phi : (X, A_1, \ldots, A_n) \to (Y, B_1, \ldots, B_n)$ is used for such maps.

The set $C(X, Y)$ is endowed with the *compact-open topology* – the topology of uniform convergence on compact sets (that is, the topology with a basis of sets of the form $C(X, A_1, \ldots, A_n; Y, B_1, \ldots, B_n)$, where $A_1, \ldots, A_n$ are compact and $B_1, \ldots, B_n$ are open). As well as $C(X, Y)$ all the sets $C(X, A_1, \ldots, A_n; Y, B_1, \ldots, B_n)$ also become topological spaces.

It is clear that if X is a point, then $C(X, Y) = Y$; if X is a discrete space with n points (that is, a collection of n isolated points), then $C(X, Y) = Y \times \cdots \times Y$ (n factors). This is the reason for denoting the space $C(X, Y)$ by Y^X.

Let X, Y, Z be topological spaces. To each continuous map $\phi : X \times Y \to Z$ there corresponds the continuous map $\phi^\vee : X \to C(X, Z)$ defined by $[\phi^\vee(x)](y) = \phi(x, y)$. It can be shown that the map $C(X \times Y, Z) \to C(X, C(Y, Z)) : \phi \mapsto \phi^\vee$ is continuous, and is a homeomorphism if Y is regular and locally compact. This relation between $C(X \times Y, Z)$ and $C(X, C(Y, Z))$ makes the notation Y^X for $C(X, Y)$ even more attractive: it takes the form of the equation $Z^{X \times Y} = (Z^Y)^X$, and is called the *exponential law*. For any topological spaces $X, Y_1, \ldots, Y_n$, there is a canonical homeomorphism $C(X, Y_1 \times \cdots \times Y_n) \to C(X, Y_1) \times \cdots \times C(X, Y_n) : f \mapsto (\text{pr}_1 \circ f, \ldots, \text{pr}_n \circ f)$.

1.5. Operations on Pointed Spaces.

In homotopy theory we often have to consider not merely topological spaces, but *pointed spaces*, that is, each space considered contains a distinguished point, or *base point*, and all maps considered take base points to base points; identical spaces with different base points are regarded as though they were different spaces. The transition to pointed spaces shows itself to a greater or lesser extent in all operations on spaces. For some operations the modification just consists in providing the resulting space with a base point. For example, the base point in the product $X \times Y$ is (x_0, y_0), where x_0, y_0 are the base points of the factors. Some operations need to be modified more significantly. Thus in the cone CX, all the points of the generator corresponding to the base point of X are identified to each other. Similar modifications are made to the suspension, join (in which the points of the generator joining the base points of the factors are identified), and the mapping cylinder and cone (where it is of course assumed that the base point is mapped to the base point). In each case the point to which the generator is shrunk is taken as the base point. With these modifications, we still have the homeomorphisms $CS^n = D^{n+1}$, $\Sigma S^n = \Sigma S^{n+1}$ and $S^p * S^q = S^{p+q+1}$, if $(1, 0, \ldots, 0)$ is taken as base point in the spheres and balls.

The space of mappings reduces to the space of mappings that take base point to base point. The base point of $C(X, Y)$ is the map taking the whole of X to the base point of Y. In homotopy theory a special role is played by the space of continuous maps of the circle with base point into a pointed space; notation: $\Omega(X, x_0) := C(S^1, (1, 0); X, x_0)$, abbreviated to ΩX; it is called the *loop space of* X *with origin at* x_0. The same terminology and notation is also used for the space $C(I, 0 \cup 1; X, x_0)$, related to $C(S^1, (1, 0); X, x_0)$ by the canonical homeomorphism $C(S^1, (1, 0); X, x_0) \to C(I, 0 \cup 1; X, x_0) : f \mapsto [t \to f(e^{2\pi i t})]$.

Finally we shall describe two operations that are specific to pointed spaces. Let $\{X_\mu\}_{\mu \in M}$ be a family of pointed topological spaces with base points x_μ. The quotient space of the sum $\coprod_{\mu \in M} X_\mu$ by the subset $\coprod_{\mu \in M} x_\mu$ is called the *bouquet of spaces* X, denoted by $\bigvee_{\mu \in M} X_\mu$, or, more precisely, $\bigvee_{\mu \in M}(X_\mu, x_\mu)$. The point $\mathrm{pr}(\coprod_{\mu \in M} x_\mu)$ is called the *centre* of the bouquet, and is naturally taken as base point. The bouquet $\bigvee_{\mu \in M} X_\mu$ is covered by copies of the spaces X_μ (usually identified with X_μ), which intersect each other only in the centre of the bouquet. Figure 1 shows a bouquet of two circles ("figure of eight").

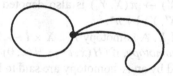

Fig. 1

Let $X_1, \ldots, X_n$ be topological spaces with base points $x_1, \ldots, x_n$. The canonical embeddings $X_1 \to X_1 \times \cdots \times X_n : x \mapsto (x, x_2, \ldots, x_n), \ldots, X_n \to X_1 \times \cdots \times X_n : x \mapsto (x_1, \ldots, x_{n-1}, x)$ determine a canonical embedding $(X_1, x_1) \vee \cdots \vee (X_n, x_n) \to X \times \cdots \times X_n$, which allows us to regard the bouquet $X_1 \vee \cdots \vee X_n$ as a subspace of $X_1 \times \cdots \times X_n$. The quotient space $X_1 \times \cdots \times X_n / X_1 \vee \cdots \vee X_n$ is called the *smash product* or *tensor product* of $X_1, \ldots, X_n$, and is denoted by $X_1 \otimes \cdots \otimes X_n$, or, more precisely, by $(X_1, x_1) \otimes \cdots \otimes (X_n, x_n)$. The notations $X_1 \wedge \cdots \wedge X_n$ and $X_1 \# \cdots \# X_n$ are also used. The point $\mathrm{pr}(X_1 \vee \cdots \vee X_n) \in X_1 \otimes \cdots \otimes X_n$ is called the *centre* of the tensor product and is taken as base point. It is not hard to see that $S^p \otimes S^q = S^{p+q}$, and that for any pointed space X, $\Sigma X = X \otimes S^1$.

If $\{X_\mu\}_{\mu \in M}$ and $\{Y_\mu\}_{\mu \in M}$ are families of pointed spaces and $\{f_\mu : X_\mu \to Y_\mu\}_{\mu \in M}$ is a family of continuous maps (taking base points to base points), then this gives rise to a continuous map $(\coprod f_\mu)/ : \bigvee_{\mu \in M} X_\mu \to \bigvee_{\mu \in M} Y_\mu$, denoted by $\bigvee_{\mu \in M} f_\mu$. The map $f_1 \otimes \cdots \otimes f_n : X_1 \otimes \cdots \otimes X_n \to Y_1 \otimes \cdots \otimes Y_n$ is defined similarly.

§2. Homotopy

2.1. Homotopies.

A. A continuous map $g : X \to Y$ is said to be *homotopic* to a continuous map $f : X \to Y$ if there exists a continuous map $H : X \times I \to Y$ such that $H(x, 0) = f(x)$, $H(x, 1) = g(x)$ for all $x \in X$. Any such mapping is called a *homotopy* connecting f and g. We also say that H is a *homotopy of f*.

Homotopy is clearly an equivalence relation. It divides the space $C(X, Y)$ of continuous maps of X into Y into equivalence classes called *homotopy classes*. The set of these classes is denoted by $\pi(X, Y)$.

B. As an example of homotopy we may take rectilinear homotopy. Let f and g be continuous maps of X into a subspace Y of $\mathbb{R}^n$. If for each point $x \in X$, the line segment joining $f(x)$ to $g(x)$ lies entirely in Y, then the formula $H(x, t) = (1 - t)f(x) + tg(x)$ defines a homotopy between f and g. Such a homotopy is called *rectilinear*. Hence any two maps of an arbitrary space into a convex subspace of a Euclidean space are homotopic.

C. If f, $f' : X \to Y$, $g : Y \to Y'$, $h : X' \to X$ are continuous maps and $F : X \times I \to Y$ is a homotopy between f and f', then $g \circ F \circ (h \times \mathrm{id}_I)$ is a homotopy between $g \circ f \circ h$ and $g \circ f' \circ h$. Hence the mapping $C(g, h) : C(X, Y) \to C(X', Y')$, induced by the maps g and h, takes homotopy classes to homotopy classes. The mapping $\pi(X, Y) \to \pi(X', Y')$ arising in this way is denoted by $\pi(g, h)$. It is determined by the homotopy classes of g and h. The mapping $\pi(g, \mathrm{id}) : \pi(X, Y) \to \pi(X, Y')$ is also denoted by g_*, and the mapping $\pi(\mathrm{id}, h) : \pi(X, Y) \to \pi(X', Y)$ by h^*.

D. Let A be a subset of X. A homotopy $H : X \times I \to Y$ is said to be *relative to A*, or, briefly, to be an *A-homotopy*, if $(H(x, t) = H(x, 0)$ for all $x \in A, t \in I$. Two maps that can be connected by an A-homotopy are said to be *A-homotopic*. Clearly, A-homotopic maps coincide on A. If we want to emphasize that a homotopy is not relative, we call it *free*.

Like ordinary homotopy, A-homotopy is an equivalence relation. The classes into which it divides the set of continuous maps $X \to Y$ that agree on A with a given map $f : A \to Y$, are called *A-homotopy classes*, or, more precisely, *homotopy classes of continuous extensions of f to X*.

Note that the rectilinear homotopy between f and g is relative to the set on which f and g coincide.

2.2. Paths.

A continuous mapping of the interval I into X is called a *path* in the space X. The points $s(0)$ and $s(1)$ are called the *origin* and *end* of the path s. If $s(0) = s(1)$ the path s is called *closed*. Closed paths are also called *loops*.

If s is a path, the path defined by $t \mapsto s(1 - t)$ is called the *inverse* of S, and denoted by s^{-1}. The path defined in terms of paths s_1 and s_2 with $s_1(1) = s_2(0)$ by the formula

$$t \mapsto \begin{cases} s_1(2t), & \text{if } t \leq \tfrac{1}{2}, \\ s_2(2t - 1), & \text{if } t \geq \tfrac{1}{2}, \end{cases}$$

is called the *product* of the paths s_1 and s_2 and denoted by $s_1 s_2$.

Being continuous maps, paths may undergo homotopies. Unfortunately, the generally accepted terminology concerning these homotopies does not entirely agree with the definitions (also generally accepted) in 2.1. Namely, when applied to paths homotopy always means a $(0 \cup 1)$-homotopy (i.e. a homotopy relative to the end-points of the interval). Also when we are speaking of loops, a free homotopy always means an ordinary free homotopy in which a loop remains a loop (i.e. a continuous map $H : I \times I \to X$ such that $H(0, t) = H(1, t)$ for each $t \in I$).

2.3. Homotopy as a Path.

A homotopy $H : X \times I \to Y$ is often interpreted as the family of continuous maps $h_t : X \to Y$, related to H by $h_t(x) = H(x, t)$ $(0 \le t \le 1)$. It follows from the continuity of H that this family is continuous as a map $I \to C(X, Y)$. In fact, H is converted into a mapping $t \to h_t$ by the mapping $C(X \times I, Y) \to C(I, C(X, Y))$. Thus a homotopy connecting continuous maps $f, g : X \to Y$ defines a path in $C(X, Y)$ joining f and g. If the space X is locally compact and regular, then, conversely, to each path in $C(X, Y)$ joining f and g, there corresponds a homotopy connecting f and g, so that in this case homotopies may be defined as paths in $C(X, Y)$.

2.4. Homotopy Equivalence.

A continuous map $g : Y \to X$ is called a *homotopy inverse* of the continuous map $f : X \to Y$ if the composition $g \circ f$ is homotopic to id_X, and $f \circ g$ is homotopic to id_Y. A continuous map having a homotopy inverse is called a *homotopy equivalence*. If there exists a homotopy equivalence $X \to Y$, the space Y is said to be *homotopy equivalent* to X.

It is clear that the identity map, as well as every homeomorphism, is a homotopy equivalence, that a homotopy inverse of a homotopy equivalence is a homotopy equivalence, and that the composition of two homotopy equivalences is a homotopy equivalence. Consequently, homotopy equivalence as a relation between topological spaces is an equivalence relation. The classes into which it divides topological spaces are called *homotopy types*.

The set $\pi(X, Y)$ is a homotopy invariant, that is, it depends only on the homotopy type of the spaces X and Y: if $g : Y \to Y'$ and $f : X' \to X$ are homotopy equivalences, then $\pi(f, g) : \pi(X, Y) \to \pi(X', Y')$ is an invertible mapping.

2.5. Retractions.

A *retraction* is a continuous map of a space onto a subspace that coincides with the identity map on the subspace. Such a subspace is called a *retract* of the space.

Any point in a topological space is a retract of the space, but a pair of points is not necessarily a retract. For example, an interval cannot be retracted onto its boundary. It is easy to show that a subset A of a topological space X is a retract iff for any topological space Y, every continuous map $A \to Y$ can be extended to a continuous map $X \to Y$.

2.6. Deformation Retractions.

A. A retraction ρ of a topological space X onto a subspace A is called a *deformation retraction* (*strong deformation retraction*) if the composite map $X \xrightarrow{\rho} A \xrightarrow{\text{in}} X$ is homotopic to id_X (A-homotopic to id_X). If X admits a deformation retraction (strong deformation retraction) onto A, then A is called a *deformation retract* (*strong deformation retract*) of A. A strong deformation retract is automatically a deformation retract. It turns out that in non-pathological situations the converse also is true. See 2.9.

It is clear that a deformation retraction $X \to A$ and the inclusion $A \to X$ are a pair of mutually inverse homotopy equivalences.

B. Let $f : X \to Y$ and $g : X_1 \to Y_1$ be continuous maps. We say that they are homotopy equivalent if there exist homotopy equivalences $\phi : X_1 \to X$ and $\psi : Y_1 \to Y$, such that the maps $\psi \circ g$ and $f \circ \phi$ are homotopic. (In this situation we also say that the diagram

$$
\begin{array}{ccc}
X_1 & \xrightarrow{g} & Y_1 \\
\phi \downarrow & & \downarrow \psi \\
X & \xrightarrow{f} & Y
\end{array}
$$

is homotopy commutative.)

Any continuous map $f : X \to Y$ is homotopy equivalent to an inclusion. This inclusion may be taken, for example, as the inclusion of X in the mapping cylinder $\text{Cyl} f$ of the map f. In fact it is clear that the natural retraction $\rho : \text{Cyl} f \to Y$ is a strong deformation retraction, and that the diagram

$$
\begin{array}{ccc}
X & \hookrightarrow & \text{Cyl} f \\
\text{id} \downarrow & & \downarrow \rho \\
X & \longrightarrow & Y
\end{array}
$$

is commutative (not just homotopy commutative, but commutative in the usual sense). It is easy to verify that X is a strong deformation retract of $\text{Cyl} f$ iff f is a homotopy equivalence. Thus every homotopy equivalence is homotopy equivalent to a strong deformation retraction, and any two homotopy equivalent spaces can be embedded as strong deformation retracts in the same space.

C. A space X is called *contractible* if the identity map id_X is homotopic to a constant map. X is contractible iff it is homotopy equivalent to a point, iff it admits a deformation retraction onto one of its points, iff each point of X is a deformation retract of X, iff any two continuous maps of an arbitrary topological space into X are homotopic, iff every continuous map of X into an arbitrary topological space is homotopic to a constant map.

The Euclidean spaces $\mathbb{R}^n$ and their convex subsets provide examples of contractible spaces.

2.7. Relative Homotopies.

Let X be a space with a distinguished sequence of subsets $A_1, \ldots, A_n$, and Y a space with a distinguished sequence of subsets $B_1, \ldots, B_n$. Recall that the notation $f : (X, A_1, \ldots, A_n) \to (Y, B_1, \ldots, B_n)$ means that f is a map $X \to Y$ such that $f(A_1) \subset B_1, \ldots, f(A_n) \subset B_n$). A map $H : (X \times I, A_1 \times I, \ldots, A_n \times I) \to (Y, B_1, \ldots, B_n)$ is called a (relative) homotopy connecting the maps $f, g : (X, A_1, \ldots, A_n) \to (Y, B_1, \ldots, B_n)$, if $H : X \times I \to Y$ is a homotopy connecting $f, g : X \to Y$. Clearly in this case the map $H| : A_i \times I \to B_i$ determined by H is a homotopy connecting the maps $f|, g| : A_i \to B_i$. It is also clear that homotopy, as defined by the homotopies $(X \times I, A_1 \times I, \ldots, A_n \times I) \to (Y, B_1, \ldots, B_n)$ is an equivalence relation. The set of homotopy classes into which it divides the set $C(X, A_1, \ldots, A_n; Y, B_1, \ldots, B_n)$ is denoted by $\pi(X, A_1, \ldots, A_n; Y, B_1, \ldots, B_n)$. The definition given in 2.1 of the map $\pi(g, h)$ can be carried over to the relative case in a corresponding way.

A continuous map $g : (Y, B_1, \ldots, B_n) \to (X, A_1, \ldots, A_n)$ is called a homotopy inverse of $f : (X, A_1, \ldots, A_n) \to (Y, B_1, \ldots, B_n)$ if the composition $g \circ f$ is homotopic to $\mathrm{id}_{(X, A_1, \ldots, A_n)}$, and $f \circ g$ is homotopic to $\mathrm{id}_{(Y, B_1, \ldots, B_n)}$. The definitions of homotopy equivalence and homotopy equivalent spaces given in 2.4 can be carried over to the relative case in a similar way.

An important special case of the situation considered in this section is that of pointed spaces. (In this case A_1 and B_1 are points, $n = 1$, and relative homotopies are relative to the base point.) The definition of contractibility in 2.6 can be carried over in an obvious way to pointed spaces (the homotopy between id_X and the constant map must be relative to the base point), as can the definition of deformation retract and deformation retraction (the base points must be the same in X and A, and the homotopy between the composite map $X \xrightarrow{\rho} A \xrightarrow{\text{in}} X$ and id_X must be relative to this base point), and the accompanying remarks there.

2.8. k-connectedness.

A. A space X is called *k-connected* $(-1 \leq k \leq \infty)$, if every continuous map $S^r \to X$ with $r \leq k$ can be extended to a continuous map $D^{r+1} \to X$; (-1)-connectedness means that the space is non-empty. For $k \geq 0$, a non-empty space X is k-connected iff every continuous map $S^r \to X$ with $r \leq k$ is homotopic relative to a point to the constant map, iff every map $S^r \to X$ with $r \leq k$ is freely homotopic to a constant map, iff any continuous maps $f, g : D^r \to X$ with $r \leq k$ that coincide on S^{r-1} are S^{r-1}-homotopic.

For non-empty spaces 0-connectedness is usually called *path-connectedness*, and sometimes (in particular in this work) just *connectedness*. A 1-connected space is usually called *simply connected*; for 0-connected spaces this is equivalent to the requirement that any two paths with common endpoints are homotopic to each other.

It follows from the homotopy invariance of the set $\pi(S^r, X)$ that a space which is homotopy equivalent to a k-connected space is itself k-connected. In particular, contractible spaces are ∞-connected.

B. The pair (X, A) is called *k-connected* $(0 \leq k \leq \infty)$ if every continuous map $f : (D^r, S^{r-1}) \to (X, A)$ with $r \leq k$ is S^{r-1}-homotopic to a map whose image is contained in A.

It is clear that the pair (X, A) is 0-connected iff each component of X intersects A. For $k > 0$, the pair (X, A) is k-connected iff every continuous map $f : (D^r, S^{r-1}) \to (X, A)$ with $r \leq k$ is homotopic to a constant map. If the inclusion $A \to X$ is a homotopy equivalence, then the pair (X, A) is ∞-connected. A pair that is homotopy equivalent to a k-connected pair is obviously k-connected.

2.9. Borsuk Pairs.

A. A topological pair (X, A) is called a Borsuk pair[1] if given any topological space Y, a continuous map $f : X \to Y$, and a homotopy $F : A \times I \to Y$ of $f|_A$, then there exists a homotopy $X \times I \to Y$ of f that extends F.

For a topological pair (X, A) to be a Borsuk pair, it is necessary, and when A is closed, also sufficient, that the set $(X \times 0) \cup (A \times I)$ be a retract of the cylinder $X \times I$. If X is Hausdorff, then the fact that A is closed follows automatically from the fact that there is a retraction of $X \times I$ onto $(X \times 0) \cup (A \times I)$, so that a topological pair (X, A) in which X is Hausdorff is a Borsuk pair iff $(X \times 0) \cup (A \times I)$ is a retract of the cylinder $X \times I$.

B. It is clear that if (X, A) and (A, B) are Borsuk pairs, then (X, B) is also a Borsuk pair. It is not hard to show that if the sets A and B form a closed covering of X and $(A, A \cap B)$ is a Borsuk pair, then (X, B) is a Borsuk pair. If (X, A) is a Borsuk pair and A is closed, then $(Z \times X, Z \times A)$ is a Borsuk pair for any topological space Z.

C. If (X, A) is a Borsuk pair, and the inclusion $A \to X$ is a homotopy equivalence, then, as is easily seen, A is a deformation retract of X. If A is a deformation retract of X and $(X \times I, (X \times 0) \cup (A \times I) \cup (X \times 1))$ is a Borsuk pair, then A is a strong deformation retract of N.

D. It can be shown that if (X, A) is a Borsuk pair, then for any neighbourhood U of A there exists a neighbourhood V of A with $V \subset U$, such that the inclusion $V \to U$ is A-homotopic to a mapping that takes V into A; in particular, A is a retract of its neighbourhood V. Conversely, in the case when X is a normal space and A is a distinguishable subset, i.e. the set of zeros of some continuous function $X \to R$ (in particular, when X is metrizable and A is closed), then for (X, A) to be a Borsuk pair it is sufficient that there exists a neighbourhood V of A, such that the inclusion $V \to X$ is A-homotopic to a mapping that takes V into A. In particular if A is a deformation retract of one of its neighbourhoods, X is normal and A is distinguishable, then (X, A) is a Borsuk pair.

E. A topological space X is called *locally contractible to the point* $x_0 \in X$ if every neighborhood u of x_0 contains a neighborhood V of x_0 such that the inclusion $V \to U$ is homotopic to the constant map taking V to x_0. X is called *locally contractible* if it is locally contractible to each of its points.

---

[1] The term *cofibration* is also used in the literature. However, strictly speaking, the word "cofibration" refers to the inclusion $A \to X$, rather than the pair (X, A).

If in these definitions homotopy is replaced by x_0-homotopy, we obtain the definitions of *strong local contractibility of X to the point x_0*, and of *strong local contractibility* of X.

The spaces $\mathbb{R}^n$, D^n and S^n provide examples of strongly locally contractible spaces.

F. In the case when A is a point, the local characterization of a Borsuk pair given above takes the following form. If a topological space forms a Borsuk pair with one of its points, then it is strongly locally contractible to that point. If a normal space is strongly locally contractible to a distinguishable point, then it forms a Borsuk pair with that point.

G. It is easy to show that if X is a normal space, (X, A) a Borsuk pair, f : $X \to Y$ a continuous map, and F a homotopy of the restriction $f|_A$ then for each neighbourhood U of A, a homotopy of f extending F may be chosen which is relative to $X \setminus U$.

2.10. CNRS Spaces.

A. A subset of a topological space is called a *neighbourhood* retract if it a retract of one of its neighbourhoods.

Retracts and open sets provide obvious examples of neighbourhood retracts. If A is a neighbourhood retract of X and B is a neighbourhood retract of A, then B is clearly a neighbourhood retract of X.

B. A topological space is called a *CNRS space* if it is compact and can be embedded as a neighbourhood retract in a Euclidean space (of a certain dimension). The expression CNRS derives from the initial letters of compact neighbourhood retract of a sphere. Obvious examples of CNRS spaces are the balls D^n and the spheres S^n. Clearly a compact neighbourhood retract of a CNRS space is a CNRS space. It is noteworthy that, as can easily be proved, the image of any embedding of a CNRS space in a normal space is a neighbourhood retract.

C. For any compact neighbourhood retract X of $\mathbb{R}^n$ there exists a positive number ε such that any continuous maps of an arbitrary space Y into X, with $\text{dist}(f(y), g(y)) < \varepsilon$ for all $y \in Y$, are homotopic. The homotopy can be chosen to be relative to the set on which f and g coincide. We may take ε to be the distance between X and the complement $\mathbb{R}^n \setminus U$ of a neighbourhood U that can be retracted onto X; then the homotopy between f and g can be constructed as the composite of the rectilinear homotopy in U and the retraction $U \to X$.

Using this result, it is easy to show that a topological pair of CNRS spaces is a Borsuk pair, and that any CNRS space is strongly locally contractible. Conversely, it can be shown that every locally contractible subspace of a Euclidean space is a neighbourhood retract of it.

2.11. Homotopy Properties of Topological Constructions.

A. It is obvious that continuous maps $f, g : Y \to X_1 \times \cdots \times X_n$ are homotopic iff the maps $\text{pr}_i \circ f$, $\text{pr}_i \circ g : Y \to X_i$ are homotopic for each i. In particular, the space $X_1 \times \cdots \times X_n$ is k-connected iff all the spaces X_i are k-connected ($-1 \leq k \leq \infty$). It is also clear that if the maps $g_1 : X_1 \to Y_1, \ldots, g_n : X_n \to Y_n$ are homotopic to

maps $f_1 : X_1 \to Y_1, \ldots, f_n : X_n \to Y_n$, then the map $g_1 \times \cdots \times g_n : X_1 \times \cdots \times X_n \to$
$Y_1 \times \cdots \times Y_n$ is homotopic to $f_1 \times \cdots \times f_n : X_1 \times \cdots \times X_n \to Y_1 \times \cdots \times Y_n$, and
that if $f_1, \ldots, f_n$ are homotopy equivalences, then $f_1 \times \cdots \times f_n$ is also a homotopy
equivalence.

Similar assertions are valid for the join $*$, the wedge $\vee$ and the tensor product
$\otimes$ (in the last two cases the maps and homotopies must of course preserve the base
points).

Similarly, if $f, g : X \to Y$ are homotopic, then the maps $\Sigma f, \Sigma g : \Sigma X \to$
ΣY are also homotopic. If $f : X \to Y$ is a homotopy equivalence, then $\Sigma f :$
$\Sigma X \to \Sigma Y$ is also a homotopy equivalence. The homotopy properties of the cone
construction are exhausted by the fact that the cone CX is obviously contractible
for any X.

B. If A is a deformation retract (strong deformation retract) of a space X, then
clearly for any Y the product $A \times Y$ is a deformation retract (strong deformation
retract) of $X \times Y$. In particular, if X is contractible, then the fibre $x \times Y$ of $X \times Y$
is a deformation retract of the product.

C. Let S and T be partitions of the spaces X and Y. If the maps $f_t : X \to Y$
form a homotopy that maps the elements of S to the elements of T, then the
map $f_t/ : X/S \to Y/T$ determined by f_t is also a homotopy. In particular, if
$f, f' : (X, A) \to (Y, B)$ are homotopic, then the maps $f/, f'/ : (X/A, \mathrm{pr}(A)) \to$
$(Y/B, \mathrm{pr}(B))$ determined by them are also homotopic. If $f : (X, A) \to (Y, B)$
is a homotopy equivalence, then $f/ : (X/A, \mathrm{pr}(A)) \to (Y/B, \mathrm{pr}(B))$ is also
a homotopy equivalence. If (X, A) is a Borsuk pair and A is contractible, then
$\mathrm{pr}(X, A) \to (X/A, \mathrm{pr}(A))$ is a homotopy equivalence.

D. If (Y, A) is a Borsuk pair and the maps $\phi, \phi' : A \to X$ are homotopic, then
the spaces $X \cup_\phi Y, X \cup_{\phi'} Y$ are homotopy equivalent; further, they nay be connected
by a homotopy equivalence f such that the diagram

$$
\begin{array}{ccc}
 & X & \\
\text{in} \swarrow & & \searrow \text{in} \\
X \cup_\phi Y & \xrightarrow{\ f\ } & X \cup_{\phi'} Y
\end{array}
$$

is commutative, where in is the natural injection.

If (Y, A) is a Borsuk pair, A is closed, $f : X \to X'$ is a homotopy equivalence,
and $\phi : A \to X$ is a continuous map, then the map $X \cup_\phi Y \to X' \cup_{f \circ \phi} Y$, determined
by id_Y and f, is a homotopy equivalence.

Thus the homotopy type of the space $X \cup_\phi Y$ is unchanged either by replacing
ϕ by a homotopic map, or by replacing the space X by a space that is homotopy
equivalent to it.

2.12. Natural Group Structures on Sets of Homotopy Classes.

Homotopy theory studies characteristics of topological spaces or continuous
maps which are naturally discrete. These characteristics usually take the same values on homotopy equivalent spaces or homotopic maps, that is, they are *homotopy
invariants*. A very widespread method of constructing them is the following: we

fix a space C, and then with each space X we associate the set $\pi(X, C)$ or the set $\pi(C, X)$.

It is much easier to study these sets if they have some kind of natural algebraic structure (this very often turns out to be a group structure). Such a structure usually also carries further useful information about the space. Before describing this in more detail, let us agree on the form of the subsequent exposition. We shall consider invariants of two kinds: C is fixed, and with each X we associate $\pi(X, C)$ or $\pi(C, X)$. The theory of each kind is developed in its own way, but these theories remain parallel, or more precisely, dual, for quite a long time. This duality, called Eckmann-Hilton duality, occupies an important place in homotopy theory. We shall not formulate it explicitly in this volume, but in order to convey some idea of it, we set out the text of this section in two columns, so that dual definitions and assertions are placed side by side.

For the rest of this section, we assume that all spaces are pointed spaces, and all maps, homotopies, etc., are understood accordingly. We fix a space C with base point c_0.

Let us assume that for each X a group structure is introduced into the set $\pi(X, C)$. Such a structure is called *natural* if for any continuous map $\phi : X' \to X''$, the map $\phi^* : \pi(X'', C) \to (X', C)$ is a homomorphism.

C is called an *H-space* if there are fixed maps
$$\mu : C \times C \to C$$
(multiplication) and
$$\nu : C \to C$$
(taking the inverse), satisfying the following three conditions:

1. (Homotopy identity)
The maps
$$C \xrightarrow{j_1} C \times C \xrightarrow{\mu} C,$$
$$C \xrightarrow{j_2} C \times C \xrightarrow{\mu} C,$$
where
$$j_1(c) = (c, c_0),$$
$$j_2(c) = (c_0, c),$$
are homotopic to the identity map $\mathrm{id} : C \to C.$

2. (Homotopy associativity)
The maps
$$C \times (C \times C) \xrightarrow{\mathrm{id} \times \mu} C \times C \xrightarrow{\mu} C,$$
$$(C \times C) \times C \xrightarrow{\mu \times \mathrm{id}} C \times C \xrightarrow{\mu} C$$
are homotopic.

Let us assume that for each X a group structure is introduced into the set $\pi(C, X)$. Such a structure is called *natural* if for any continuous map $\phi : X' \to X''$, the map $\phi_* : \pi(C, X') \to \pi(C, X'')$ is a homomorphism.

C is called an *H'-space* if there are fixed maps
$$\mu : C \to C \vee C$$
(comultiplication) and
$$\nu : C \to C$$
(taking the co-inverse), satisfying the following three conditions:

1. (Homotopy co-identity)
The maps
$$C \xrightarrow{\mu} C \vee C \xrightarrow{\pi_1} C,$$
$$C \xrightarrow{\mu} C \vee C \xrightarrow{\pi_2} C,$$
where π_1 is the identity on the first (summand) C and maps the second C to c_0, and π_2 is the identity on the second C, and maps the first C to c_0, are homotopic to the identity map $\mathrm{id} : C \to C.$

2. (Homotopy co-associativity)
The maps
$$C \xrightarrow{\mu} C \vee C \xrightarrow{\mathrm{id} \vee \mu} C \vee (C \vee C),$$
$$C \xrightarrow{\mu} C \vee C \xrightarrow{\mu \vee \mathrm{id}} (C \vee C) \vee C$$
are homotopic.

3. (Homotopy inverse)
The maps
$$C \xrightarrow{\mathrm{id} \times \nu} C \times C \xrightarrow{\mu} C,$$
$$C \xrightarrow{\nu \times \mathrm{id}} C \times C \xrightarrow{\mu} C$$
are homotopic to the constant map
(that is, the map to the point c_0).

An important example of an H-space is the loop space ΩZ of any space Z. The map
$$\mu : \Omega Z \times \Omega Z \to \Omega Z$$
is defined by
$$\mu(f, g)(t) = \begin{cases} f(2t), & \text{if } t \in [0, 1/2], \\ g(2t - 1), & \text{if } t \in [1/2, 1], \end{cases}$$
that is, with two loops we associate the loop formed by successively tracing both loops:

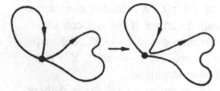

The map $\nu : \Omega Z \to \Omega Z$ is defined by
$$\nu(f)(t) = f(1 - t),$$
that is, a loop is mapped to the same loop traced out in the reverse direction.

Another important example of an H-space is a topological group.

Theorem. *The set $\pi(X, C)$ has a natural (with respect to X) group structure iff C is an H-space.* $\square$

3. (Homotopy co-inverse)
The maps
$$C \xrightarrow{\mu} C \vee C \xrightarrow{\mathrm{id} \vee \nu} C,$$
$$C \xrightarrow{\mu} C \vee C \xrightarrow{\nu \vee \mathrm{id}} C$$
are homotopic to the constant map
(that is, the map to the point c_0).

An important example of an H'-space is the suspension ΣZ over any space Z. The map
$$\mu : \Sigma Z \to \Sigma Z \vee \Sigma Z$$
is defined by
$$\mu \mathrm{pr}(z, t) = \begin{cases} \mathrm{pr}_I(z, 2t), & \text{if } t \in [0, 1/2], \\ \mathrm{pr}_{II}(z, 2t - 1), & \text{if } t \in [1/2, 1], \end{cases}$$
the Roman figures indicating in which of the two ΣX that form $\Sigma X \vee \Sigma X$ the point in question lies:

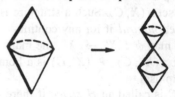

The map $\nu : \Sigma Z \to \Sigma Z$ is defined by
$$\nu \mathrm{pr}(z, t) = \mathrm{pr}(z, 1 - t).$$
There is no parallel to this statement.

Theorem. *The set $\pi(C, X)$ has a natural (with respect to X) group structure iff C is an H'-space.* $\square$

Omitting details, we confine ourselves to a description of the constructions involved in the proof of these theorems.

If there is a group structure in $\pi(X, C)$, natural with respect to X, then $\mu : C \times C \to C$ can be obtained as a mapping from the homotopy class resulting from the multiplication of homotopy classes
$$[\mathrm{pr}_1], [\mathrm{pr}_2] \in \pi(C \times C, C)$$

If there is a group structure in $\pi(C, X)$, natural with respect to X, then $\mu : C \to C \vee C$ can be obtained as a mapping from the homotopy class resulting from the multiplication of homotopy classes
$$[\mathrm{in}_1], [\mathrm{in}_2] \in \pi(C, C \vee C)$$

of the projections of the product $C \times C$ onto its factors, and $\nu : C \to C$ as the representative of the homotopy class inverse in the group $\pi(C, C)$ to the class [id] of the identity map. It is easy to verify that these maps μ and ν define the structure of an H-space on C.

If C is an H-space with multiplication $\mu : C \times C \to C$, and X is any space, then the composition of the natural bijection

$$\pi(X, C) \times \pi(X, C) \to \pi(X, C \times C)$$

with the map

$$\mu_* : \pi(X, C \times C) \to \pi(X, C),$$

induced by μ, is a group operation on $\pi(X, C)$, natural with respect to X. The map $\nu : C \to C$ induces a map ν_*, relating each element of the group $\pi(X, C)$ to its inverse.

It can be shown that the group $\pi(X, \Omega\Omega Z)$ is abelian for any X and Z.

For any $n \geq 1$ there is a space K_n, unique up to homotopy equivalence, such that

$$\pi(S^n, K_n) = \begin{cases} 0, & i \neq n, \\ \mathbb{Z}, & i = n. \end{cases}$$

The construction of such spaces K^n is discussed below in §11. It is easy to show that K^n is homotopy equivalent to K^{n+1}. Hence $\pi(X, K_n)$ has a natural group structure. The group $\pi(X, K_n)$ is called the n-dimensional (integral) cohomology group of X, denoted by $H^n(X; \mathbb{Z})$ or $H^n(X)$. Since K_n is homotopy equivalent to $\Omega\Omega K_{n+2}$, this group is abelian for any X and n.

$$H^i(S^n) = \begin{cases} 0, & i \neq n, \\ \mathbb{Z}, & i = n. \end{cases}$$

The computation of the groups $H^i(K_n)$ has proved to be a very important problem, solved in the 1950s by the efforts of A. Borel, H. Cartan, and J.-P. Serre.

of the natural embeddings of C in $C \vee C$, and $\nu : C \to C$ as the representative of the homotopy class inverse in the group $\pi(C, C)$ to the class [id] of the identity map. It is easy to verify that these maps μ and ν define the structure of an H'-space on C.

If C is an H'-space with comultiplication $\mu : C \to C \vee C$, and X is any space, then the composition of the natural bijection

$$\pi(C, X) \times \pi(C, X) \to \pi(C \vee C, X)$$

with the map

$$\mu^* : \pi(C \vee C, X) \to \pi(C, X),$$

induced by μ, is a group operation on $\pi(C, X)$, natural with respect to X. The map $\nu : C \to C$ induces a map ν^* relating each element of the group $\pi(C, X)$ to its inverse.

It can be shown that the group $\pi(\Sigma\Sigma Z, X)$ is abelian for any X and Z.

The dual object is the sphere S^n.

The dual fact is $S^{n+1} = \Sigma S^n$.

It follows from this fact that $\pi(S^n, X)$ has a natural group structure for any $n \geq 1$. The group $\pi(S^n, X)$ is called the n-dimensional homotopy group of X, and denoted by $\pi_n(X)$, or, more precisely, by $\pi_n(X, x_0)$. Since $S^n = \Sigma\Sigma S^{n-2}$, this group is abelian for any $n \geq 2$ and any X.

$$\pi_i(K_n) = \begin{cases} 0, & i \neq n, \\ \mathbb{Z}, & i = n. \end{cases}$$

The computation of the groups $\pi_i(S^n)$ proved to be a very important problem at one time. It has remained unsolved to this day.

Cohomology groups are one of the central objects to be studied in the second part of this volume.

Homotopy groups are one of the central objects to be studied in this part of the volume.

§3. Homotopy Groups

3.1. Absolute Homotopy Groups.

A. We shall discuss in more detail the definition of the homotopy groups recalled in the previous section. Let (X, x_0) be a pointed space and r a non-negative integer. In the previous section $\pi_r(X, x_0)$ denoted the set of classes of maps $S^r \to X$ taking the base point of the sphere S^r (usually the point $(1, 0, \ldots, 0) \in S^r$) to x_0. Two other equivalent definitions are often used: $\pi_r(X, x_0)$ is defined as $\pi(I^r, \partial I^r; X, x_0)$ or as $\pi(D^r, S^{r-1}; X, x_0)$. To establish the equivalence of these definitions we need only fix a homeomorphism $I^r, \partial I^r \to D^r, S^{r-1}$ and a continuous map $I^r \to S^r$, taking I^r to $(1, 0, \ldots, 0)$ and mapping $I^r \setminus \partial I^r$ bijectively onto $S^r \setminus (1, 0, \ldots, 0)$. This gives natural bijections $C(D^r, S^{r-1}; X, x_0) \to C(I^r, \partial I^r; X, x_0)$ and $C(S^r, (1, 0, \ldots, 0); X, x_0) \to C(I^r, \partial I^r; X, x_0)$, and induced bijections $\pi(D^r, S^{r-1}; X, x_0) \to \pi(I^r, \partial I^r; X, x_0)$ and $\pi(S^r, (1, 0, \ldots, 0); X, x_0) \to \pi(I^r, \partial I^r; X, x_0)$. We shall regard the definition with I^r as the basic one. The set $C(I^r, \partial I^r; X, x_0)$ will be denoted by $\mathrm{Sph}_r(X, x_0)$, and its elements will be called r-*dimensional spheroids of X with origin* x_0.

B. For $r > 0$ we define the *product* $\phi\psi$ of spheroids $\phi, \psi \in \mathrm{Sph}_r(X, x_0)$ as the spheroid in $\mathrm{Sph}_r(X, x_0)$ given by

$$\phi\psi(t_1, t_2, \ldots, t_n) = \begin{cases} \phi(2t_1, t_2, \ldots, t_n), & \text{if } 0 \leq t_1 \leq 1/2, \\ \psi(2t_1 - 1, t_2, \ldots, t_n), & \text{if } 1/2 \leq t_1 \leq 1, \end{cases}$$

and the spheroid ϕ^{-1}, the *inverse* of $\phi \in \mathrm{Sph}_r(X, x_0)$, as the spheroid in $\mathrm{Sph}_r(X, x_0)$ given by $\phi^{-1}(t_1, t_2, \ldots, t_n) = \phi(1 - t_1, t_2, \ldots, t_n)$. Clearly if ϕ_1 is homotopic to ϕ and ψ_1 homotopic to ψ, then $\phi_1\psi_1$ is homotopic to $\phi\psi$. Consequently multiplication of spheroids defines a multiplication in $\pi_r(X, x_0)$. This multiplication is associative, and the homotopy class of the constant spheroid (mapping I^r to x_0) acts as a two-sided identity. The homotopy classes of ϕ and ϕ^{-1} are mutually inverse. The set $\pi_r(X, x_0)$ made into a group in this way for $r > 0$ is called the r-*dimensional homotopy group of X at* x_0.

C. If $r > 0$, then each r-spheroid maps I^r to the component X_0 of X that contains x_0. Hence, for $r > 0$, the group $\pi_r(X, x_0)$ is canonically isomorphic to $\pi_r(X_0, x_0)$.

D. Since I^0 is a point, and $\partial I^0 = \varnothing$, $\mathrm{Sph}_0(X, x_0)$ may be identified with X, and $\pi_0(X, x_0)$ with the set of components of X. The set $\pi_0(X, x_0)$ has no natural group structure, but merely has a distinguished element, which is called the identity, as in higher dimensions: it is the homotopy class of the constant spheroid, that is, the component of X containing x_0.

In order to give a unified formulation for the cases $r > 0$ and $r = 0$, the set $\pi_0(X, x_0)$ is called the *zero-dimensional homotopy group of X at* x_0, and the group-theoretic terminology is extended to sets with a distinguished element. In particular, a homomorphism means a map taking the distinguished element to the distinguished element, the kernel of a homomorphism is the inverse image of the distinguished element, and an isomorphism is an invertible homomorphism.

E. One-dimensional spheroids are just closed paths, and the definitions of multiplication, inverse and homotopy of spheroids agree with the corresponding definitions for paths given in 1.2. The one-dimensional homotopy group is also called the *fundamental group*. It was defined several decades earlier than the higher homotopy groups, and, as we shall see, occupies a special position among the homotopy groups.[2]

F. Let X be a space with base point x_0. The formula

$$[\psi(t_1, \ldots, t_r)](u_1, \ldots, u_s) = \phi(t_1, \ldots, t_r, u_1, \ldots, u_s)$$

associates with the spheroid $[\phi : (I^{r+s}, \partial I^{r+s}) \to (X, x_0)] \in \mathrm{Sph}_{r+s}(X, x_0)$ an r-spheroid of the space $\mathrm{Sph}_s(X, x_0)$ $(= C(I^s, \partial I^s; X, x_0))$, in which the constant spheroid is taken as base point. Thus we obtain a map $\mathrm{Sph}_{r+s}(X, x_0) \to \mathrm{Sph}_r(\mathrm{Sph}_s(X, x_0), \mathrm{const})$, which is clearly invertible. Both this map and its inverse take homotopic spheroids to homotopic spheroids. For $r > 0$ this map takes the multiplication in $\mathrm{Sph}_{r+s}(X, x_0)$ to the multiplication in $\mathrm{Sph}_r(\mathrm{Sph}_s(X, x_0), \mathrm{const})$. Hence for $r > 0$ it defines a group isomorphism

$$\pi_{r+s}(X, x_0) \to \pi_r(\mathrm{Sph}_s(X, x_0), \mathrm{const}),$$

so that the group $\pi_r(X, x_0)$ may be described as $\pi_{r-q}(\mathrm{Sph}_q(X, x_0), \mathrm{const})$, and in particular as $\pi_{r-1}(\mathrm{Sph}_1(X, x_0), \mathrm{const}) = \pi_{r-1}(\Omega(X, x_0))$.

G. For $r > 1$, $\pi_r(X, x_0)$ is commutative. This means that for any spheroids $\phi, \psi \in \mathrm{Sph}_r(X, x_0)$ with $r > 1$, the products $\phi\psi$ and $\psi\phi$ are homotopic.

H. If $f : (X, x_0) \to (Y, y_0)$ is a continuous map from one pointed space to another, then to each spheroid $\phi : (I^r, \partial I^r) \to (X, x_0)$ there corresponds the spheroid $f \circ \phi : (I^r, \partial I^r) \to (Y, y_0)$, so that we obtain a map $f_\# : \mathrm{Sph}_r(X, x_0) \to \mathrm{Sph}_r(Y, y_0)$. Clearly $f_\#$ takes homotopic spheroids to homotopic spheroids, and the constant spheroid to the constant spheroid; also $f_\#(\phi\psi) = f_\#(\phi)f_\#(\psi)$ for $r > 0$. Hence for each $r \geq 0$, $f_\#$ determines a homomorphism $\pi_r(X, x_0) \to \pi_r(Y, y_0)$, called the *homomorphism induced by* f, and denoted by f_* or, more precisely, f_{*r}.

For any continuous maps $f : (X, x_0) \to (Y, y_0)$, $g : (Y, y_0) \to (Z, z_0)$ and any $r \geq 0$, $(g \circ f)_{*r} = g_{*r} \circ f_{*r}$. If $f = \mathrm{id}_{(X, x_0)}$, then $f_{*r} = \mathrm{id}_{\pi_r(X, x_0)}$. If maps $f, f' : (X, x_0) \to (Y, y_0)$ are homotopic, then $f'_{*r} = f_{*r}$. If $f : (X, x_0) \to (Y, y_0)$ is a homotopy equivalence, then f_{*r} is an isomorphism.

I. For any pointed spaces (X, x_0), (Y, y_0) and any r, the group $\pi_r(X \times Y, (x_0, y_0))$ is canonically isomorphic to the direct product $\pi_r(X, x_0) \times \pi_r(Y, y_0)$: the canonical isomorphism $\pi_r(X \times Y, (x_0, y_0)) \to \pi_r(X, x_0) \times \pi_r(Y, y_0)$ is defined by the formula $\alpha \to (\mathrm{pr}_{1*}(\alpha), \mathrm{pr}_{2*}(\alpha))$. If (X', x_0'), (Y', y_0') are other pointed spaces and $f : (X, x_0) + (X', x_0')$, $g : (Y, y_0) \to (Y', y_0')$ are continuous maps, then the diagram

---

[2] For these two reasons it is sometimes denoted simply by $\pi(X, x_0)$.

$$\pi_r(X \times Y, (x_0, y_0)) \quad \longrightarrow \quad \pi_r(X, x_0) \times \pi_r(Y, y_0)$$

$$\downarrow (f \times g)_* \qquad\qquad\qquad \downarrow f_* \times g_*$$

$$\pi_r(X' \times Y', (x_0', y_0')) \quad \longrightarrow \quad \pi_r(X', x_0') \times \pi_r(Y', y_0')$$

where the horizontal arrows are canonical isomorphisms, is commutative.

J. Let X be a T_1-space, and let the subsets $X \subset X_1 \subset \cdots$ form a fundamental cover of X, and $x_0 \in X$. It is not hard to show that if for some r all the inclusion homomorphisms $\pi_r(X_k, x_0) \to \pi_r(X_{k+1}, x_0)$ are isomorphisms, then the inclusion homomorphisms $\pi_r(X_k, x_0) \to \pi_r(X, x_0)$ are also all isomorphisms.

3.2. Digression: Local Systems.

A. We say that a *local system of groups*[3] is given on a space X, if for each point $x \in X$, there is a group G_x and for each path $s : I \to X$, there is a homomorphism $T_s : G_{s(0)} \to G_{s(1)}$, satisfying the three conditions:

(i) if $s_1(0) = s(1)$, then $T_{ss_1} = T_{s_1} \circ T_s$;

(ii) if s is a constant path, then T_s is the identity automorphism;

(iii) if the paths s, s_1 are homotopic, then $T_s = T_{s_1}$.

Condition (iii) shows that we may write T_σ instead of T_s, where σ is the homotopy class of the path s. It also follows from conditions (i)–(iii) that all the homomorphisms T_s are isomorphisms and $T^{-1} = T_{s^{-1}}$. The isomorphism T_s is called the *translation along s*.

If s is a loop representing an element σ in $\pi_1(X, x_0)$, then $T_s = T_\sigma$ is an automorphism of the group G_x. The formula $\sigma \mapsto T_\sigma$ defines a right group action of $\pi_1(X, x_0)$ on G_x. For each path $s : I \to X$ there is a natural isomorphism $t_s : i\pi_1(X, s(0)) \to \pi_1(X, s(1))$, defined by $t_s\omega = \sigma^{-1}\omega\sigma$, where σ is the homotopy class of s; it is also denoted by t_σ. An obvious verification shows that $T_s(g\omega) = T_s(g)t_s(\omega)$ for $g \in G_{s(0)}$, $\omega \in \pi_1(X, s(0))$, that is, $T_s : G_{s(0)} \to G_{s(1)}$ is a t_s-map.

B. Let $(X, \{G_x\}, \{T_s\})$, $(X', \{G'_{x'}\}, \{T'_{s'}\})$ be local systems of groups on spaces X, X' and let $f : X \to X'$ be a continuous map, and suppose that a homomorphism $h_x : G_x \to G'_{f(x)}$ is given for each $x \in X$. We say that the homomorphisms f_x and f form a *homomorphism of the first local system into the second* if $h_{s(1)} \circ T_s = T'_{f \circ s} \circ h_{s(0)}$ for any path $s : I \to X$. A homomorphism $(f, \{h_x\})$ is an *isomorphism* if f is a homeomorphism, and every h_x is an isomorphism, and an *equivalence* if in addition $X' = X$ and $f = \mathrm{id}_X$.

C. If $(X', \{G'_{x'}\}, \{T'_{s'}\})$ is a local system of groups and f is a continuous map from X to X', then there is an *induced local system* $(X, \{G_x\}, \{T_s\})$ on X, in which G_x is defined to be $G'_{f(x)}$ and T_s to be $T'_{f \circ s}$. Clearly $(f, \{\mathrm{id}_{G_x}\})$ is a homomorphism of the induced local system to the original one.

Two local systems of groups $(X, \{G_x\}, \{T_s\})$ and $(X, \{G'_x\}, \{T'_s\})$ on a connected space X with base point x_0 are equivalent iff the actions of the group $\pi_1(X, x_0)$ on G_{x_0} and G'_{x_0} that they define are isomorphic in the sense that there exists a group isomorphism $G_{x_0} \to G'_{x_0}$ that is a $\pi_1(X, x_0)$-map.

---

[3] Other terms are found in the literature: local system of coefficients, locally constant sheaf of groups, flat bundle with fibre a group, etc.

D. A local system of groups on a space X is *simple* if it is equivalent to a *canonical simple local system* $(X, \{G_x\}, \{T_s\})$ in which all the G_x are equal to a certain group G and all the homomorphisms T_s are the identity automorphism of this group.

A local system of groups on a connected space X with base point x_0 is simple iff the action of $\pi_1(X, x_0)$ on G_x that it defines is the identity action. In particular, it is automatically simple if $\pi_1(X, x_0)$ is trivial, or if the groups G_x are all isomorphic to $\mathbb{Z}_2$.

E. Everything above can be transferred in an obvious way from local systems of groups to local systems of other algebraic structures, for example, vector spaces or rings. In this section we shall meet local systems of sets with an identity (a distinguished element), as well as local systems of groups.

3.3. Local Systems of Homotopy Groups of a Topological Space.

A. Two spheroids $\phi_0 \in \mathrm{Sph}_r(X, x_0)$, $\phi_1 \in \mathrm{Sph}_r(X, x_1)$ are called *freely homotopic* if they can be joined by a homotopy consisting of spheroids. In more detail: spheroids ϕ_0, ϕ_1 are freely homotopic if there is a continuous map $h : I^r \times I \to X$, constant on each of the sets $\partial I^r \times t$ $(t \in I)$, and such that $h(y, 0) = \phi_0(y)$, $h(y, 1) = \phi_1(y)$ for each $y \in I^r$. An essential element of such a homotopy h is the path described by the origin of the spheroid, defined by $t \mapsto h(\partial I^r \times t)$. We say that h is a *free homotopy joining* ϕ_0 *and* ϕ_1 *along this path*.

Every spheroid with origin x_0 admits a free homotopy along any path with origin x_0. Free homotopies of homotopic spheroids along homotopic paths lead to homotopic spheroids. Hence the free homotopies along a path $s : I \to X$ determine a map $T_s : \pi_r(X, s(0)) \to \pi_r(X, s(1))$ for any $r \geq 0$. We thus obtain a local system $(X, \{\pi_r(X, x)\}, \{T_s\})$, which is a local system of groups for $r \geq 1$, and of sets with a distinguished point for $r = 0$. It is called the *local system of r-dimensional homotopy groups of* X. In particular, for any $x \in X$ and $r \geq 1$, there is a natural right group action of $\pi_1(X, x)$ on $\pi_r(X, x)$.

For $r = 1$ the isomorphism T_s acts according to the formula $T_s \omega = \sigma^{-1} \omega \sigma$, where σ is the homotopy class of the path s (that is, it is the same as the isomorphism t_s in 3.2). In particular, the right action of $\pi_1(X, x_0)$ on $\pi_1(X, x_0)$ is the inner right action.

Thanks to the existence of the isomorphisms T_s, in the case of a connected space X all the homotopy groups $\pi_r(X, x)$ are isomorphic to each other for each r.

B. A space X is *r-simple* if it is connected and the local system of its r-dimensional homotopy groups is simple. In this case the groups $\pi_r(X, x)$ are not only isomorphic, but related by canonical isomorphisms, allowing them to be identified with a single group $\pi_r(X)$, the *homotopy group of* X *without base point*, whose elements are classes of freely homotopic spheroids. A space is *simple* if it is r-simple for all r.

If a space X is not r-simple, then the isomorphisms T_s do not allow us to identify the groups $\pi_r(X, x)$ with different x. In this case we can only speak of the group $\pi_r(X)$ of X as an abstract group.

Obviously the local system of 0-dimensional homotopy groups of a space is always simple, and if the space is connected, it reduces to a local system of one-point sets. Clearly a space is 1-simple iff it is connected and its fundamental group is commutative.

C. By 3.1, every continuous map f of X into X' defines a homomorphism $f_* = (f_*)_x : \pi_r(X, x) \to \pi_r(X', f(x))$ at each point $x \in X$. Clearly, if h is a free homotopy joining spheroids ϕ_0, ϕ_1 along a path s, then $f \circ h$ is a free homotopy joining the spheroids $f \circ \phi_0$, $f \circ \phi_1$ along the path $f \circ s$, so that $(f_*)_{s(1)} \circ T_s = T_{f \circ s} \circ (f_*)_{s(0)}$. Hence for any $r \geq 0$, the induced homomorphisms $(f_{*r})_x$ and f form a homomorphism of the local system of r-dimensional homotopy groups of X into the local system of r-dimensional homotopy groups of X'. In the case when X and X' are r-simple, the local systems of r-dimensional homotopy groups of X and X' reduce to groups $\pi_r(X)$, $\pi_r(X')$, and the homomorphisms $(f_{*r})_x$ to a single homomorphism $f_{*r} : \pi_r(X) \to \pi_r(X')$.

If $f : X \to X'$ is a homotopy equivalence, then all the induced homomorphisms $(f_{*r})_x : \pi_r(X, x) \to \pi_r(X', f(x))$ are isomorphisms. A space X is k-connected iff all the groups $\pi_r(X, x_0)$ with $r \leq k$ ($0 \leq k < \infty$) are trivial. X is ∞-connected iff all the groups $\pi_r(X, x_0)$ are trivial.

D. The influence of a group structure. Let X be a topological group (that is, X is both a topological space and a group, and the group operations are continuous: the maps $X \times X \to X$ and $X \to X$ defined by $(g, h) \to gh$ and $g \to g^{-1}$ are continuous). Then for each path $s : I \to X$, the translation $\pi_r(X, s(0)) \to \pi_r(X, s(1))$ coincides for each $r \geq 0$ with the isomorphism induced by the left group translation by $[s(1)][s(0)]^{-1}$. There is even a canonical free homotopy $I^r \times I \to X$ joining a spheroid $\phi : (I^r, \partial I^r) \to (X, s(0))$ and $[s(1)][s(0)]^{-1}\phi$ along s: it is defined by the formula

$$((t_1, \ldots, t_r), t) \mapsto [s(t)][s(0)]^{-1}\phi(t_1, \ldots, t_r).$$

Corollary. *The components of a topological group are simple. In particular their fundamental groups are commutative.*

When X is a topological group, then beside the multiplication on the sets $\mathrm{Sph}_r(X, 1)$ defined in 3.1.B, there is another defined by the multiplication in X (the product of spheroids $\phi, \psi \in \mathrm{Sph}_r(X, 1)$ is given by $y \mapsto \phi(y)\psi(y)$); this second multiplication is still meaningful for $r = 0$, when the first one is not defined. Clearly, the new multiplication makes the set $\mathrm{Sph}_r(X, 1)$ into a group for any $r \geq 0$, the spheroids homotopic to a constant form a normal subgroup, and the corresponding factor group coincides as a set with $\pi_r(X, 1)$. For $r = 0$ the resulting group $\pi_0(X, 1)$ coincides with the factor group X/X_0, where X_0 is the component of the identity. For $r \geq 1$, the new group structure on $\pi_r(X, 1)$ coincides with the third: for any two spheroids $\phi, \psi \in \mathrm{Sph}_r(X, 1)$, the homotopy $I^r \times I \to X$, defined by the formula

$$((t_1, \ldots, t_r), t) \mapsto \phi(\min(1, \tfrac{2t_1}{1+t}), t_2, \ldots, t_r)\psi(\max(0, \tfrac{2t_1+t-1}{1+t}), t_2, \ldots, t_r),$$

relates the previous product of these spheroids to the new one.

In addition, the multiplication in X also makes the set $\bigcup_{x \in X} \mathrm{Sph}_r(X, x)$ of all r-dimensional spheroids into a group. Spheroids homotopic to a constant form a normal subgroup, and the factor group is canonically isomorphic to $\pi_r(X, 1)$ for $r \geq 1$.

Every inner automorphism of a topological group X induces an automorphism of the groups $\pi_r(X, 1)$, and so the inner right action of X defines a right group action on all the groups $\pi_r(X, 1)$. The transformations induced by the elements of the subgroup X_0 are all the identity: if $w : I \to X$ is a path from 1 to x, then for any spheroid $\phi \in \mathrm{Sph}_r(X, 1)$, the formula $(y, t) \mapsto [w(t)]^{-1}\phi(y)w(t)$ defines a homotopy $I^r \times I \to X$ between ϕ and the spheroid $y \mapsto x^{-1}\phi(y)x$. Thus there is a natural right group action of $\pi_0(X, 1) = X/X_0$ on $\pi_r(X, 1)$.

E. The case of H-spaces. Like topological groups, every connected H-space is simple, and, in particular, has a commutative fundamental group. This is proved in the same way as for topological groups.

The second description of the homotopy groups $\pi_r(X, e)$ given for groups in 3.3.D can be transferred in an obvious way to an H-space X with identity e. Although the resulting multiplication on the sets $\mathrm{Sph}_r(X, e)$ does not make them into groups in the general case, it defines a multiplication on $\pi_r(X, e)$ which is the same as the usual one for $r \geq 1$. The set $\pi_0(X, e)$ turns out to be a group if X is homotopy associative and has a homotopy inverse.

Spaces of spheroids are an important class of H-spaces. For any topological space X with base point x_0, the sets $\mathrm{Sph}_r(X, x_0)$ with $r \geq 1$, topologized as subsets of $C(I^r, X)$, and with the constant spheroid as base point and the usual multiplication, are H-spaces. They are homotopy associative and have a homotopy inverse for $r \geq 1$ and are homotopy commutative for $r \geq 2$.

F. The case of homogeneous spaces. Let G be a topological group and H a connected subgroup. If $(G, \mathrm{pr}, X = G/H)$ is a Serre bundle (see 4.3 below), then for each path $s : I \to X$ and any $r \geq 0$, the translation $\pi_r(X, s(0)) \to \pi_r(X, s(1))$ coincides with the isomorphism induced by any transformation of X produced (under the canonical action $G \times X \to X$) by an element of G and taking $s(0)$ to $s(1)$. (Cf. 3.4.D).

Corollary. *If H is a connected subgroup of a topological group G and $(G, \mathrm{pr}, G/H)$ is a Serre bundle, then the components of the space G/H are simple spaces. In particular the fundamental groups of such spaces are commutative.*

The condition that $(G, \mathrm{pr}, G/H)$ is a Serre bundle is usually automatically satisfied, since the group G and the quotient space G/H are usually smooth manifolds and the projection $G \to G/H$ is a submersion.

3.4. Relative Homotopy Groups.

A. One of the faces of the r-cube I^r is the $(r-1)$-cube $I^{r-1} \subset \mathbb{R}^{r-1}$. Let J^{r-1} denote the complement of the interior of this face in the boundary ∂I^r of I^r. For any topological pair (X, A) with base point $x_0 \in A$ and any positive integer r, the set $C(I^r, \partial I^r, J^{r-1}; X, A, x_0)$ of all continuous maps $I^r, \partial I^r, J^{r-1} \to X, A, x_0$ will be denoted by $\mathrm{Sph}_r(X, A, x_0)$. The elements of $\mathrm{Sph}_r(X, A, x_0)$ are called *n-dimensional spheroids of the pair (X, A) with origin*

x_0. The set $\pi(I^r, \partial I^r, J^{r-1}; X, A, x_0)$ of their homotopy classes is denoted by $\pi_r(X, A, x_0)$.

As in the case of $\pi_r(X, x_0)$, the set $\pi_r(X, A, x_0)$ has another equivalent description in which the triple $D^r, S^{r-1}, (1, 0, \ldots, 0)$ is used instead of $I^r, \partial I^r, J^{r-1}$. The equivalence of these two approaches is established by fixing any continuous map $(I^r, \partial I^r, J^{r-1}) \to (D^r, S^{r-1}, (1, 0, \ldots, 0))$ that induces a homeomorphism $I^r \setminus J^{r-1} \to D^r \setminus (1, 0, \ldots, 0)$.

Note that any spheroid $\phi \in \text{Sph}_r(X, A, x_0)$ with $\phi(I^r) \subset A$ is homotopic to the constant spheroid: a standard homotopy $I^r \times I \to X$ between ϕ and the constant spheroid is given by

$$((t_1, \ldots, t_{r-1}, t_r), t) \mapsto \phi(t_1, \ldots, t_{r-1}, (1 - t)t_r + t).$$

A one-dimensional spheroid of (X, A) with origin x_0 is obviously a path with origin in A and end-point x_0.

Warning. A homotopy of such a spheroid is stationary at the point 1, but not necessarily stationary at 0 if A is not reduced to x_0.

For $r \geq 2$, the formula which defined a multiplication in $\text{Sph}_r(X, x_0)$ (see 2.1) also defines a multiplication in $\text{Sph}_r(X, A, x_0)$. This multiplication goes over to $\pi_r(X, A, x_0)$ and makes it into a group for $r \geq 2$. The identity element of this group is the homotopy class of the constant spheroid. The class of ϕ^{-1}, defined by

$$\phi^{-1}(t_1, t_2, \ldots, t_r) = \phi(1 - t_1, t_2, \ldots, t_r),$$

is the inverse of the class of ϕ. The group $\pi_r(X, A, x_0)$, $r \geq 2$, is called the *r-dimensional homotopy group of the pair* (X, A) *at the point* x_0. The homotopy groups of a pair are also called *relative* homotopy groups (in contrast to the *absolute* homotopy groups of a space). The 1-dimensional homotopy group of (X, A) at x_0 is defined to be the set $\pi_1(X, A, x_0)$ with the class of the constant spheroid as identity (distinguished element).

If $A = x_0$ then $\pi_r(X, A, x_0)$ coincides with $\pi_r(X, x_0)$ (for $r \geq 2$ they coincide as groups, and for $r = 1$ as sets with distinguished elements).

For $r \geq 2$, $\pi_r(X, A, x_0)$ is canonically isomorphic to $\pi_r(X_0, A_0, x_0)$, where X_0 and A_0 are the components of X and A containing x_0.

For $r \geq 3$ $\pi_r(X, A, x_0)$ is commutative.

B. For each continuous map $f : (X, A, x_0) \to (X', A', x_0')$ there is an *induced homomorphism* $f_* : \pi_r(X, A, x_0) \to \pi_r(X', A', x_0')$ for $r \geq 1$, defined in the same way as in the absolute case, and coinciding with the absolute induced homomorphism $f_* : \pi_r(X, x_0) \to \pi_r(X', x_0')$ when $A = x_0$, $A' = x_0'$. As in the absolute case, $(g \circ f)_* = g_* \circ f_*$, and $\text{id}_* = \text{id}$. If f and f' are homotopic, then $f'_* = f_*$, and if f is a homotopy equivalence then f_* is an isomorphism.

C. For a spheroid $\phi \in \text{Sph}_r(X, A, x_0)$, the map $\phi' : (I^{r-1}, \partial I^{r-1}) \to (A, x_0)$ is a spheroid in $\text{Sph}_{r-1}(A, x_0)$ and is called the *boundary of the spheroid* ϕ, denoted by $\partial \phi$. The resulting map $\partial : \text{Sph}_r(X, A, x_0) \to \text{Sph}_{r-1}(A, x_0)$ obviously takes homotopic spheroids to homotopic spheroids, and the constant spheroid to the constant spheroid. Hence for $r \geq 1$, it defines a homomorphism

$\pi_r(X, A, x_0) \to \pi_{r-1}(A, x_0)$, called the *boundary homomorphism*, also denoted by ∂. For any continuous map $f : (X, A, x_0) \to (X', A', x_0')$ the following diagram is commutative for $r \geq 1$.

$$
\begin{array}{ccc}
\pi_r(X, A, x_0) & \xrightarrow{\ \partial\ } & \pi_{r-1}(A, x_0) \\
{\scriptstyle f_*}\downarrow & & \downarrow{\scriptstyle (f|)_*} \\
\pi_r(X', A', x_0') & \xrightarrow{\ \partial\ } & \pi_{r-1}(A', x_0')
\end{array}
$$

In fact the corresponding diagram with π replaced by Sph is already commutative.

D. Two spheroids $\phi_0 \in \mathrm{Sph}_r(X, A, x_0)$, $\phi_1 \in \mathrm{Sph}_r(X, A, x_1)$ of (X, A) are called *freely homotopic* if, regarded as maps of $(I^r, \partial I^r)$ into (X, A), they can be joined by a homotopy consisting of spheroids of (X, A), that is, there is a map $h : I^r \times I \to X$ with $h(\partial I^r \times I) \subset A$, constant on each of the sets $J^{r-1} \times t$ $(t \in I)$, and such that $h(y, 0) = \phi_0(y)$, $h(y, 1) = \phi_1(y)$ for any $y \in I^r$. We say that h is a *free homotopy* joining ϕ_0 and ϕ_1 along the path $t \mapsto h(J^{r-1} \times t)$.

Every spheroid of (X, A) with origin x_0 admits a free homotopy along any path in A with origin x_0. Free homotopies of homotopic spheroids along homotopic paths in A produce homotopic spheroids. Hence free homotopies along a path $s : I \to A$ determine a map $T_s : \pi_r(X, A, s(0)) \to \pi_r(X, A, s(1))$ for any $r \geq 1$. As in the absolute case, the maps T_s are homomorphisms and have properties (i)–(iii) in the definition of local systems, so that we obtain a local system $(A, \{\pi_r(X, A, x)\}, \{T_s\})$ on A, consisting of groups for $r \geq 2$, and sets with a distinguished element for $r = 1$. It is called the *local system of r-dimensional homotopy groups of the pair* (X, A). In particular, for any $x \in A$ and $r \geq 1$, there is a natural right action of the group $\pi_r(A, x)$ on $\pi_r(X, A, x)$; this is a group action for $r \geq 2$, and fixes the identity for $r = 1$.

The existence of this local system shows that for any $r \geq 1$, all the r-dimensional homotopy groups $\pi_r(X, A, x)$ of (X, A) with A connected are isomorphic to each other.

A pair (X, A) with A connected is called *r-simple* if the local system of its r-dimensional homotopy groups is simple. In this case all the groups $\pi_r(X, A, x)$ with $x \in A$ can be identified with the *r-dimensional homotopy group* $\pi_r(X, A)$ *of the pair* (X, A) *without base point*, whose elements are classes of freely homotopic spheroids. A pair is called *simple* if it is r-simple for all $r \geq 1$. For example, a space with a base point is a simple pair.

Clearly for any pair (X, A) and any path $s : I \to A$, the diagram

$$
\begin{array}{ccc}
\pi_r(X, A, s(0)) & \xrightarrow{\ \partial\ } & \pi_{r-1}(A, s(0)) \\
{\scriptstyle T_s}\downarrow & & \downarrow{\scriptstyle T_s} \\
\pi_r(X, A, s(1)) & \xrightarrow{\ \partial\ } & \pi_{r-1}(A, s(1))
\end{array}
$$

is commutative. Hence the boundary homomorphisms $\partial = \partial_x : \pi_r(X, A, x) \to \pi_{r-1}(A, x)$ and $\mathrm{id}_A A$ form a homomorphism of the local system of r-dimensional homotopy groups of (X, A) into the local system of $(r-1)$-dimensional homotopy groups of A.

For any continuous map $f : (X, A) \to (Y, B)$, the homomorphisms $f_* = (f_*)_x : \pi_r(X, A, x) \to \pi_r(Y, B, f(x))$ and f form a homomorphism of the local system of r-dimensional homotopy groups of (X, A) into the local system of r-dimensional homotopy groups of (Y, B). If f is a homotopy equivalence then all the homomorphisms $(f_*)_x$ are isomorphisms.

E. Let X and A be connected spaces; then (X, A) is ∞-connected iff all the groups $\pi_r(X, A, x_0)$ are trivial, and (X, A) is k-connected iff $\pi_r(X, A, x_0)$ is trivial for all r with $1 \le r \le k$.

F. The group action of $\pi_1(A, x_0)$ on $\pi_2(X, A, x_0)$ is related to the group operation in $\pi_2(X, A, x_0)$ in the following way: if $\alpha, \beta \in \pi_2(X, A, x_0)$, then $\alpha^{-1}\beta\alpha = T_{\partial\alpha}\beta$.

If w is a spheroid in $\mathrm{Sph}_1(X, A, x_0)$ and s is a loop in $\mathrm{Sph}(X, x_0)$, then the product ws is defined, and is clearly a spheroid in $\mathrm{Sph}_1(X, A, x_0)$, whose class is determined by the classes of w and s; thus the product $\omega\sigma$ is defined for $\omega \in \pi_1(X, A, x_0)$ and $\sigma \in \pi_1(X, x_0)$. The formula $(\omega, \sigma) \mapsto \omega\sigma$ defines a right group action of $\pi_1(X, x_0)$ on $\pi_1(X, A, x_0)$. If $A = x_0$ this action obviously coincides with the canonical right action of $\pi_1(X, x_0)$ on itself.

G. For any path $s : I \to A$ the translation $T_s : \pi_1(X, A, s(0)) \to \pi_1(X, A, s(1))$ is a $[T_{\mathrm{inos}} : \pi_1(X, s(0)) \to \pi_1(X, s(1))]$-map; that is, $T_s(\omega\sigma) = T_s(\omega)T_{\mathrm{inos}}(\sigma)$ for $\sigma \in \pi_1(X, s(0))$, $\omega \in \pi_1(X, A, s(0))$. It is also clear that for any continuous map $f : (X, A, x_0) \to (X', A', x_0')$, the homomorphism $f_* : \pi_1(X, A, x_0) \to \pi_1(X', A', x_0)$ is a $[f_* : \pi_1(X, x_0) \to \pi_1(X', x_0)]$-map; that is, $f_*(\omega\sigma) = f_*(\omega)f_*(\sigma)$ for $\omega \in \pi_1(X, A, x_0)$, $\sigma \in \pi_1(X, x_0)$. Applied to the inclusion $\mathrm{rel} : (X, x_0, x_0) \to (X, A, x_0)$, the last formula shows that $(\mathrm{rel}_*\omega)\sigma = \mathrm{rel}_*(\omega\sigma)$. Finally, for any $\omega \in \pi_1(X, A, x_0)$, $\sigma \in \pi_1(A, x_0)$,

$$T_\sigma\omega = \omega(\mathrm{in}_*\sigma),$$

where in is the inclusion $(A, x_0) \to (X, x_0)$.

3.5. The Homotopy Sequence of a Pair.

A. Let (X, A) be a topological pair with base point $x_0 \in A$. As we know, the homotopy groups $\pi_r(X, x_0)$, $\pi_r(A, x_0)$ are defined for $r \ge 0$, and the groups $\pi_r(X, A, x_0)$ and homomorphisms $\partial : \pi_r(X, A, x_0) \to \pi_{r-1}(A, x_0)$ are defined for $r \ge 1$. Together with these we consider the homomorphisms $\mathrm{in}_* : \pi_r(A, x_0) \to \pi_r(X, x_0)$ and $\mathrm{rel}_* : \pi_r(X, x_0) \to \pi_r(X, A, x_0)$, induced by the inclusions in $: (A, x_0) \to (X, x_0)$ and $\mathrm{rel} : (X, x_0, x_0) \to (X, A, x_0)$. These three series of homotopy groups and homomorphisms can be combined into a left-infinite sequence

$$\cdots \xrightarrow{\partial} \pi_2(A, x_0) \xrightarrow{\mathrm{in}_*} \pi_2(X, x_0) \xrightarrow{\mathrm{rel}_*} \pi_2(X, A, x_0) \xrightarrow{\partial} \pi_1(A, x_0) \xrightarrow{\mathrm{in}_*}$$

$$\longrightarrow \pi_1(X, x_0) \xrightarrow{\mathrm{rel}_*} \pi_1(X, A, x_0) \xrightarrow{\partial} \pi_0(A, x_0) \xrightarrow{\mathrm{in}_*} \pi_0(X, x_0).$$

Here all the terms except the last six are commutative groups, all the terms except the last three are groups, and the last three terms are sets with a distinguished element. All the maps except the last three are group homomorphisms, and the last three maps are homomorphisms of sets with distinguished elements. By 3.3. and 3.4, $\pi_1(X, x_0)$ has a right group action on the groups $\pi_r(X, x_0)$ for $r \geq 1$, and on the set $\pi_1(X, A, x_0)$, and the group $\pi_1(A, x_0)$ has a right group action on the groups $\pi_r(A, x_0)$ for $r \geq 1$, and $\pi_r(X, A, x_0)$ for $r \geq 2$. The homomorphisms in_*, rel_* and ∂ are compatible with these actions in the following sense:

(i) $\text{in}_{*r} : \pi_r(A, x_0) \to \pi_r(X, x_0)$ is an in_{*1}-homomorphism, that is, $\text{in}_{*r}(T_\sigma \alpha) = T_{\text{in}_{*1}\sigma}(\text{in}_{*r}\alpha)$ for $\sigma \in \pi_1(A, x_0)$, $\alpha \in \pi_r(A, x_0)$;

(ii) $\partial : \pi_{r+1}(X, A, x_0) \to \pi_r(A, x_0)$ is a $\pi_1(A, x_0)$-homomorphism, that is, $\partial(T_\sigma(\alpha)) = T_\sigma(\partial\alpha)$ for $\sigma \in \pi_1(A, x_0)$, $\alpha \in \pi_{r+1}(X, A, x_0)$;

(iii) $\text{rel}_{*r} : \pi_r(X, x_0) \to \pi_r(X, A, x_0)$ is a $\pi_1(A, x_0)$-homomorphism relative to the right group action of $\pi_1(A, x_0)$ on $\pi_r(X, x_0)$ induced by the existing action of $\pi_1(X, x_0)$ by means of the homomorphism $\text{in}_{*1} : \pi_1(A, x_0) \to \pi_1(X, x_0)$, that is, $\text{rel}_{*r} T_{\text{in}_{*}\sigma} = T_\sigma \text{rel}_{*r}\alpha$ for $\sigma \in \pi_1(A, x_0)$, $\sigma \in \pi_r(X, x_0)$;

(iv) the transformation $T_{\partial\alpha}$ of $\pi_2(X, A, x_0)$, produced by the image $\partial\alpha$ of an element α of $\pi_2(X, A, x_0)$ is the inner automorphism $\beta \mapsto \alpha^{-1}\beta\alpha$;

(v) on the set $\text{rel}_*(\pi_1(X, x_0))$ the transformation of the set $\pi_1(X, A, x_0)$ produced by the element σ in $\pi_1(X, x_0)$ is defined by $\text{rel}_*(\omega)\sigma = \text{rel}_*(\omega\sigma)$.

The sequence

$$\cdots \xrightarrow{\partial} \pi_2(A, x_0) \xrightarrow{\text{in}_*} \pi_2(X, x_0) \xrightarrow{\text{rel}_*} \pi_2(X, A, x_0) \xrightarrow{\partial}$$

$$\pi_1(A, x_0) \xrightarrow{\text{in}_*} \pi_1(X, x_0) \xrightarrow{\text{rel}_*} \pi_1(X, A, x_0) \xrightarrow{\partial} \pi_0(A, x_0) \xrightarrow{\text{in}_*} \pi_0(X, x_0)$$

together with the actions listed above is called the *homotopy sequence of the pair* (X, A).

B. From the algebraic point of view this is a very unwieldy object. However, since sequences algebraically related to it arise in various geometric situations, it makes sense to give the corresponding purely algebraic definitions.

A left-infinite sequence

$$\longrightarrow \Pi_7 \xrightarrow{\rho_6} \Pi_6 \xrightarrow{\rho_5} \Pi_5 \xrightarrow{\rho_4} \Pi_4 \xrightarrow{\rho_3} \Pi_3 \xrightarrow{\rho_2} \Pi_2 \xrightarrow{\rho_1} \Pi_1 \xrightarrow{\rho_0} \Pi_0,$$

in which Π_0, Π_1, Π_2 are sets with identity (distinguished element), Π_3, Π_4, Π_5 are groups, Π_6, $\Pi_7, \ldots$ are commutative groups, ρ_0, ρ_1, ρ_2 are homomorphisms in the sense explained in 2.1, ρ_3, $\rho_4, \ldots$ are group homomorphisms, is called a π-*sequence* if we are given: a right group action of Π_3 on the groups Π_{3k} with $k \geq 2$, and on the set Π_2, and a right group action of Π_4 on the groups Π_{3k+1} and Π_{3k-1} with $k \geq 2$, such that:

(i) ρ_{3k} with $k \geq 2$ is a Π_3-homomorphism;

(ii) ρ_{3k+1} with $k \geq 2$ is a Π_4-homomorphism, ρ_4 is a Π_4-homomorphism relative to the inner right action of Π_4;

(iii) ρ_{3k-1} with $k \geq 2$ is a Π_4-homomorphism relative to the right action of Π_4 on Π_{3k}, induced by the existing action of Π_3 on Π_{3k} by means of the homomorphism ρ_3;

(iv) the transformation of Π_5 produced by the image $\rho_4(\alpha)$ of an element α of Π_5 is the inner automorphism $\beta \mapsto \alpha^{-1}\beta\alpha$;

(v) on $\rho_2(\Pi_3)$ the transformation of the set Π_2 produced by an element σ of Π_3 is defined by $\rho_2(\omega)\sigma = \rho_2(\omega\sigma)$.

A *homomorphism* of the π-sequence $\{\Pi_i, \rho_i\}_{i=0}^{\infty}$ into the π-sequence $\{\Pi_i', \rho_i'\}_{i=0}^{\infty}$ is a sequence of homomorphisms $\{h_i : \Pi_i \to \Pi_i'\}_{i=0}^{\infty}$ such that $\rho_i \circ h_{i+1} = h_i \circ \rho_i$ for any $i \geq 0$; $h_{3k}, h_{3k+1}, h_{3k-1}$ with $k \geq 2$ are (respectively) Π_3, Π_4, Π_5-homomorphisms; $h_2(\omega)h_3(\sigma) = h_2(\omega\sigma)$ for any $\omega \in \Pi_2, \sigma \in \Pi_3$. An *isomorphism* is defined to be a homomorphism in which the constituent homomorphisms are isomorphisms.

C. The vertical homomorphisms

$$\cdots \pi_r(A, x_0) \xrightarrow{\text{in}_*} \pi_r(X, x_0) \xrightarrow{\text{rel}_*} \pi_r(X, A, x_0) \xrightarrow{\partial} \pi_{r-1}(A, x_0) \longrightarrow \cdots$$
$$\downarrow{(f|)_*} \qquad\qquad \downarrow{f_*} \qquad\qquad \downarrow{f_*} \qquad\qquad \downarrow{(f|)_*}$$
$$\cdots \pi_r(A', x_0') \xrightarrow{\text{in}_*} \pi_r(X', x_0') \xrightarrow{\text{rel}_*} \pi_r(X', A', x_0') \xrightarrow{\partial} \pi_{r-1}(A', x_0') \longrightarrow \cdots$$

induced by a continuous map $f : (X, A, x_0) \to (X', A', x_0')$ give a homomorphism of the first sequence into the second. For any path $s : I \to A$, the vertical isomorphisms

$$\cdots \pi_r(A, s(0)) \xrightarrow{\text{in}_*} \pi_r(X, s(0)) \xrightarrow{\text{rel}_*} \pi_r(X, A, s(0)) \xrightarrow{\partial} \pi_{r-1}(A, s(0)) \longrightarrow \cdots$$
$$\downarrow{T_s} \qquad\qquad \downarrow{T_{\text{in}\circ s}} \qquad\qquad \downarrow{T_s} \qquad\qquad \downarrow{T_s}$$
$$\cdots \pi_r(A, s(1)) \xrightarrow{\text{in}_*} \pi_r(X, s(1)) \xrightarrow{\text{rel}_*} \pi_r(X, A, s(1)) \xrightarrow{\partial} \pi_{r-1}(A, s(1)) \longrightarrow \cdots$$

give an isomorphism of the first sequence onto the second.

D. From the definition of a π-sequence it follows that (i) Ker ρ_4 is contained in the centre of the group Π_5, (ii) if Π_4 acts identically on Π_5, then Im ρ_4 is contained in the centre of the group Π_4 and the group Π_5 is commutative; (iii) if the group Π_5 is commutative and ρ_4 is an epimorphism, then Π_4 acts identically on Π_5.

We say that a π-sequence is *exact* if Ker $\rho_i =$ Im ρ_{i+1} for each $i \geq 0$, and in addition the inverse images of the elements of the set Π_1 under ρ_1 coincide with the orbits of the action of the group Π_3 on Π_2.

In an exact π-sequence, clearly for any $i \geq 0$, the homomorphism ρ_i is trivial iff ρ_{i+1} is an epimorphism, and Ker ρ_i is trivial iff the homomorphism ρ_{i+1} is trivial. For $i \geq 2$, if Ker ρ_i is trivial then ρ_i is one-one. But if Ker ρ_0 is trivial, it does not follow that ρ_0 is one-one, and the same holds for ρ_1. However, if the π-sequence is exact, then ρ_1 is guaranteed to be one-one if the group Π_3 is trivial or it acts identically on Π_2.

It follows from what has been said that for an exact π-sequence with $i \geq 1$, (i) ρ_i and ρ_{i+2} are trivial iff ρ_{i+1} is invertible, (ii) if Π_i and Π_{i+2} are trivial then Π_{i+1} is trivial, (iii) if Π_{i-1} and Π_{i+2} are trivial then ρ_i is invertible.

It is easy to show that in an exact π-sequence, if the action of Π_4 on Π_2 induced by the action of Π_3 by means of ρ_3, is the identity action, then Im ρ_3 is a normal subgroup of Π_3, and that the converse is true if ρ_2 is an epimorphism.

E. It is easy to verify that the homotopy sequence of any pair is exact.

F. Let us list some simple consequences of the exactness of the homotopy sequence of a pair. If X is ∞-connected then all the homomorphisms

∂ : $\pi_r(X, A, x_0) \to \pi_{r-1}(A, x_0)$ are isomorphisms. If X is k-connected and $k < \infty$, then ∂ : $\pi_r(X, A, x_0) \to \pi_{r-1}(A, x_0)$ is an isomorphism for $r \le k$, and ∂ : $\pi_{k+1}(X, A, x_0) \to \pi_k(A, x_0)$ is an epimorphism. In both cases the converse is true if X is connected.

If A is ∞-connected then all the homomorphisms rel$_*$: $\pi_r(X, x_0) \to \pi_r(X, A, x_0)$ are isomorphisms. If A is k-connected and $k < \infty$, then rel$_*$: $\pi_r(X, x_0) \to \pi_r(X, A, x_0)$ is an isomorphism for $r \le k$, and rel$_*$: $\pi_{k+1}(X, x_0) \to \pi_{k+1}(X, A, x_0)$ is an epimorphism. In both cases the converse is true if one of the spaces X, A is connected.

If the pair (X, A) is ∞-connected, then all the homomorphisms in$_*$: $\pi_r(A, x_0)$ $\to \pi_r(X, x_0)$ are isomorphisms. If the pair (X, A) is k-connected with $k < \infty$, then in$_*$: $\pi_r(A, x_0) \to \pi_r(X, x_0)$ is an isomorphism for $r < k$, and in$_*$: $\pi_k(A, x_0) \to \pi_k(X, x_0)$ is an epimorphism. In both cases the converse is true without any extra conditions. In particular, if in : $A \to X$ is a homotopy equivalence, then (X, A) is ∞-connected (cf. 1.6).

3.6. Splitting

A. We say that the π-sequence $\ldots \Pi_3 \xrightarrow{\rho_2} \Pi_2 \xrightarrow{\rho_1} \Pi_1 \xrightarrow{\rho_0} \Pi_0$ is *split on the right at the term* Π_α *by the homomorphism* ζ : $\Pi_{\alpha-1} \to \Pi_\alpha$, if $\rho_{\alpha-1} \circ \zeta = \mathrm{id}_{\Pi_{\alpha-1}}$, and that the splitting is *normal* if $\alpha = 0, 1, 2$ or Im ζ is a normal subgroup of Π_α. The π-sequence is *split on the left at the term* Π_α *by the homomorphism* ζ : $\Pi_\alpha \to \Pi_{\alpha+1}$ if $\zeta \circ \rho_\alpha = \mathrm{id}_{\Pi_{\alpha+1}}$. (Homomorphism means group homomorphism if this makes sense, and homomorphism of a set with identity otherwise.)

At terms Π_α with $\alpha > 5$ a right splitting of a π-sequence is always normal. At the term Π_4 a right splitting is also normal if the sequence is exact and Π_4 acts identically on Π_5.

B. Suppose a π-sequence is exact, splits normally on the right at Π_α and splits on the right at $\Pi_{\alpha+3}$. Then, as is easily seen, it also splits on the left at Π_α and if $\alpha \ge 4$, then Π_α splits into the direct product of the subgroup Im ρ_α (canonically isomorphic to $\Pi_{\alpha+1}$) and a subgroup isomorphic to $\Pi_{\alpha-1}$. For $\alpha = 1, 2, 3$, the isomorphism $\Pi_\alpha \simeq \Pi_{\alpha+1} \times \Pi_{\alpha-1}$ need not hold, as obvious examples show.

Now suppose the π-sequence is exact and splits on the left at the terms Π_α, $\Pi_{\alpha-3}$. Then, as is easily shown, it also splits on the right at Π_α, and if $\alpha \ge 4$, this splitting is normal and Π_α is the direct product of the subgroup Im ρ_α (canonically isomorphic to $\Pi_{\alpha+1}$) and a subgroup isomorphic to $\Pi_{\alpha-1}$.

C. Later we shall often meet the situation where a π-sequence is exact and splits at every third term. It follows from what has been said above that if a π-sequence is exact and splits normally on the right at terms Π_{i_0+3k} with $i_0 + 3k \ge 1$, then it also splits on the left at these terms; if it is exact and splits on the left at the terms Π_{i_0+3k}, then it splits on the right at terms Π_{i_0+3k} with $i_0 + 3k \ge 3$, and splits normally on the right at terms Π_{i_0+3k} with $i_0 + 3k \ge 4$.

D. If A is a retract of X, the homotopy sequence of (X, A) clearly splits on the left at the terms $\pi_r(X, x_0)$: a splitting homomorphism is ρ_* : $\pi_r(X, x_0) \to \pi_r(A, x_0)$ induced by any retraction ρ : $X \to A$. In particular, $\pi_r(X, x_0) \simeq \pi_r(A, x_0) \times \pi_r(X, A, x_0)$ for $r \ge 1$.

E. If (X, x_0) can be contracted to (A, x_0), that is, if id_X is x_0-homotopic to a map $h : X \to X$ with $h(X) \subset A$, then the homotopy sequence of (X, A) with base point x_0 splits on the right at the terms $\pi_r(A, x_0)$. The homomorphisms induced by $h| : X \to A$ are splitting homomorphisms. In particular, $\pi_r(A, x_0) \simeq \pi_r(X, x_0) \times \pi_{r+1}(X, A, x_0)$ for $r \geq 1$.

F. If (A, x_0) is contractible in (X, x_0) that is, if the inclusion $A \to X$ is x_0-homotopic to a constant, then the homotopy sequence of (X, A) with base point x_0 splits on the right at the terms $\pi_r(X, A, x_0)$. Splitting homomorphisms $\pi_r(A, x_0) \to \pi_{r+1}(X, A, x_0)$ are induced by the maps $\gamma_r : \mathrm{Sph}_r(A, x_0) \to \mathrm{Sph}_{r+1}(X, A, x_0)$ defined by $[\gamma_r(\phi)](t_1, \ldots, t_{r+1}) = h(\phi(t_1, \ldots, t_r), t_{r+1})[\phi \in \mathrm{Sph}_r(A, x_0)]$, where h is any x_0-homotopy $h : A \times I \to X$ from in $: A \to X$ to the constant map. Hence $\pi_{r+1}(X, A, x_0) \simeq \pi_{r+1}(X, x_0) \times \pi_r(A, x_0)$.

3.7. The Homotopy Sequence of a Triple.

Let (X, A, B) be a topological triple (that is, X is a topological space and $B \subset A \subset X$) with base point $x_0 \in B$. By 3.4, the homotopy groups $\pi_r(X, A, x_0)$, $\pi_r(X, B, x_0)$, $\pi_r(A, B, x_0)$ and homomorphisms $\mathrm{in}_* : \pi_r(A, B, x_0) \to \pi_r(X, B, x_0)$, $\mathrm{rel}_* : \pi_r(X, B, x_0) \to \pi_r(X, A, x_0)$ induced by the inclusions in $: (A, B) \to (X, B)$, rel $: (X, B) \to (X, A)$ are defined for $r \geq 1$. For $r \geq 2$ we again define a homomorphism $\partial : \pi_r(X, A, x_0) \to \pi_{r-1}(A, B, x_0)$ as the composition of the boundary homomorphism $\pi_r(X, A, x_0) \to \pi_{r-1}(A, x_0)$ and the homomorphism $\pi_{r-1}(A, x_0) \to \pi_{r-1}(A, B, x_0)$ induced by the inclusion $(A, x_0, x_0) \to (A, B, x_0)$. These three series of groups and homomorphisms can be joined into a left-infinite sequence which, like the homotopy sequence of a pair, is a π-sequence: the right

$$\cdots \overset{\partial}{\longrightarrow} \pi_2(A, B, x_0) \overset{\mathrm{in}_*}{\longrightarrow} \pi_2(X, B, x_0) \overset{\mathrm{rel}_*}{\longrightarrow} \pi_2(X, A, x_0) \overset{\partial}{\longrightarrow} \pi_1(A, B, x_0) \overset{\mathrm{in}_*}{\longrightarrow}$$
$$\longrightarrow \pi_1(X, B, x_0) \overset{\mathrm{rel}_*}{\longrightarrow} \pi_1(X, A, x_0),$$

group actions of $\pi_2(A, X, x_0)$ on the groups $\pi_r(X, A, x_0)$ and the right group actions of $\pi_2(X, B, x_0)$ on $\pi_r(A, B, x_0)$ and $\pi_r(X, B, x_0)$ are induced by the actions of $\pi_1(A, x_0)$ and $\pi_1(B, x_0)$ by means of the homomorphisms $\partial : \pi_2(X, A, x_0) \to \pi_1(A, x_0)$ and $\partial : \pi_2(X, B, x_0) \to \pi_1(B, x_0)$; the right group action of $\pi_2(X, A, x_0)$ on $\pi_1(A, B, x_0)$ is induced by the action of $\pi_1(A, x_0)$ by means of $\partial : \pi_2(X, A, x_0) \to \pi_1(A, x_0)$. This sequence is called the *homotopy sequence of the triple* (X, A, B) *with base point* x_0. It can easily be verified that it is exact. For any path $s : I \to B$ the translations $\pi_r(X, A, s(0)) \to \pi_r(X, A, s(1))$, $\pi_r(X, B, s(0)) \to \pi_r(X, B, s(1))$ and $\pi_r(A, B, s(0)) \to \pi_r(A, B, s(1))$ form an isomorphism of the homotopy sequence of (X, A, B) with base point $s(0)$ onto the homotopy sequence of (X, A, B) with base point $s(1)$. For any continuous map f of (X, A, B) with base point $x_0 \in B$ into (X', A', B') with base point $x_0' \in B'$, the homomorphisms $f_* : \pi_r(X, A, x_0) \to \pi_r(X', A', x_0')$, $f_*(X, B, x_0) \to \pi_r(X', B', x_0')$ and $(f|)_* : \pi_r(A, B, x_0) \to \pi_r(A', B', x_0')$ form a homomorphism of the homotopy sequence of the first triple into the homotopy sequence of the second triple.

Chapter 2
Bundle Techniques

§4. Bundles

4.1. General Definitions.

A. A *bundle* is a triple of the form (E, p, B), where E and B are topological spaces and $p : E \to B$ is a continuous map. E is called the *total space*, B the *base*, and p the *projection* of the bundle (E, p, B). The inverse image $p^{-1}(b)$ of a point $b \in B$ is called the *fibre of the bundle over the point b*.

A *section* of the bundle $\xi = (E, p, B)$ is a continuous map $s : B \to E$ such that $p \circ s = \mathrm{id}_B$. Two sections of ξ are *homotopic* if they can be joined by a homotopy consisting of sections, that is, a homotopy $h : B \times I \to E$ such that $p \circ h$ coincides with $\mathrm{pr}_1 : B \times I \to B$.

The *restriction of the bundle* $\xi = (E, p, B)$ *to a subspace* A of the base B is the bundle with base A, total space $p^{-1}(A)$ and projection $p|$. It is denoted by $\xi|_A$

The *product of the bundles* $\xi_1 = (E_1, p_1, B_1)$, $\xi_2 = (E_2, p_2, B_2)$ is the bundle with total space $E_1 \times E_2$, base $B_1 \times B_2$ and projection $p_1 \times p_2$. It is denoted by $\xi_1 \times \xi_2$. Its fibre over the point (b_1, b_2) is equal to the product of the fibres $p_1^{-1}(b_1) \times p_2^{-1}(b_2)$.

B. A *map of the bundle* $\xi' = (E', p', B')$ *into the bundle* $\xi = (E, p, B)$ is a pair of continuous maps $F : E' \to E$, $f : B' \to B$ such that the diagram

$$
\begin{array}{ccc}
E' & \xrightarrow{F} & E \\
{\scriptstyle p'}\downarrow & & \downarrow{\scriptstyle p} \\
B' & \xrightarrow{f} & B
\end{array}
\tag{1}
$$

is commutative.

If $\phi = (F, f)$ is such a pair, we write $\phi : \xi' \to \xi$. The total space of ξ is denoted by $E(\xi)$, the base by $B(\xi)$, the projection by $p(\xi)$, and the fibre over $b \in B$ by $F_b(\xi)$. In conformity with these notations the maps constituting a bundle map ϕ are denoted by $E(\phi)$ and $B(\phi)$.

A map $\phi : \xi' \to \xi$ is called an *isomorphism* if $E(\phi)$ and $B(\phi)$ are homeomorphisms, and an *equivalence* if in addition $B(\xi) = B(\xi')$ and $B(\phi) = \mathrm{id}_{B(\xi)}$. Bundles are called *isomorphic* (*equivalent*) if there is an isomorphism (equivalence) between them.

A map $\phi : \xi' \to \xi$ is called an *inclusion* if $E(\phi)$ and $B(\phi)$ are inclusions For example, the inclusions $p(\xi)^{-1}(A) \to E(\xi)$, $A \to B(\xi)$ constitute an inclusion of the bundle $\xi|_A$ in ξ for any subset A of $B(\xi)$.

The commutativity of diagram (1) shows that the map is a *fibre map*, that is, it maps each fibre of ξ' into a fibre of ξ. Clearly, if $p(\xi)$ is surjective, then for each fibre map $F : E(\xi') \to E(\xi)$ there is a unique map $f : B(\xi') \to B(\xi)$ such that the diagram

$$E(\xi') \xrightarrow{F} E(\xi)$$

$$p(\xi') \downarrow \qquad \downarrow p(\xi)$$

$$B(\xi') \xrightarrow{f} B(\xi)$$

is commutative. If the map $p(\xi')$ is an identification (or quotient) map then the continuity of F implies the continuity of f. Thus if ξ' is a bundle whose projection is an identification map, then for each continuous fibre map $F : E(\xi') \to E(\xi)$ there is a unique continuous map $\phi : \xi' \to \xi$ with $E(\phi) = F$.

C. Let f be a continuous map of a topological space B into the base of a bundle ξ. Then there is a bundle with base B, called the *bundle induced from ξ by f*, denoted by $f^!\xi$ (the notation $f^*(\xi)$ is also found in the literature). Its total space is the subspace of $B \times E(\xi)$ consisting of the points (k, x) with $f(b) = p(\xi)(x)$, and the projection is the restriction of $\mathrm{pr}_1 : B \times E(\xi) \to B$.

It is clear that the restriction of $\mathrm{pr}_2 : B \times E(\xi) \to E(\xi)$ is a map $E(f^!\xi) \to E(\xi)$ that maps the fibre of $f^!\xi$ over any $b \in B(f^!(\xi)) = B$ homeomorphically onto the fibre of ξ over the point $f(b)$. This map, together with f, defines a map of $f^!\xi$ into ξ, called the *adjoint* of f.

The space $E(f^!\xi)$ can usefully be pictured as follows: over each point $b \in B$ we place the fibre of ξ over the point $f(b)$; $E(f^!\xi)$ consists of these fibres. If several points of B are mapped to the same point of $B(\xi)$, then the fibre of ξ over this point is taken correspondingly many times.

If f is a homeomorphism, then the adjoint map $f^!\xi \to \xi$ is a homeomorphism, and if $f = \mathrm{id}_{B(\xi)}$ it is an equivalence. If f is an inclusion (and hence $B \subset B(\xi)$), then the adjoint map $f^!\xi \to \xi$ establishes an equivalence between $f^!\xi$ and $\xi|_B$. For any continuous maps $f : B \to B(\xi)$ and $g : B' \to B$, the bundles $(f \circ g)^!\xi$ and $g^!(f^!\xi)$ are clearly canonically equivalent.

If $\phi = (F, f)$ is a map of ξ' to ξ, then the map $E(\xi') \to E(f^!\xi)$ defined by $x \mapsto (p(\xi')(x), F(x))$ is continuous. This map together with $\mathrm{id}_{B(\xi')}$ defines a map of ξ' into $f^!\xi$; the composition of this map with the adjoint of f gives ϕ. Thus any map $\phi = (F, f)$ of a bundle ξ' to ξ can be factored through the induced bundle $f^!\xi$.

4.2. Locally Trivial Bundles.

A. An obvious example of a bundle with given base B and fibres homeomorphic to a given space F is the *standard trivial bundle* $(B \times F, \mathrm{pr}_1, B)$. Its fibres are the fibres $b \times F$ of the product $B \times F$, which are of course canonically homeomorphic to F.

B. Note that the sections $B \to B \times F$ of the standard trivial bundle with base B and fibre F are in one-one correspondence with the continuous maps $B \to F$: a map $f : B \to F$ corresponds to the section $B \to B \times F : b \mapsto (b, f(b))$, and a section $s : B \to B \times F$ corresponds to the map $\mathrm{pr}_2 \circ s$.

C. A bundle ξ is called *trivial* or, more precisely, *topologically trivial*, if it is equivalent to a standard trivial bundle. Any equivalence between a standard trivial bundle and the bundle ξ is called a *trivialization* of ξ. To construct a trivialization of a bundle (E, p, B) it is sufficient to find a bundle (E, p', F) such that each of its fibres intersects each fibre of the original bundle in a single point: then $E \to B \times F : x \mapsto (p(x), p'(x))$ is a trivialization. Thus to specify a trivialization of a bundle is essentially to specify homeomorphisms between its fibres and a certain space, so that the homeomorphisms depend continuously on the fibre.

D. A bundle ξ is called *locally trivial*, or, more precisely, *locally topologically trivial*, if each point of its base has a neighbourhood U such that the restriction $\xi|_U$ is trivial.

A trivial bundle is obviously locally trivial. Perhaps the most popular example of a non-trivial but locally trivial bundle is the projection of a Möbius band onto its centre line (a circle). See Fig. 2.

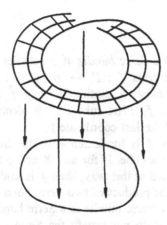

Fig. 2

The fibre of this bundle is an interval. Another famous example of a bundle is the *Hopf bundle* with total space S^3, base S^2, and projection defined by $(z_1, z_2) \mapsto (z_1 : z_2)$ (here S^3 is regarded as the subset $|z_1|^2 + |z_2|^2 = 1$ in $\mathbb{C}^2$, and S^2 as $\mathbb{C}P^1$). The Hopf bundle is also non-trivial but locally trivial. It is the first in a series of similarly defined bundles also called *Hopf bundles*: $S^{2n+1} \to \mathbb{C}P^n$, $S^{4n+3} \to \mathbb{H}P^n$, and $S^{15} \to \mathrm{Ca}P^1$. The most important of these, along with $S^3 \to \mathbb{C}P^1 = S^2$, are $S^7 \to \mathbb{H}P^1 = S^4$ and $S^{15} \to \mathrm{Ca}P^1 = S^8$.

E. A locally trivial bundle is a *covering in the wide sense* if its fibre is a discrete space. Non-trivial bundles of this type include, for example, the bundle with total space $\mathbb{R}$, base S^1 and projection $x \mapsto e^{2\pi i x}$, and the bundle with total space and base S^1 and projection $z \mapsto z^m$, where m is any non-zero integer.

F. Since the projection of a product space onto a factor is an open map, the projection of a trivial bundle, and hence of a locally trivial bundle, is an open map. An obvious verification shows that the product of two trivial bundles is a trivial bundle, and the product of two locally trivial bundles is locally trivial. It is also clear that the bundle induced by a trivial bundle is trivial, and that induced by a locally trivial bundle is locally trivial. Also, if $f : B \to B(\xi)$ is a constant map, then the induced bundle $f^!\xi$ is trivial for any ξ.

G. Like the fibres of a standard trivial bundle, the fibres of a trivial bundle are homeomorphic to each other, but in the non-standard case the homeomorphism is no longer canonical (see above). It is easy to show that the fibres of a locally trivial bundle are also homeomorphic to each other if the base is connected. On the other hand, the example of the bundle $((B \times F) \coprod (B' \times F'), \mathrm{pr}_1 \coprod \mathrm{pr}_2, B \coprod B')$, where B, F, B', F' are arbitrary topological spaces, shows that the fibres of a locally trivial bundle over points of different components of the base need not be homeomorphic.

H. It is easy to show that if X and Y are closed smooth manifolds then every submersion $p : X \to Y$ (that is, a map whose differential is an epimorphism at every point of X) defines a locally trivial bundle.

4.3. Serre Bundles.

A. A bundle ξ is called a *Serre bundle* if it satisfies the *Serre condition*: for any $r \geq 1$ and continuous maps $f : I^r \to B(\xi)$, $f_0 : I^{r-1} \to E(\xi)$ such that $p(\xi) \circ f_0 = f|_{I^{r-1}}$, there exists a continuous map $\tilde{f} : I^r \to E(\xi)$ such that $p(\xi) \circ \tilde{f} = f$ and $\tilde{f}|_{I^{r-1}} = f_0$. (The cube I^{r-1} is identified with the face of I^r defined by the vanishing of the last coordinate.)

The relation $p(\xi) \circ \tilde{f} = f$ is fundamental in the theory of bundles, and the space mapped is not always a cube. If for any X and ξ the maps $\tilde{f} : X \to E(\xi)$ and $f : X \to B(\xi)$ are related in this way, then f is said to *cover* f.

It is easy to check that the product of two Serre bundles is a Serre bundle, and that the bundle induced by a Serre bundle is a Serre bundle.

Examples of bundles that do not satisfy the Serre condition are the bundles (I, p, I) with $p(x) = x/2$, and with $p(x) = 4x(1-x)$. In the first bundle, there is no map $\tilde{f}$ with $p \circ \tilde{f} = f$ and $\tilde{f}|_{I^{r-1}} = f_0$ if $r = 1$, $f = \mathrm{id}_I$, $f_0(0) = 0$; and the same holds for the second bundle if $r = 2$, f is defined by $f(x, y) = 4x(1 - x)(1 - y)$ and $f_0 = \mathrm{id}_I$. Note that the first bundle has both empty and non-empty fibres, while in the second bundle one fibre is connected, and the rest are not connected. As will be seen later (see 4.3), such peculiarities of a bundle are incompatible with the Serre condition when the base is connected.

B. It is easy to show that the Serre condition is local in the following sense. If each point of the base of ξ has a neighbourhood U such that the restriction $\xi|_U$ is a Serre bundle, then ξ is a Serre bundle.

It is easy to see that trivial bundles satisfy the Serre condition, and hence so does every locally trivial bundle.

The following example shows that a Serre bundle need not be locally trivial. Let T be the triangle in $\mathbb{R}^2$ with vertices $(0, 0)$, $(0, 1)$, $(1, 0)$ and $p : T \to I$ the map defined by $p(x, y) = x$ (see Fig. 3). Then the bundle (T, p, I) is not locally trivial (the fibres over the points 0 and 1 are not homeomorphic, although the base is connected). However, it obviously satisfies Serre's condition. This example also shows that a Serre bundle with connected base may have non-homeomorphic fibres. In fact, the fibres of a Serre bundle with connected base need not even be homotopy equivalent (they are only weakly homotopy equivalent; see 9.4 below).

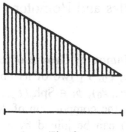

Fig. 3

C. A bundle ξ satisfies the *strong Serre condition* or *Hurewicz condition* if for any topological space X, any continuous map $f : X \to E(\xi)$, and any homotopy F of the map $p(\xi) \circ f$, there exists a homotopy of f that covers F.

If the arbitrary space X in this condition is replaced by a cube, we obtain the usual Serre condition. At first sight the distance between the Serre condition and the strong Serre condition may seem much greater than it really is. In fact if the space X in the strong Serre condition is restricted to be a cellular space (see §7), then the two conditions become equivalent (see 9.6.**B**).

It can be shown that a bundle with a connected base satisfying the strong Serre condition has fibres that are pairwise homotopy equivalent.

4.4. Bundles of Spaces of Maps.

A. Let (X, A) be a Borsuk pair and Y a topological space. Consider the bundle $(C(X, Y), C(\text{in}, \text{id}), C(A, Y))$, in which the projection $C(\text{in}, \text{id})$ takes a map $f :$ $X \to Y$ to its restriction to A. It is easy to show that it satisfies the strong Serre condition if X is locally compact and regular.

In particular, the space of paths $C(I, X)$ in X is a bundle over $C(0 \cup 1, X) =$ $X \times X$. Since the fibres of a bundle with connected base satisfying the strong Serre condition are homotopy equivalent, it follows that if points x_0, x_1, y_0, y_1 belong to the same component of X, then the spaces of paths $C(I, 0, 1; X, x_0, x_1)$ and $C(I, 0, 1; X, y_0, y_1)$, joining x_0 to x_1 and y_0 to y_1 respectively, are homotopy equivalent.

B. Any continuous map $f : X \to Y$ is homotopy equivalent to the projection of a strong Serre bundle. For the total space of this bundle we may take the subspace E of $X \times C(I, Y)$ consisting of the pairs (x, s) with $s(0) = f(x)$; the base is Y and the projection is the map $(x, s) \mapsto s(1)$. It can easily be proved that this is a strong Serre bundle. Mutually homotopy inverse homotopy equivalences $X \to E$ and $E \to X$ can be defined by $x \mapsto (x, u_x)$ and $(x, s) \mapsto x$, where u_x is the constant path in Y with origin $f(x)$ In terms of Eckmann-Hilton duality (discussed in 2.12), the construction of this bundle is dual to the construction of the mapping cylinder. The space E is therefore sometimes called the *cocylinder* of f.

§5. Bundles and Homotopy Groups

5.1. The Local System of Homotopy Groups of the Fibres of a Serre Bundle.

A. Let ξ be a Serre bundle, F_0, F_1 two of its fibres, and x_0, x_1 points in F_0, F_1. Two spheroids $\phi_0 \in \text{Sph}_r(F_0, x_0)$, $\phi_1 \in \text{Sph}_r(F_1, x_1)$ are called *fibre homotopic* if the spheroids resulting from the composition of ϕ_0 and ϕ_1 with the inclusions $F_0 \hookrightarrow E(\xi)$ and $F_1 \hookrightarrow E(\xi)$ can be joined by a free homotopy consisting of spheroids of $E(\xi)$, mapping I^r into fibres of ξ; in other words, there exists a map $h : I^r \times I \to E(\xi)$, constant on each of the sets $\partial I^r \times t$, $t \in I$, and such that $h(y, 0) = \phi_0(y)$, $h(y, 1) = \phi_1(y)$, $y \in I^r$, and the map $p(\xi) \circ h$ is constant on each of the sets $I^r \times t$, $t \in I$. We say that h is a *fibre homotopy between* ϕ_0 and ϕ_1 *along the path* $t \mapsto h(\partial I^r \times t)$.

It is easy to show that every spheroid of F_0 with origin x_0 admits a fibre homotopy along any path in $E(\xi)$ starting at x_0. Fibre homotopies of homotopic (in the fibre) spheroids along homotopic paths in $E(\xi)$ always lead to homotopic (in the fibre) spheroids. Fibre homotopies of freely homotopic (in the fibre) spheroids along paths covering homotopic paths in $B(\xi)$ lead to freely homotopic (in the fibre) spheroids. Thus, fibre homotopies along the path $s : I \to E(\xi)$ determine (for any $r \geq 0$) a mapping $T_s : \pi_r(F_0, s(0)) \to \pi_r(F_1, s(1))$, where $F_i = F_{p(\xi)(s(i))}(\xi)$ $(= p(\xi)^{-1}(p(\xi)s(i)))$, i. These maps are clearly homomorphisms and have properties (i)–(iii) of 3.2, so that on $E(\xi)$ we have a local system $(E(\xi), \{\pi_r(F_{p(\xi)(x)}(\xi), x)\}, \{T_s\})$, which is a local system of groups for $r \geq 1$, and of sets with identity for $r = 0$. It is called the *upper local system of r-dimensional homotopy groups of the fibres of the bundle* ξ. In particular, for any $x \in E(\xi)$ and $r \geq 1$, there is a natural right group action of $\pi_r(E(\xi), x)$ on $\pi_r(F_{p(\xi)(x)}(\xi), x)$.

The restriction of this local system to any fibre of ξ clearly coincides with the local system of r-dimensional homotopy groups of this fibre, and the inclusion homomorphisms $\pi_r(F_{p(\xi)(x)}(\xi), x) \to \pi_r(E(\xi), x)$ together with $\text{id}_{E(\xi)}$ form a homomorphism of this local system into the local system of r-dimensional homotopy groups of $E(\xi)$.

B. If the fibres of ξ are r-simple, then for each $b \in B(\xi)$ all the groups $\pi_r(F_b(\xi), x)$ with $x \in p(\xi)^{-1}(b)$ can be identified with the homotopy group $\pi_r(F_b(\xi))$ (see 3.3). In this case, given a path $s : I \to B(\xi)$, we define the map $T_s : \pi_r(F_{s(0)}(\xi)) \to \pi_r(F_{s(1)}(\xi))$ as the translation $T_{\tilde{s}} : \pi_r(F_{s(0)}(\xi), \tilde{s}(0)) \to \pi_r(F_{s(1)}(\xi), \tilde{s}(1))$ along any path $\tilde{s} : I \to E(\xi)$ covering s. From what was said above it is clear that T_s does not depend on the choice of $\tilde{s}$. The maps T_s are obviously homomorphisms and have properties (i)–(iii) of 3.2, so that on $B(\xi)$ we have a local system $(B(\xi), \{\pi_r(F_b(\xi))\}, \{T_s\})$, which is a local system of groups for $r \geq 1$, and of one-point sets for $r = 0$. It is called the *lower local system of r-dimensional homotopy groups of the fibres of* ξ. In particular, for $r \geq 1$ and each point $b \in B(\xi)$ there is a natural right group action of $\pi_1(B(\xi), b)$ on $\pi_r(F_b(\xi))$.

The local system that this system induces on $E(\xi)$ by means of $p(\xi)$ is clearly none other than the upper local system of r-dimensional homotopy groups of the fibres of ξ.

If ϕ is a map from the Serre bundle ξ to a Serre bundle ξ_1, then the homomorphisms $(E(\phi)|)_* : \pi_r(F_{p(\xi)(x)}(\xi), x) \to \pi_r(F_{p(\xi_1)E(\phi)(x)}(\xi_1), E(\phi)(x))$, $x \in E(\xi)$, together with $E(\phi)$ form a homomorphism of the upper local system of r-dimensional homotopy groups of the fibres of ξ into the corresponding local system for ξ_1. If the fibres of ξ and ξ_1 are r-simple, then the homomorphisms $(E(\phi)|)_* : \pi_r(F_b(\xi)) \to \pi_r(F_{B(\phi)(b)}(\xi_1))$, $b \in B(\xi)$, together with $B(\phi)$ form a homomorphism of the lower local system of r-dimensional homotopy groups of the fibres of ξ into the corresponding system for ξ_1.

5.2. The Homotopy Sequence of a Serre Bundle.

A. Let $\xi = (E, p, B)$ be a Serre bundle, $x_0 \in E$, $b_0 = p(x_0)$, $F_0 = p^{-1}(x_0)$. An immediate application of the Serre condition shows that if A is a subset of the base B containing b_0, then the homomorphism

$$p_* : \pi_r(E, p^{-1}(A), x_0) \to \pi_r(B, A, b_0)$$

and in particular the homomorphism

$$p_* : \pi_r(E, F_0, x_0) \to \pi_r(B, b_0)$$

are isomorphisms for any $r \geq 1$.

B. This allows us to transform the homotopy sequence of the pair (E, F_0) by replacing for each $r \geq 1$ the group $\pi_r(E, F_0, x_0)$ by $\pi_r(B, b_0)$, the homomorphism $\mathrm{rel}_* : \pi_r(E, x_0) \to \pi_r(E, F_0, x_0)$ by its composition with the isomorphism $p_* : \pi_r(E, F_0, x_0) \to \pi_r(B, b_0)$, and the homomorphism $\partial : \pi_r(E, F_0, x_0) \to \pi_{r-1}(F_0, x_0)$ by the composition $\Delta = \partial \circ p_*^{-1} : \pi_r(B, b_0) \to \pi_{r-1}(F_0, x_0)$. Note that $p_* \circ \mathrm{rel}_* : \pi_r(E, x_0) \to \pi_r(B, b_0)$ is just the homomorphism induced by the projection $(E, x_0) \to (B, b_0)$. If we add the group $\pi_0(B, b_0)$ and the homomorphism $p_* : \pi_0(E, b_0) \to \pi_0(B, b_0)$ to the right-hand end of the resulting sequence we arrive at the sequence

$$\ldots \longrightarrow \pi_2(F_0, x_0) \xrightarrow{\mathrm{in}_*} \pi_2(E, x_0) \xrightarrow{p_*} \pi_2(B, b_0) \xrightarrow{\Delta} \pi_1(F_0, x_0) \xrightarrow{\mathrm{in}_*} \pi_1(E, x_0) \longrightarrow$$
$$\xrightarrow{p_*} \pi_1(B, b_0) \xrightarrow{\Delta} \pi_0(F_0, x_0) \xrightarrow{\mathrm{in}_*} \pi_0(E, x_0) \xrightarrow{p_*} \pi_0(B, b_0). \qquad (1)$$

C. There is a group action of $\pi_1(B, b_0)$ on the set $\pi_0(F_0, x_0)$ defined as follows: an element σ of $\pi_1(B, b_0)$ acts on a component C of F_0 sending it to the component $C\sigma$ of F_0 containing the origins of paths ending in C and covering the loops of σ. This action is compatible with the action of the fundamental group of E on the homotopy groups of the fibres (see 5.1) in the sense that $Cp_*(\sigma) = T_\sigma C$ for any component C of F_0 and any $\sigma \in \pi_1(E, x_0)$ and any $x_0 \in F_0$.

Also, according to 3.6 and 5.1, right group actions are defined of $\pi_1(B, b_0)$ on $\pi_r(F_0, x_0)$ and of $\pi_1(E, x_0)$ on $\pi_r(E, x_0)$ and $\pi_r(F_0, x_0)$. The homomorphisms in_*, p_* and Δ are compatible with these actions, as required by the definition of a π-sequence (see 3.5), so that the sequence (1) above is a π-sequence. It is called the *homotopy sequence of the bundle* ξ with base point x_0. This sequence too is exact, as can easily be shown from the exactness of the homotopy sequence of the pair (E, F_0).

D. It is also easy to see that for any map of ξ into another Serre bundle $\xi' = (E', p', B')$, the vertical homomorphisms

$$\cdots \pi_r(F_0, x_0) \xrightarrow{\text{in}_*} \pi_r(E, x_0) \xrightarrow{p_*} \pi_r(B, b_0) \xrightarrow{\Delta} \pi_{r-1}(F_0, x_0) \longrightarrow \cdots$$

$$\downarrow (E(f)|)_* \qquad\qquad \downarrow E(f)_* \qquad\qquad \downarrow B(f)_* \qquad\qquad \downarrow (E(f)|)_*$$

$$\cdots \pi_r(F_0', x_0') \xrightarrow{\text{in}_*} \pi_r(E', x_0') \xrightarrow{p_*} \pi_r(B', b_0') \xrightarrow{\Delta} \pi_{r-1}(F_0', x_0') \longrightarrow \cdots$$

where $x_0' = E(f)(x_0)$, $b_0' = B(f)(b_0)$, $F_0' = (p')^{-1}(b_0')$, constitute a homomorphism of the first sequence into the second. (This means that the homotopy sequence of a Serre bundle is a functor from the category of Serre bundles to the category of π-sequences.)

5.3. Important Special Cases.[4]

A. As before, let $\xi = (E, p, B)$ be a Serre bundle, $x_0 \in E$, $b_0 = p(x_0)$, $F_0 = p^{-1}(b_0)$. If E is ∞-connected then all the homomorphisms $\Delta : \pi_r(B, b_0) \to \pi_{r-1}(F_0, x_0)$ are isomorphisms. If E is k-connected with $k < \infty$, then for $r \leq k$ the homomorphisms $\Delta : \pi_r(B, b_0) \to \pi_{r-1}(F_0, x_0)$ are isomorphisms, and $\Delta : \pi_{k+1}(B, b_0) \to \pi_k(F_0, x_0)$ is an epimorphism. In both cases the converse is true when E is connected.

B. If B is ∞-connected then all the homomorphisms $\text{in}_* : \pi_r(F_0, x_0) \to \pi_r(E, x_0)$ are isomorphisms. If B is k-connected with $k < \infty$, then for $r \leq k$ the homomorphisms $\text{in}_* : \pi_r(F_0, x_0) \to \pi_r(E, x_0)$ are isomorphisms, and $\text{in}_* : \pi_{k+1}(F_0, x_0) \to \pi_{k+1}(B, b_0)$ is an epimorphism. In both cases the converse is true when B is connected.

C. If F is ∞-connected then all the homomorphisms $p_* : \pi_r(E, x_0) \to \pi_r(B, b_0)$ are isomorphisms. If F_0 is k-connected with $k < \infty$, then for $r \leq k$ the homomorphisms $p_* : \pi_r(E, x_0) \to \pi_r(B, b_0)$ are isomorphisms, and $p_* : \pi_{k+1}(E, x_0) \to \pi_{k+1}(B, b_0)$ is an epimorphism. In both cases the converse is true without any additional assumptions.

D. If ξ has a section taking b_0 to x_0, then the sequence of splits on the right at the term $\pi_r(E, x_0)$. The homomorphism $s_* : \pi_r(B, b_0) \to \pi_r(E, x_0)$ induced by any section $s : (B, b_0) \to (E, x_0)$ is a splitting homomorphism.

E. If F_0 is a retract of E, then the sequence of ξ splits on the left at the term $\pi_r(E, x_0)$. The homomorphism $\rho_* : \pi_r(E, x_0) \to \pi_r(F_0, x_0)$ induced by any retraction $\rho : E \to F_0$ is a splitting homomorphism.

F. If the inclusion in : $F_0 \to E$ is x_0-homotopic to a constant, then the sequence of ξ splits on the right at the terms $\pi_r(B, b_0)$. For any x_0-homotopy $h : F_0 \times I \to E$ between in : $F \to E$ and the constant map, define maps $\gamma_r : \text{Sph}_r(F_0, x_0) \to \text{Sph}_{r+1}(B, b_0)$ by the formula

$$[\gamma_r(\phi)](t_1, \ldots, t_{r+1}) = p \circ h(\phi(t_1, \ldots, t_r), t_{r+1}), \quad [\phi \in \text{Sph}_r(F_0, x_0)].$$

Then the induced homomorphisms $\pi_r(F_0, x_0) \to \pi_{r+1}(B, b_0)$ are splitting homomorphisms.

---

[4] For applications of these results see 10.2, 10.5, 10.7, and 10.8.

G. If the projection p is x_0-homotopic to the constant map, then the sequence of ξ splits on the left at the terms $\pi_r(F_0, x_0)$. For any x_0-homotopy $h : E \times I \to B$ between p and the constant map, define maps $\gamma_r : \mathrm{Sph}_r(F_0, x_0) \to \mathrm{Sph}_{r+1}(B, b_0)$ by the formula

$$[\gamma_r(\phi)](t_1, \ldots, t_{r+1}) = h(\phi(t_1, \ldots, t_r), t_{r+1}).$$

Then the induced homomorphisms $\pi_r(F_0, x_0) \to \pi_{r+1}(B, b_0)$ are splitting homomorphisms.

H. If ξ is a covering in the wide sense then $p_* : \pi_r(E, x_0) \to \pi_r(B, b_0)$ is an isomorphism for $r \geq 2$, and a monomorphism for $r = 1$. If in addition E is connected, then the map $\Delta / : \pi_1(B, b_0)/\mathrm{Im}\, p_* \to F_0$, induced by $\Delta : \pi_1(B, b_0) \to \pi_0(F_0, x_0) = F_0$, is invertible.

I. Suppose that f is a map of the Serre bundle $\xi = (E, p, B)$ to the Serre bundle $\xi' = (E', p', B')$, and let $x_0 \in E$, $b_0 = p(x_0)$, $F_0 = p^{-1}(b_0)$, $x_0' = E(f)(x_0)$, $b_0' = B(f)(b_0)$, $F_0' = (p')^{-1}(b_0')$. The following results follow from the functorial property of the homotopy sequence of a Serre bundle and the 5-lemma:

(i) if all the homomorphisms $B(f)_* : \pi_r(B, b_0) \to \pi_r(B', b_0')$ for $r \geq 1$, and all the homomorphisms $(E(f)|)_* : \pi_r(F_0, x_0) \to \pi_r(F_0', x_0')$ for $r \geq 0$ are isomorphisms, then all the homomorphisms $E(f)_* : \pi_r(E, x_0) \to \pi_r(E', x_0')$ for $r \geq 1$ are isomorphisms;

(ii) if all the homomorphisms $E(f)_* : \pi_r(E, x_0) \to \pi_r(E', x_0')$ for $r \geq 0$, and all the homomorphisms $(E(f)|)_* : \pi_r(F_0, x_0) \to \pi_r(F_0', x_0')$ for $r \geq 0$ are isomorphisms, then all the homomorphisms $B(f)_* : \pi_r(B, b_0) \to \pi_r(B', b_0')$ for $r \geq 0$ are isomorphisms;

(iii) if all the homomorphisms $B(f)_* : \pi_r(B, b_0) \to \pi_r(B', b_0')$ for $r \geq 0$ and all the homomorphisms $E(f)_* : \pi_r(E, x_0) \to \pi_r(E', x_0')$ for $r \geq 0$ are isomorphisms, then all the homomorphisms $(E(f)|)_* : \pi_r(F_0, x_0) \to \pi_r(F_0', x_0')$ are isomorphisms for $r \geq 1$, and $(E(f)|)_* : \pi_0(F_0, x_0) \to \pi_0(F_0', x_0')$ is an epimorphism with zero kernel, and is an isomorphism when all the homomorphisms $E(f)_* : \pi_r(E, x) \to \pi_r(E', E(f)(x))$ with $x \in F_0$ are epimorphisms.

§6. The Theory of Coverings

6.1. Coverings.

A. Recall from 4.2 that a covering in the wide sense is a locally trivial bundle with discrete fibre. The total space of such a bundle is usually called a *covering space*. Clearly each point of a covering space has a neighbourhood that is mapped homeomorphically by the projection onto its image in the base.

B. A covering in the wide sense is said to be a *covering in the narrow sense* or simply a *covering* if the covering space and the base are connected and non-empty. All the fibres of a covering clearly have the same cardinality, called the *number of sheets of the covering*. The projection of a one-sheeted covering is obviously a homeomorphism, but a covering (in the narrow sense) with more than one sheet cannot be trivial.

C. The examples in 4.2 of coverings in the wide sense (the bundle with projections $S^1 \to S^1 : z \mapsto z^m$, $m \neq 0$, and $\mathbb{R} \to S^1 : x \mapsto e^{2\pi i x}$) are in fact coverings in the narrow sense. The bundle with total space $G_+(n, k)$ (= the space of oriented k-dimensional subspaces of $\mathbb{R}^{n+k}$) and base $G(n, k)$ (= the space of k-dimensional subspaces of $\mathbb{R}^{n+k}$), with the map that "forgets" orientation as the projection, is a two-sheeted covering for $k \neq 0$. A special case of this is the bundle $(S^n, \text{pr}, \mathbb{R}P^n)$, $n \geq 1$. By taking products further examples can be obtained from these. In this way we get a covering with covering space and base the torus $S^1 \times S^1$, and projection $(z, w) \mapsto (z^p, z^q)$ and a covering with a torus as base covered by $\mathbb{R}^2$ with projection $(x, y) \mapsto (e^{2\pi i x}, e^{2\pi i y})$.

6.2. The Group of a Covering.

The principal result in this section is an effective method of enumerating the coverings of a connected non-pathological space, together with a condition for two given coverings to be equivalent. The tool used is the fundamental group, and its application depends on the following two theorems, in which $\xi = (E, p, B)$ is a given covering with base point $x \in E$.

(i) each path in B with origin $p(x)$ is covered by exactly one path (in E) with origin x;

(ii) if two paths in B with origin $p(x)$ are homotopic, then the covering paths (in E) are also homotopic, and in particular, have the same end-point.

The existence of the covering path in (i) is a consequence of the Serre property of a covering, and the uniqueness follows from the following general theorem: if two continuous maps $f, g : X \to E$ of a connected space X into a covering space coincide at some point and $p \circ f = p \circ g$, then $f = g$. However, theorems (i) and (ii) can easily be proved by themselves.

The theory of coverings is based on the concept of a covering group. Recall that the homomorphism $p_* : \pi_1(E, x) \to \pi_1(B, p(x))$ is a monomorphism for any covering $\xi = (E, p, B)$ with base point $x \in E$ (see 5.3). (This also follows directly from the theorems just stated on covering paths.) The image of this monomorphism is called the *covering group of ξ with base point x*, and is denoted by $\text{gr}\,\xi(x)$. It is easy to see how this group depends on x. If $p(x_1) = p(x_0)$ and s is a path in E from x_0 to x_1, then $\text{gr}\,\xi(x_1) = \sigma[\text{gr}\,\xi(x_0)]\sigma^{-1}$, where σ is the class of the loop $p \circ s$. In particular, if $p(x_1) = p(x_0)$, then $\text{gr}\,\xi(x_0)$ and $\text{gr}\,\xi(x_1)$ are conjugate in $\pi_1(B, p(x_0))$. The converse is also true: the groups $\text{gr}\,\xi(x)$ with $x \in p^{-1}(p(x_0))$ consist of all the subgroups conjugate to $\text{gr}\,\xi(x_0)$ in $\pi_1(B, p(x_0))$.

6.3. Hierarchies of Coverings.

A. A covering $\xi = (E, p, B)$ with base point $x_0 \in E$ is said to be *subordinate* to a covering $\xi' = (E', p', B)$ with base point $x_0' \in E'$ and the same base B, if there is a map $\phi : \xi' \to \xi$ such that $B(\phi) = \text{id}_B$ and $E(\phi)(x_0') = x_0$. In this case the map ϕ is called a *subordination*. If $\phi : \xi' \to \xi$ is a subordination then clearly $(E', E(\phi), B)$ is a covering. By using the uniqueness of the covering path it is easy to show that if a subordination exists it is unique. It follows from this that if two coverings with base points are each subordinate to the other, then the subordinations are mutually inverse equivalences.

B. Let $\xi = (E, p, b)$ and $\xi' = (E', p', B)$ be two coverings with the same base B, and base points $x_0 \in E$, $x_0' \in E'$ such that $p'(x_0') = p(x_0)$. Then, if ξ is subordinate to ξ', clearly $\mathrm{gr}\,\xi'(x_0') \subset \mathrm{gr}\,\xi(x_0)$; if $\mathrm{gr}\,\xi'(x_0) \subset \mathrm{gr}\,\xi(x_0)$ and B is locally connected, then ξ is subordinate to ξ'. Under this subordination a point $x' \in E'$ is the image of the common end-point of all paths in E that begin at x_0 and cover a path of the form $p' \circ u'$, where u' is a path in E' joining x_0' to x'.

C. From what has been said it is clear that the coverings $\xi = (E, p, B)$, $\xi' = (E', p', B)$ with the same locally connected base are equivalent iff for points $x_0 \in E$ and $x_0' \in E'$ with $p(x_0) = p'(x_0')$, the groups $\mathrm{gr}\,\xi(x_0)$ and $\mathrm{gr}\,\xi'(x_0')$ are conjugate in $\pi_1(B, p(x_0))$.

6.4. The Existence of Coverings.

A. A topological space X is called *semilocally simply connected* if each point $x \in X$ has a neighbourhood U such that the homomorphism $\mathrm{in}_* : \pi_1(U, x) \to \pi_1(X, x)$ is trivial.

The class of semilocally simply connected spaces obviously contains all simply connected spaces and all locally contractible spaces (the latter include all locally Euclidean spaces and all CNRS spaces (see 2.10)). It is not hard to show that a space that has a simply connected covering is semilocally simply connected. The significance of the concept of semilocal simple connection is shown by the following theorem.

B. If B is connected, locally connected and semilocally simply connected, then for each $b_0 \in B$ and each subgroup π of $\pi_1(B, b_0)$ there is a covering $\xi = (E, p, B)$ with base B and base point $x_0 \in E$ such that $p(x_0) = b_0$ and $\mathrm{gr}\,\xi(x_0) = \pi$. Such a covering can be obtained in the following way: as the space E we take with an appropriate topology the quotient set of the set of paths $C(I, 0; B, b_0)$ in B with origin b_0, where two paths s_1, s_2 are identified if they have the same end-point and the class of the loop $s_1 s_2^{-1}$ is in π; the projection $p : E \to B$ maps each point of E to the common end-point of the paths that represent it.

This theorem completes the classification of coverings with a fixed base. With the previous results it establishes a one-one correspondence between classes of equivalent coverings over a connected, locally connected and semilocally simply connected base B with base point b_0 and the classes of conjugate subgroups of $\pi_1(B, b_0)$. It takes the hierarchy of coverings into the usual set-theoretic hierarchy of subgroups. The trivial subgroup corresponds to a covering with a simply connected covering space. Since this latter covering is subordinate to any other covering, it is called the *universal covering*. (Warning: this meaning of universal should not be confused with that in the theory of Steenrod bundles. However, universal Steenrod bundles include the universal coverings: they are the universal principal bundles corresponding to discrete groups; see Fuchs (1971), Chap. 4, §4.)

6.5. Automorphisms of a Covering.

Like the automorphisms (self-equivalences) of any bundle ξ, the automorphisms of a covering ξ form a group Aut. An automorphism ϕ of the covering $\xi = (E, p, B)$ is uniquely determined by the image $E(\phi)(x_0)$ of an arbitrary point $x_0 \in E$. For ϕ can be regarded as a subordination of ξ with base point $E(\phi)(x_0)$ to ξ with base point x_0, and subordinations are unique (see 6.3). The images of x_0 under automorphisms of ξ of course belong to the fibre $p^{-1}(p(x_0))$ but they do not necessarily fill out this fibre. By using the homotopy sequence of ξ, the fibre $p^{-1}(p(x_0)) = \pi_0(p^{-1}(p(x_0)), x_0)$ can be identified with the set of right cosets of $\operatorname{gr}\xi(x_0) = p_*\pi_1(E, x_0)$ in $\pi_1(B, p(x_0))$. Under this identification the image of the orbits of x_0 (under the action of Autξ) is contained in the subset of the set of cosets $\pi_1(B, p(x_0))/\operatorname{gr}\xi(x_0)$ that inherits a natural group structure from $\pi_1(B, p(x_0))$, namely, the quotient of the normalizer of $\operatorname{gr}\xi(x_0)$ in $\pi_1(B, p(x_0))$ by $\operatorname{gr}\xi(x_0)$. (Recall that the normalizer $\operatorname{Nr}(H)$ of a subgroup H of a given group G is the set of $g \in G$ with $gHg^{-1} = H$; it is a subgroup of G containing H as a normal subgroup, so that $\operatorname{Nr}(H)/H$ is a group.) Thus there is an injective map of Autξ into the group $\operatorname{Nr}(\operatorname{gr}\xi(x_0))/\operatorname{gr}\xi(x_0)$. It is not difficult to verify that this map is an antihomomorphism, and that it is bijective if the base B is locally connected. Hence for a locally connected base, Autξ is anti-isomorphic to $\operatorname{Nr}(\operatorname{gr}\xi(x_0))/\operatorname{gr}\xi(x_0)$.

6.6. Regular Coverings.

A covering $\xi = (E, p, B)$ is *regular* if for some $x_0 \in F$, $\operatorname{gr}\xi(x_0)$ ($= p_*\pi_1(E, x_0)$) is a normal subgroup of $\rho_1(B, p(x_0))$. It follows from 6.2 that in this case $\operatorname{gr}\xi(x)$ is a normal subgroup of $\pi_1(B, p(x))$ for any $x \in E$.

If ξ is regular, then for any $b \in B$, the groups $\operatorname{gr}\xi(x)$ are the same for any $x \in p^{-1}(b)$.

Note that all two-sheeted coverings are regular (a subgroup of index 2 is always normal). Further examples of regular coverings are $(\mathbb{R}, x \mapsto e^{2\pi i x}, S^1)$ and $(S^1, x \mapsto x^m, S^1)$. An example of a non-regular covering is shown in Fig. 4 (where the two points marked A are identified, as are the two points marked B). The following easily proved criterion for regularity shows that this example is not regular.

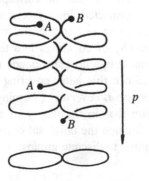

Fig. 4

A covering $\xi = (E, p, B)$ is regular iff for some point $x \in E$, any loop s : $I \to E$ with origin x_0, and any $x \in p^{-1}(p(x_0))$, the path with origin x covering the loop $p \circ s$ is closed.

6.7. Covering Maps.

Let $\xi = (E, p, B)$ and $\xi' = (E', p', B')$ be coverings with base points $x_0 \in E$ and $x_0' \in E'$, and let $f : (B', p'(x_0')) \to (B, p(x_0))$ be a continuous map. It can be shown that the condition $f_*(\text{gr}\,\xi'(x_0')) \subset \text{gr}\,\xi(x_0)$ is necessary, and, if B' is locally connected, also sufficient for the existence of a map $F : E' \to E$ with $F(x_0') = x_0$, such that $p \circ F = F \circ p'$ (so that F and f define a map $\xi' \to \xi$). If such a map F exists it is easily seen to be unique.

By applying this result to the situation where $\xi' = (Y, \text{id}, Y)$, we obtain the following criterion for the existence of a map covering a given map. Let $\xi = (E, p, B)$ be a covering with base point $x_0 \in E$, let Y be a locally connected space with base point y_0, and $f : (Y, y_0) \to (B, p(x_0))$ be a continuous map. If $f_*(\pi_1(Y, y_0)) \subset \text{gr}\,\xi(x_0)$ (in particular, if Y is simply connected), then there is a map $F : (Y, y_0) \to (E, x_0)$ covering f.

Chapter 3
Cellular Techniques

§7. Cellular Spaces

7.1. Basic Concepts.

A. A decomposition S of a topological space X is said to be *cellular* if there is a function d that maps the elements of S to the non-negative integers, such that for each $e \in S$ there is a continuous map $D^{d(e)} \to X$ with the following two properties:

(i) it maps $\text{Int}\,D^{d(e)}$ homeomorphically onto e;

(ii) it maps $S^{d(e)-1}$ to a union of elements of S on which d takes values less than $d(e)$.

The elements of a cellular decomposition are called *cells*, and their closures *closed cells*. The number $d(e)$ is the *dimension of the cell e*, usually denoted by $\dim e$. Any continuous map $D^{d(e)} \to X$ with properties (i), (ii) is said to be *characteristic* for e. Its image is obviously contained in the closure $\text{Cl}\,e$, and if X is Hausdorff the image is equal to $\text{Cl}\,e$. In particular, the closed cells of a cellular decomposition of a Hausdorff space are compact. It is also clear that in the Hausdorff case, $\text{Cl}\,e \setminus e$ is covered by cells of lower dimension, for any cell e.

B. A cellular decomposition is said to be *rigged* (or *equipped*) if a characteristic map is fixed for each of its cells. The family $\{\chi_e : D^{\dim e} \to X\}$ arising in this way is called a *rigging* of the decomposition, and the map $\coprod_{e \in X/S} \chi_e : \coprod_{e \in X/S} D^{\dim e} \to X$ is the *total characteristic map*.

C. The covering of X by the closed cells of a cellular decomposition S enables us to define a new topology on X. The closed sets in this topology are specified by the rule: a set is closed if its intersection with the closed cells of the decomposition are closed in the original topology. This new topology is called the *weak* or *cellular* topology, and the passage from the original topology to the new topology is called *cellular weakening* of the topology. Weakening a topology can only increase the supply of open and closed sets, and in particular a Hausdorff space remains Hausdorff. In all cases the topology of the closed cells is unchanged, so that the decomposition S is still cellular and retains the characteristic maps.

When X is Hausdorff and S is provided with a rigging $\{\chi_e\}$, the cellular topology can be effectively described by means of the total characteristic map $\chi = \bigsqcup_{e \in X/S} \chi_e$: a set A is open (closed) iff its inverse image $\chi^{-1}(A)$ is open (closed). In other words, the cellular topology is the same as the topology induced on X by the one-one quotient map of χ from the quotient space of the sum $\bigsqcup_{e \in X/S} D^{\dim e}$ factored by the decomposition into the inverse images of single points under χ. The equivalence of these two descriptions of the cellular topology follows because the maps χ_e are closed.

D. A *cellular space* is a Hausdorff topological space with a cellular decomposition having the two properties:

(C) each closed cell intersects only finitely many cells;

(W) the topology of the space is the same as the cellular topology.

The notations (C) and (W) are generally accepted, and come from the terms "Closure finite" and "Weak topology". They were introduced by J.H.C. Whitehead (as were cellular spaces themselves, which he called *CW-complexes*).

Clearly property (C) is preserved under cellular weakening of the topology. Hence a Hausdorff space with a cellular decomposition satisfying (C) remains a cellular space after cellular weakening of its topology.

The terminology relating to cellular decompositions is usually applied to cellular spaces as well. In particular, a cellular space may be rigged, and finite or countable. A finite cellular space is one with a finite number of cells, not a finite number of points.

E. The *dimension of a cellular space* is the upper bound of the dimensions of its cells; for the empty space (which is not excluded as a cellular space) it is taken as -1. The dimension, finite or infinite, of a cellular space X is denoted by $\dim X$.

F. The simplest cellular spaces are the discrete spaces, decomposed into 0-dimensional cells (isolated points). Clearly all 0-dimensional cellular spaces are of this type: a decomposition of a Hausdorff space with a non-discrete topology into 0-cells does not satisfy condition (W).

An example of a cellular decomposition satisfying (W) but not (C) is the decomposition of the ball D^n, $n > 1$, into the n-cell $\text{Int}\, D^n$ and the 0-cells covering S^{n-1}.

G. A cellular decomposition is called *locally finite* if each point of the space has a neighbourhood that intersects only finitely many cells, or, equivalently, each point has a neighbourhood that intersects only finitely many closed cells.

Clearly any compact subset of a space with a locally finite cellular decomposition has a neighbourhood that meets only finitely many cells. It follows that a locally finite cellular decomposition of a Hausdorff space satisfies (C). As for condition (W), it is easily shown that it is satisfied in general for any locally finite cellular decomposition. Thus a Hausdorff space with a finite or locally finite cellular decomposition is a cellular space.

The local finiteness condition can be easily reformulated in terms of cells: a cellular space is locally finite iff each cell intersects only finitely many closed cells iff each closed cell intersects only finitely many closed cells.

H. A subset of a cellular space that contains together with each point the closure of the cell containing that point is called a *subspace* of the given cellular space. Subspaces are themselves cellular spaces: the cellular decomposition of the space induces a cellular decomposition on each subspace, which clearly satisfies (C) and (W).

By condition (W) the subspaces of a cellular space are closed. Note that the union and intersection of any collection of subspaces are subspaces and that every covering of a cellular space by subspaces is fundamental. (Recall that a covering of a space X is called *fundamental* if a subset $A \subset X$ is open (closed) in X whenever $A \cap U$ is open (closed) in U for every $U \in \Gamma$. In particular, condition (W) may be stated thus: the closed cells form a fundamental covering of the space.)

A pair consisting of a cellular space and a subspace is called a *cellular pair*. A *cellular triple* and *cellular triad* are defined similarly.

I. Warning: a closed cell need not be a subspace. An example is the bouquet $(D^1, 0) \vee (S^2, (1, 0, 0))$, decomposed into four cells: the 0-cells -1, $1 \in D^1$, the 1-cell $\text{Int } D^1$, and the 2-cell $S^2 \setminus (1, 0, 0)$; see Fig. 5. This is clearly a cellular space, but the closure of the 2-cell touches the 1-cell without containing it.

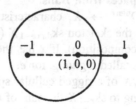

Fig. 5

J. Important subspaces of a cellular space X are its *skeletons* $\text{sk}_0 X, \text{sk}_1 X, \ldots$, defined by $\text{sk}_r X = \bigcup_{\dim e \leq r} e$. They are all nonempty if X is nonempty (since the presence of cells of some positive dimension implies the presence of cells of lower dimension); for formal reasons, we add the empty skeleton sk_{-1} and $\text{sk}_\infty X - X$. The skeletons form a filtration $\{\text{sk}_r X\}_{0 \leq r < \infty}$ of X.

K. It follows from condition (C) that each cell of a cellular space is contained in a finite subspace. It is easily deduced from condition (W) that a compact subset of a cellular space intersects only finitely many cells, and hence is contained in a finite subspace. Similar arguments show that a compact subset of a locally finite cellular space is contained in the interior of a finite subspace.

L. A map from a cellular space X to a cellular space Y is called *cellular* if it is continuous and maps the skeleton $sk_r X$ into $sk_r Y$ for any r.

A cellular map obviously maps each 0-cell to a 0-cell. A cell of positive dimension need not be mapped to a single cell by a cellular map; for example, the identity map of D^1, decomposed into 0-cells $-1, 1$, and the 1-cell $(-1, 1)$, into the same interval decomposed into 0-cells $-1, 0, 1$ and 1-cells $(-1, 0), (0, 1)$, maps the 1-cell to the union of a 0-cell and two 1-cells; see Fig. 6. A cellular map is a *cellular equivalence* if it is invertible and its inverse is also a cellular map; equivalently: a cellular equivalence is a homeomorphism which exactly transforms one cellular decomposition into another. Two cellular spaces that can be related by a cellular equivalence are *cellularly equivalent*. Two rigged cellular spaces related by a cellular equivalence that takes one rigging into the other are called *rigged-equivalent*.

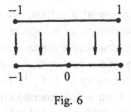

Fig. 6

A map f of a cellular space X into a cellular space Y is a *cellular inclusion* if $f(x)$ is a subspace of Y and $f| : X \to f(X)$ is a cellular equivalence.

Warning: a cellular homeomorphism need not be a cellular equivalence. Example: the homeomorphism described above and shown in Fig. 6.

7.2. Gluing of Cellular Spaces from Balls.

A. Note that each map $D^{\dim e} \to X$, characteristic for the cell e, maps the boundary sphere $S^{\dim e-1}$ into the skeleton $sk_{\dim e-1} X$ (this is in fact property (i) in the definition of a cellular decomposition). If x has the rigging $\{\chi_e\}$, then the map $\chi_e| : S^{\dim e-1} \to sk_{\dim e-1} X$ is called a *gluing* for e.

For $r \geq 0$, the skeleton $sk_r X$ of a rigged cellular space X is canonically homeomorphic to the result of gluing to $sk_{r-1} X$ the sum of r-balls $\coprod_{e \in M_r} D_e$, where M_r is the set of r-cells of X, by means of the sum of the gluing maps of the r-cells. The canonical homeomorphism $(sk_{r-1} X) \cup_\phi (\coprod_{e \in M_r} D_e) \to sk_r X$ is the one-one quotient map of the map $(sk_{r-1} X) \coprod (\coprod_{e \in M_r} D_e) \to sk_r X$ defined by the inclusion $sk_{r-1} X \to sk_r X$ and the maps $\chi_e| : D_e \to sk_r X$.

The description above of the weak topology in terms of the total characteristic map shows that every cellular space can be glued together from balls, in a "nice" way. This gluing consists of a sequence of attachings: the rth attaching turns $sk_{r-1} X$ into $sk_r X$ ($r = 0, 1, \ldots$) and X is the inductive limit of the $sk_r X$.

B. The following formalization transforms this description of cellular spaces into a useful inductive method for their construction. Note first that if A is a topological space provided with a rigged cellular decomposition in which all the cells have dimension less than q, and we attach to A a sum of q-dimensional balls

$\coprod_{\mu \in M}(D_\mu = D^q)$ by means of any continuous map $\phi : \coprod_{\mu \in M}(S = S^{q-1}) \to A$, then the resulting space is provided with an obvious rigged cellular decomposition in which all the cells have dimension less than $q + 1$. This space satisfies condition (W), and if A is normal, the new space is also normal. From the fact that a compact subset of a cellular space intersects only finitely many cells, it follows that the new space satisfies condition (C) if A is cellular. Hence if A is a normal rigged cellular space, then so is $A \cup_\phi (\coprod_{\mu \in M} D_\mu)$. These remarks form the basis of our inductive construction. We start with $q = 0$, that is, with A empty, and at the rth step we attach the space $\coprod_{\mu \in M_r}(D_\mu = D^r)$ by a continuous map ϕ_r : $\coprod_{\mu \in M_r}(S_\mu = S^{r-1}) \to X_{r-1}$ to the normal rigged cellular space X_{r-1} already obtained, where $\dim X_{r-1} \leq r - 1$. The result of the rth step is a normal rigged cellular space $X_r = X_{r-1} \cup_\phi (\coprod_{\mu \in M} D_\mu)$ with $\dim X_r \leq r$, and the whole process produces a sequence $x_0, x_1, \ldots$, with natural cellular inclusions $X_r \to X_{r+1}$ and limit space $X = \lim X_r$. This last space is normal (being the limit of normal spaces each embedded as a closed subset in the next) and has an obvious cellular decomposition with properties (C) and (W). Thus X is a normal rigged cellular space; obviously $\mathrm{sk}_r X = X_r$.

X is called an *inductively glued cellular space*. From what has been said it follows that every rigged cellular space is rigged-equivalent to an inductively glued cellular space. From this it follows incidentally that every cellular space is normal.

7.3. Examples of Cellular Decompositions.

Some very important spaces such as spheres, balls, projective spaces, and Grassmann manifolds have canonical cell decompositions that make them into cellular spaces. We shall describe some of them in this section; the fact that they satisfy conditions (C) and (W) is obvious in all cases.

A. The most economical cell decomposition of the sphere S^n ($0 \leq n < \infty$) consists of one 0-cell (say $(1, 0, \ldots, 0)$) and one n-cell.

B. The most economical cell decomposition of the ball D^n ($1 \leq n < \infty$) consists of three cells: the 1-cell $(1, 0, \ldots, 0)$, the $(n-1)$-cell $S^{n-1} \setminus (1, 0, \ldots, 0)$ and the n-cell $\mathrm{Int}\, D^n$.

C. Cell decompositions of S^∞ and D^∞ obviously cannot consist of a finite number of cells. One of the simplest cell decompositions of S^∞ contains two cells of each dimension: the two hemispheres that are the components of $S^r \setminus S^{r-1}$. In this decomposition $\mathrm{sk}_r S^\infty = S^r$. A similar decomposition of D^∞ contains three 0-cells and four cells of each dimension > 0; here $\mathrm{sk}_r D^\infty = D^r \cup S^r$ for any $r \geq 0$. Figure 7 shows $\mathrm{sk}_2 D^\infty$. With these cell decompositions of S^∞ and D^∞, the finite-dimensional spheres and balls S^n and D^n are cellular subspaces, but the cell decompositions induced on S^n and D^n are different from those described in 7.3.A and **B**.

D. A canonical cell decomposition of real projective space $\mathbb{R}P^n$ $0 \leq n \leq \infty$) consists of cells $e_r = \mathbb{R}P^r \setminus \mathbb{R}P^{r-1}$ with $\dim e_r = r$, where $0 \leq r \leq n$ for $n < \infty$, and $0 \leq r < \infty$ for $n = \infty$. The characteristic map for e_r may be taken as the composite map

$$D^r \xrightarrow{\mathrm{pr}} \mathbb{R}P^r \xrightarrow{\mathrm{in}} \mathbb{R}P^n,$$

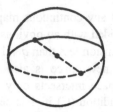

Fig. 7

where pr is the quotient map when the ball is factored by the partition consisting of the one-point subsets of its interior, and the pairs of antipodal points of its boundary. In this decomposition $\text{sk}_r \mathbb{R}P^n = \mathbb{R}P^r$, and the natural projection $S^{r-1} \to \mathbb{R}P^{r-1}$ may be used as the attaching map for e_r.

E. A canonical cell decomposition of complex projective space $\mathbb{C}P^n$ ($0 \leq n \leq \infty$) consists of cells $e_r = \mathbb{C}P^r \setminus \mathbb{C}P^{r-1}$ with $\dim e_r = 2r$, where $0 \leq r \leq n$ for $n < \infty$, and $0 \leq r < \infty$ for $n = \infty$. The characteristic map for e_r may be taken as the composite map

$$D^{2r} \longrightarrow \mathbb{C}P^r \xrightarrow{\text{in}} \mathbb{C}P^n.$$

Clearly $\text{sk}_r \mathbb{C}P^n = \mathbb{C}P^{[r/2]}$ if $r \leq 2n$, and the Hopf bundle $S^{2r-1} \to \mathbb{C}P^{r-1}$ is an attaching map for e_r.

F. (Rigged) cell decompositions of $\mathbb{H}P^n$ ($0 \leq n \leq \infty$) and $\text{Ca}P^n$ ($0 \leq n \leq 2$) can be defined similarly. The decomposition of $\mathbb{H}P^n$ consists of cells $e_r = \mathbb{H}P^r \setminus \mathbb{H}P^{r-1}$ with $\dim e_r = 4r$, where $0 \leq r \leq n$ for $n < \infty$, and $0 \leq r < \infty$ for $n = \infty$; the decomposition of $\text{Ca}P^n$ has cells $e_r = \text{Ca}P^{r-1} \setminus \text{Ca}P^{r-1}$, $0 \leq r \leq n$, where $\dim e_r = 8r$. Note that when $n = 1$ the cell decompositions of $\mathbb{R}P^n$, $\mathbb{C}P^n$, $\mathbb{H}P^n$, and $\text{Ca}P^n$ described here consist of two cells, and are the canonical decompositions of the spheres $S^1 = \mathbb{R}P^1$, $S^2 = \mathbb{C}P^1$, $S^4 = \mathbb{H}P^1$, and $S^8 = \text{Ca}P^1$.

G. It is well known that every connected closed two-dimensional manifold is homeomorphic to a sphere, or a sphere with handles, or a sphere with crosscaps. [Recall that a sphere with g handles is obtained from a sphere with g holes (that is, a sphere S^2 from which the interiors of g pairwise disjoint spherical segments have been removed; see Fig. 8) by attaching g handles (a handle is a torus $S^1 \times S^1$ from which the interior of an embedded disc has been removed; see Fig. 9) by homeomorphisms of the boundary circles of the handles to the boundary circles of the holes; see Fig. 10.

Fig. 8

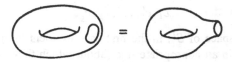

Fig. 9

Fig. 10

A sphere with g crosscaps is obtained from a sphere with g holes by attaching g Möbius bands by homeomorphisms of the boundary circle of the Möbius bands to the boundary circles of the holes. Recall also that a sphere with one handle is homeomorphic to the torus, a sphere with one crosscap to the projective plane, and a sphere with two crosscaps to the Klein bottle. A sphere with g handles and h crosscaps (defined in an obvious way) is homeomorphic to a sphere with $2g + h$ crosscaps.

Spheres with handles and spheres with crosscaps have standard rigged cell decompositions, generalizing the canonical two-cell decomposition of S^2, the canonical three-cell decomposition of the projective plane, and the canonical four-cell decomposition of the torus $S^1 \times S^1$ (obtained by taking the product of the canonical decompositions of two copies of S^1; see 7.5.A below). Each of these standard decompositions contains a single 0-cell and a single 2-cell; there are $2g$ 1-cells for the sphere with g handles, and g 1-cells for the sphere with g crosscaps. Thus the 1-skeleton for a sphere with g handles is a bouquet of $2g$ circles, while for a sphere with g crosscaps it is a bouquet of g circles. The description of the whole rigged cell decomposition reduces to describing the attaching map for the 2-cell, that is, describing a map of S^1 to the bouquet above.

Leaving aside the case $g = 1$ already considered, for a sphere with g handles we regard S^1 as the perimeter of a regular polygon with $4g$ sides labelled successively

$$a_1, b_1, a_1', b_1', \ldots, a_g, b_g, a_g', b_g',$$

and for a sphere with g crosscaps as the perimeter of a regular polygon with $2g$ sides labelled successively

$$c_1, c_1', \ldots, c_g, c_g'.$$

We form a quotient space of S^1, in the first case by identifying a_i with a_i' and b_i with b_i' by reflexion in an appropriate line (about which they are symmetric), and in the second case by identifying c_i with c_i' by a rotation of the polygon (about its centre). In both cases the quotient space is a bouquet of circles, containing $2g$ circles in the first case and g in the second. The projection of S^1 onto this quotient space gives the desired attaching map.

It can easily be verified that the resulting cellular space really is homeomorphic to a sphere with handles or a sphere with crosscaps by noting that the pentagon bounded by the sides a_i, a_i', b_i, b_i' and the diagonal that cuts these sides off from the $4g$-gon becomes a handle after the identifications, while the triangle bounded by c_i, c_i' and the diagonal that cuts these sides off from the $2g$-gon becomes a Möbius band after the identification.

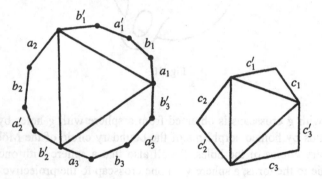

H. For a description of cell decompositions of Grassmann manifolds and flag manifolds, see Fuchs (1986).

7.4. Topological Properties of Cellular Spaces.

A. It is easy to see that a cellular space is compact iff it is finite (that is, has a finite number of cells), and is locally compact iff it is locally finite.

B. Any finite cellular space can be embedded in a Euclidean space of sufficiently high dimension. The simplest general construction of such an embedding is inductive: given an embedding $sk_{n-1}X \to \mathbb{R}^q$ an embedding $sk_n X \to \mathbb{R}^{q+n+1}$ is constructed. This construction gives an embedding in a space whose dimension depends quadratically on the dimension of the embedded space; however, for any finite cellular space X there exists an embedding $X \to \mathbb{R}^{2\dim X+1}$.

Every countable locally finite cellular space admits an embedding in $\mathbb{R}^\infty$ and every finite-dimensional countable locally finite cellular space admits a topological embedding in $\mathbb{R}^q$ for sufficiently large q.

C. A finite cellular space is a CNRS space. This can be proved using a concrete construction for embedding it in a Euclidean space.

D. The components of a cellular space are open subspaces. If $r \geq 1$, the r-skeleton of a component of a cellular space is a component of its r-skeleton. In particular, a cellular space is connected iff its 1-skeleton is connected.

E. A connected locally finite cellular space is countable.

F. A cellular space has a countable basis iff it is countable and locally finite. A cellular space is metrizable iff it satisfies the first axiom of countability iff it is locally finite.

7.5. Cellular Constructions.

When applied to cellular spaces the constructions in 1.4 and 1.5 are naturally modified. For some constructions the modification consists in endowing the resulting space with a cellular decomposition so that it becomes a cellular space; an obvious example is the sum of spaces. In other constructions, for example in products, the modification affects the actual topology of the resulting space. Modifications of both types are described below. We stress that when they are applied to rigged cellular spaces they all give rigged cellular spaces.

A. The *cellular product* $X \times Y$ of topological spaces X, Y endowed with cell decompositions S, T has a natural cell decomposition, namely, the decomposition $S \times T = \{e_1 \times e_2 | e_1 \in S, e_2 \in T\}$ with $\dim e_1 \times e_2 = \dim e_1 + \dim e_2$. As a characteristic map for $e_1 \times e_2$ we may take the composition of a homeomorphism $D^{\dim e_1 + \dim e_2} \to D^{\dim e_1} \times D^{\dim e_2}$ and the product $\chi_{e_1} \times \chi_{e_2} : D^{\dim e_1} \times D^{\dim e_2} \to X \times Y$ of arbitrary characteristic maps $\chi_{e_1} : D^{\dim e_1} \to X$, $\chi_{e_2} : D^{\dim e_2} \to Y$. If S and T are rigged this gives a canonical rigging for $S \times T$.

If S, T have property (C), then obviously so does $S \times T$. In contrast to this, $S \times T$ need not satisfy property (W) even when X and Y are cellular spaces. Example: $X = Y$ is a bouquet $\vee_{t \in \mathbb{R}} (I_t = (1, 0))$ of a continuous family of intervals (regarded as a cellular space; see **F** below). The cellular space which is obtained from the product of cellular spaces X, Y after cellular weakening of the topology is called the *cellular product*, denoted by $X \times_c Y$.

Note that cellular weakening of the topology does not change the topology of compact subsets of $X \times Y$. It can also be shown that if X is locally finite, then $X \times_c Y = X \times Y$ for any cellular space Y, and that this is also true if X and Y are locally countable, that is, if each point in both spaces has a neighbourhood that intersects only countably many cells.

B. *Gluing.* Let X, Y be cellular spaces, A a subspace of Y, and $\phi : A \to X$ a cellular map. The topological space $X \cup_\phi Y$ is defined as in 1.4.**D**. The decompositions of $Y \setminus A$ and X into cells define its cell decomposition. The natural maps $X \to X \cup_\phi Y$ and $Y \to X \cup_\phi Y$ are cellular, and the first one is a cellular embedding.

If $X = D^0$ then $\phi : A \to X$ is cellular for any cellular pair (Y, A) and $X \cup_\phi Y = Y/A$. Hence the previous definition makes the quotient space of a cellular space by a subspace into a cellular space.

C. Since the decomposition of the interval I into the cells $0, 1, \text{Int } I$ makes it a finite cellular space, the cylinder $X \times I$ is a cellular space for any cellular space X (see **A**). Since the bases of this cylinder are cellular subspaces of it, then the quotient operation that makes it into the cone CX, and the second quotient operation that makes CX into ΣX are included in the scheme **B**, so that the cone and suspension over a cellular space are also cellular spaces.

If $f : X \to Y$ is a cellular map, then the gluings that turn $X \times I$ into $\mathrm{Cyl}\,f$ and CX into $\mathrm{Con}\,f$ are also included in the scheme of **B**, so that the mapping cylinder and cone of a cellular map are cellular spaces.

D. We define the *cellular join* $X *_c Y$ of cellular spaces X, Y by

$$X *_c Y = \left(X \coprod Y\right) \cup_\phi \left[(X \times Y) \times I\right],$$

where ϕ is the map of the union $[(X \times_c Y) \times 0] \cup [(X \times_c Y) \times 1]$ into $X \times Y$ defined by $\phi(x, y, 0) = \mathrm{in}_1(x)$, $\phi(x, y, 1) = \mathrm{in}_2(y)$. Since ϕ is a cellular map, $X *_c Y$ is a cellular space.

If X is locally finite, then $X *_c Y$ coincides as a topological space with $X * Y$ (see **A**). In the general case the cell decomposition of $X *_c Y$ is obviously also a cell decomposition for $X * Y$, so that $X *_c Y$ is obtained from $X * Y$ by cellular weakening of the topology, which does not change the topology of compact subsets of $X * Y$.

E. Spaces of maps in non-trivial cases are too large for cell decompositions to be possible. However, there is the following theorem of Milnor (1959): if X, Y are cellular spaces, then the space $C(X, Y)$ is homotopy equivalent to a cellular space.

F. *The case of pointed spaces.* If X is a pointed cellular space with base point x_0, then the cone $C(X, x_0)$ and the suspension $\Sigma(X, x_0)$ are obtained from the cone CX and the suspension ΣX by factoring by a subspace, and so are cellular spaces. Similarly the bouquet of a family of pointed spaces with 0-cells as base points is obtained from their sum by factoring by a subspace, and hence is a cellular space.

Finally we define the *cellular tensor product* of cellular spaces X, Y with base points the 0-cells x_0, y_0 as the quotient space $(X \times_c Y)/[(X \times y_0) \cup (x_0 \times Y)]$ and their *cellular join* as $(X *_c Y)/(x_0 * y_0)$. These are cellular spaces, denoted by $(X, x_0) \otimes_c (Y, y_0)$ and $(X, x_0) *_c (Y, y_0)$. If X is locally finite then they are equal as topological spaces to $(X, x_0) \otimes (Y, y_0)$ and $(X, x_0) * (Y, y_0)$; in the general case their cell decompositions are also cellular for $(X, x_0) \otimes (Y, y_0)$, $(X, x_0) * (Y, y_0)$ since they are obtained from $(X, x_0) \otimes (Y, y_0)$, $(X, x_0) * (Y, y_0)$ by cellular weakening of the topology.

§8. Simplicial Spaces

8.1. Basic Concepts.

A. Let A be a subset of $\mathbb{R}^n$ consisting of $r + 1$ ($r \geq 0$) points not lying in a $(r - 1)$-dimensional plane. The convex hull $\mathrm{conv}\,A$ of A (the smallest convex set containing A) is called the *Euclidean simplex spanned by* A. The points of A are the *vertices* of this simplex, and the number r its *dimension*.

Obviously a point of a simplex is a vertex iff there is no non-degenerate segment in the simplex having the given point as its mid-point. Thus the set of vertices of a simplex is determined by the simplex.

The simplexes spanned by the subsets of A are called the *faces* of the simplex spanned by A. Faces spanned by two mutually complementary subsets of A are called *opposite*. If A_1, A_2 are such subsets, then the map

$$\operatorname{conv} A_1 * \operatorname{conv} A_2 \to \operatorname{conv} A : \operatorname{pr}(x_1, x_2, t) \to (1 - t)x_1 + tx_2$$

is a homeomorphism. Hence a Euclidean simplex is canonically homeomorphic to the join of any two of its opposite faces.

Since the ball D^r is homeomorphic to the join $D^p * D^q$ with $p + q = r - 1$, an obvious induction shows that any r-simplex is homeomorphic to D^r.

The boundary of an r-simplex in the r-dimensional plane determined by it is clearly the union of its $(r-1)$-dimensional faces. This boundary and its complement in the simplex are usually called simply the boundary and interior of the simplex.

The simplex spanned by A can also be described as the set of sums $\sum_{a \in A} t_a a$, where the t_a are non-negative real numbers with sum 1. Since A does not lie in any $(r - 1)$-dimensional plane, the numbers t_a are uniquely determined by the point $x = \sum_{a \in A} t_a a$; the number t_a is called the ath *barycentric coordinate* of the point x. We shall denote it by $b_a(x)$.

If $B \subset A$ then the face spanned by B of the simplex spanned by A is obviously determined in barycentric coordinates by the equations $b_a(x) = 0$ for $a \in A \setminus B$. It is also clear that for $x \in \operatorname{conv} B$ the coordinates $b_a(x)$ with $a \in B$ relative to the simplex $\operatorname{conv} A$ are the same as the coordinates $b_a(x)$ relative to $\operatorname{conv} B$.

The point of a simplex with all its barycentric coordinates equal (that is, all equal to $1/(r + 1)$, where r is the dimension of the simplex) is called its *centre*.

B. A map from the simplex spanned by A to the simplex spanned by B is *simplicial* if it is affine and maps A into B. Such a map maps each face of the first simplex simplicially onto some face of the second simplex, and maps the interior of the first simplex onto the interior of the simplex that is its image.

Clearly any map $A \to B$ can be uniquely extended to a simplicial map from the simplex spanned by A to the simplex spanned by B. If the map $A \to B$ is one-one, then its simplicial extension is an embedding, and if $A \to B$ is invertible, then its simplicial extension is a homeomorphism.

A simplex is *ordered* if the set of its vertices is (linearly) ordered. Since a subset of an ordered set has a natural order, the faces of an ordered simplex are ordered simplexes.

There is a unique bijection compatible with the orderings between the sets of vertices of two ordered simplexes of the same dimension. Hence all ordered Euclidean simplexes of the same dimension are canonically simplicially homeomorphic to each other.

The simplex spanned by the points $(1, 0, \ldots, 0)$, $(0, 1, 0, \ldots, 0)$, $\ldots$, $(0, 0, \ldots, 1)$ in $\mathbb{R}^{r+1}$ is called the *unit simplex, denoted by T^r*. It is convenient because the barycentric coordinates of its points are the same as their ordinary coordinates in $\mathbb{R}^{r+1}$. The order of the vertices listed above makes it into an ordered simplex, and hence any ordered Euclidean simplex of dimension r is canonically simplicially homeomorphic to T^r.

C. A topological space X is called an *ordered topological simplex of dimension r* if it is endowed with a homeomorphism $T^r \to X$, called the *characteristic homeomorphism* of the simplex; the set X is sometimes called the *support* of the simplex. For example, an ordered Euclidean simplex of dimension r is an ordered topological simplex of dimension r.

A standard method of suppressing an ordering is to consider all possible orderings simultaneously. In accordance with this we call a topological space X a *topological simplex of dimension r*, or *r-simplex*, if it is endowed with $(r + 1)!$ homeomorphisms $T^r \rightarrow X$ that can be transformed into each other by simplicial homeomorphisms $T^r \rightarrow T^r$. The terms "characteristic homeomorphism" and "support" are used in this situation also, but instead of one characteristic homeomorphism there are $(r + 1)!$ equally valid characteristic homeomorphisms. The Euclidean simplexes are examples of topological simplexes.

If X is a topological r-simplex (ordered topological r-simplex) and Y a topological space, then any homeomorphism $X \rightarrow Y$ turns Y into a topological r-simplex (ordered topological r-simplex). Hence any homeomorphic image of a Euclidean r-simplex (ordered Euclidean r-simplex) is a topological r-simplex (ordered topological r-simplex).

The vertices, faces, boundary, interior, barycentric coordinates, centre, and simplicial map are all defined in an obvious way for topological simplexes. The faces of an (ordered) topological simplex are (ordered) topological simplexes. As for a Euclidean simplex, a topological simplex becomes ordered by fixing an order of its vertices.

D. A *triangulation* of a set is a covering of it by topological simplexes satisfying the three conditions:

(i) the faces of any simplex in the covering are simplexes in the covering;

(ii) if a simplex in the covering is a subset of another, then the first is a face of the second;

(iii) the intersection of the supports of two overlapping simplexes of the covering is the support of a simplex of the covering.

A set endowed with a triangulation is called a *simplicial space*. The simplexes of the triangulation are called *simplexes of the space* and the 0-simplexes its *vertices*.

A triangulation of a set makes it into a topological space (in which a subset is open iff its intersection with each simplex of the triangulation is open in that simplex). The supports of the simplexes of a triangulation form a fundamental covering of the space.

If a is a vertex of a simplicial space X, then the points of the simplexes with vertex a have ath barycentric coordinate b_a (see **A** and **C**), and we get a continuous function $b_a : X \rightarrow \mathbb{R}$ if we put $b_a(x) = 0$ for any point x not in a simplex with vertex a. This function is called the ath *barycentric function*. Clearly for any two distinct points x, y of X there is a vertex a such that $b_a(x) \neq b_a(y)$. Consequently a simplicial space is Hausdorff.

In the case when the set X endowed with a triangulation already has a topology, and this topology is the same as that defined by the triangulation, we say that the triangulation is a triangulation of the original topological space X. For example, the covering of a topological simplex consisting of all its faces is a triangulation of it.

A simplicial space is *ordered* if its simplexes are ordered in such a way that the ordering of the faces of each simplex is compatible with the ordering of the simplex itself. This condition is satisfied, for example, if the ordering of the simplexes is induced by some ordering of the set of all the vertices of the space; hence we see that a simplicial space can always be ordered.

E. A fundamental class of simplicial spaces is the following. For any nonempty set A, let Si A denote the set of all non-negative finitely supported functions ϕ : $A \to R$ with $\sum_{a \in A}(A) = 1$. If $B \subset A$, we shall identify Si B with a subset of Si A by regarding the functions in Si B as equal to 0 on $A \setminus B$. If A is finite and consists of $r + 1$ elements, then Si A is a topological simplex in a natural sense: it is a subset of the $(r+1)$-dimensional Euclidean space of all functions $A \to R$, and the $(r+1)!$ orderings of A correspond to the $(r+1)!$ homeomorphisms $T^r \to$ Si A taking the point $(x_1, \ldots, x_{r+1})$ to the function with values $x_1, \ldots, x_{r+1}$, which can be transformed into each other by simplicial homeomorphisms $T^r \to T^r$. In the general case, Si A is covered by the topological simplexes Si B corresponding to all possible finite subsets of A, and this is clearly a triangulation of the set Si A. Hence Si A is a simplicial space, called the *simplex spanned by* A. An ordering of it is equivalent to an ordering of the set A.

F. The interiors of the simplexes of a simplicial space X form a decomposition of the set X, and it is clear that this is a cell decomposition if we define the dimension of the interior e of a simplex s by $\dim e = \dim s$: a characteristic map for e may be taken as the composite map

$$D^{\dim s} \longrightarrow T^{\dim s} \xrightarrow{\phi} s \xrightarrow{\text{in}} X,$$

where the first arrow is an arbitrary homeomorphism and ϕ is some characteristic homeomorphism of the simplex s. Conditions (C) and (W) are obviously satisfied here, and since X is Hausdorff (by D), this cell decomposition turns X into a cellular space. Hence a simplicial space decomposed into the interiors of its simplexes is a cellular space.

Clearly the skeleton $\text{sk}_r X$ of a simplicial space X is the union of its simplexes of dimension not greater than r, and $\dim(\text{sk}_r X) = r$, if $r \leq \dim X$ (for an arbitrary cellular space X with $\dim X = r$ we may have $\dim \text{sk}_r X < r$). In particular $\text{sk}_0 X$ is the set of vertices of X. This set is finite iff the space X is finite. It follows from what was said in 7.1.G that a simplicial space is locally finite iff each vertex belongs to only finitely many simplexes.

G. A *subspace* of a simplicial space is a subset consisting of whole simplexes of the triangulation. It has a natural triangulation and hence is a simplicial space. Subspaces of an ordered simplicial space are ordered simplicial spaces. Obviously a subset of a simplicial space is a subspace iff it is a subspace in the cellular sense.

A subspace of a simplicial space is *complete* if its intersection with the support of any simplex of the space is either the support of a simplex of the space or is empty; equivalently: a subspace is complete if it contains all the simplexes whose vertices lie in it.

The simplexes of a simplicial space are obviously complete subspaces. The complete subspaces of the space Si A (see **E** above) are precisely the spaces Si B with $B \subset A$.

H. A map of a simplicial space X to a simplicial space Y is said to be *simplicial* if it maps each simplex of X simplicially to a simplex of Y. Such a map is clearly cellular and maps X onto a subspace of Y.

The following facts are also obvious. If a simplicial map is invertible, it is a homeomorphism and the inverse map is simplicial. An injective simplicial map is a topological embedding. A simplicial map $f : X \to Y$ is determined by the submap $f| : \mathrm{sk}_0 X \to \mathrm{sk}_0 Y$ of the set of vertices of X to the set of vertices of Y. A map $\mathrm{sk}_0 X \to \mathrm{sk}_0 Y$ can be extended to a simplicial map $X \to Y$ iff it takes the vertices of each simplex of X to the vertices of some simplex of Y. A simplicial map $f : X \to Y$ is injective iff the submap $f| : \mathrm{sk}_0 X \to \mathrm{sk}_0 Y$ is injective, and is invertible iff the submap is invertible. Simplicial spaces that can be mapped into each other by simplicial homeomorphisms are said to be *simplicially homeomorphic*.

A simplicial map f of an ordered simplicial space X to an ordered simplicial space Y is *monotonic* if $f(a) \le f(b)$ for any two vertices a, b in X belonging to the same simplex with $a \le b$. A simplicial map of one simplicial space into another may always be made monotonic by ordering the spaces appropriately. Further, if X is a simplicial space and Y an ordered simplicial space, then a simplicial map $f : X \to Y$ can be made monotonic by an appropriate ordering of X: it is sufficient to order the inverse image of each vertex of Y arbitrarily, and to define an order in the simplexes of X by the rule: $a < b$ if $f(a) < f(b)$ or $f(a) = f(b)$ and $a < b$ in $f^{-1}(f(a))$.

I. A *polyhedron* is a subset of Euclidean space with a finite triangulation in which all the simplexes are Euclidean. A Euclidean simplex is the simplest example of a polyhedron.

The subspaces of a polyhedron are obviously polyhedra. From what has been said in **H** it is clear that any finite simplicial space can be simplicially embedded in a Euclidean simplex with the same number of vertices. Thus any finite simplicial space is simplicially homeomorphic to a polyhedron. Moreover it is not hard to show that a finite simplicial space of dimension n is simplicially homeomorphic to a polyhedron in $\mathbb{R}^{2n+1}$. This result is best possible: for any n there is an n-dimensional polyhedron that cannot be topologically embedded in $\mathbb{R}^{2n}$. An example is $\mathrm{sk}_n T^{2n+2}$.

8.2. Simplicial Schemes.

In this section we shall show that simplicial spaces may be regarded as purely combinatorial objects.

A. A *simplicial scheme* is a pair whose first element is a set M, and whose second element is a covering s of this set by finite subsets of it, containing along with any set all its subsets.

A *map of a simplicial scheme* (M, S) *to a simplicial scheme* (M', S') is a pair of maps $\phi : M \to M'$, $\Phi : S \to S'$, such that $\Phi(A) = \phi(A)$ for $A \in S$. From the last condition a map (ϕ, Φ) of the scheme (M, S) to the scheme (M', S') is determined by the map ϕ, and it is clear that a map $\phi : M \to M'$ determines a map of (M, S) into (M', S') iff $\phi(A) \in S'$ for each $A \in S$. If the maps ϕ and Φ are invertible (that is, ϕ is invertible and $\Phi(S) = S'$) then (ϕ, Φ) is called an *isomorphism*. If there is an isomorphism between two simplicial schemes they are said to be *isomorphic*.

A simplicial scheme (M, S) is a *subscheme* of (M', S') if $M \subset M'$ and $S \subset S'$. A subscheme (M, S) is *complete* if S contains all the sets of S' lying in M.

B. The simplicial scheme consisting of the 0-dimensional skeleton of a simplicial space X and the covering of this skeleton by the 0-skeletons of the simplexes of X is called the *scheme of* X. For example, the scheme of the space Si A (see 8.1.E) consists of the set A and its covering by all finite subsets.

Since the submap $sk_0 X \to sk_0 Y$ of a simplicial map $f : X \to Y$ takes the 0-skeleton of any simplex of X into the 0-skeleton of some simplex of Y, it defines a map of the scheme of X into the scheme of Y. This last map is called the *scheme of the map* f. Clearly a simplicial map is determined by its scheme, each map of the scheme of X into the scheme of Y is the scheme of some simplicial map $X \to Y$ and a simplicial map is invertible iff its scheme is an isomorphism. In particular, simplicial spaces are simplicially homeomorphic iff their schemes are isomorphic.

If X is a subspace of a simplicial space Y, then evidently the scheme of X is a subscheme of the scheme of Y, complete if X is complete. Also every subscheme of the scheme of a simplicial space is the scheme of some subspace.

In particular, for an arbitrary simplicial scheme (M, S) consider the simplex Si M spanned by M. (M, S) is obviously a subscheme of the scheme of this simplex, and hence (M, S) is the scheme of a certain subspace of Si M. Thus every simplicial scheme is the scheme of some simplicial space. If we take the scheme of an arbitrary given simplicial space as (M, S) we see that every simplicial space X can be simplicially embedded in Si $sk_0 X$.

C. A simplicial scheme (M, S) is *ordered* if the sets of S are ordered and the ordering of these sets is compatible with the ordering of their subsets. A map (ϕ, Φ) of an ordered simplicial scheme (M, S) to an ordered simplicial scheme (M', S') is *monotonic* if $a \leq b$ implies $\phi(a) \leq \phi(b)$. Thus to order the scheme of a simplicial space is the same as to order the space, and the scheme of a simplicial map from one ordered simplicial space to another is monotonic iff the map itself is monotonic.

8.3. Simplicial Constructions.

Many of the topological and cellular constructions described in 1.4, 1.5 and 7.5 can be supplemented by analogous constructions that convert simplicial spaces into simplicial spaces. The simplest examples are the operations $\coprod$ and $\vee$: the sum of simplicial spaces and the bouquet of simplicial spaces with a vertex as a base point in each are simplicial spaces in an obvious sense. The most important of the more complicated constructions are considered below. First we describe the construction of the barycentric subdivision which enables us to refine a triangulation, and has no analogue in 1.4, 1.5 or 7.5.

A. The following construction produces a simplicial space ba X from any simplicial space X which is identical with X as a topological space, but has a finer triangulation, called the *barycentric subdivision* of the original space.

First let X be a Euclidean simplex. For an arbitrary enumeration $a_0, \ldots, a_r$ of the vertices of X consider the set $\{x \in X | b_{a_0(x)} \le b_{a_1(x)} \le \cdots \le b_{a_r(x)}\}$. This is obviously a Euclidean simplex with vertices at the centres of the simplexes spanned by the sets $\{a_0\}, \{a_0, a_1\}, \ldots, \{a_0, \ldots, a_r\}$, and the simplexes of this form corresponding to all possible enumerations of the vertices of X together with their faces determine a triangulation of X. This is the barycentric subdivision of the standard triangulation of the simplex X, converting it into ba X (see Fig. 11, where $\dim X = 2$). This construction obviously has the property that if X is a face of a simplex Y, then the inclusion ba $X \to$ ba Y is a simplicial embedding.

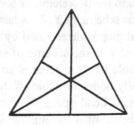

Fig. 11

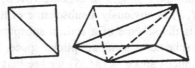

Fig. 12

The barycentric subdivision of the standard triangulation of a topological simplex X is defined as the image of the barycentric subdivision of the unit simplex T', $r = \dim X$, under the simplicial homeomorphism $T' \to X$. This is clearly well-defined, that is, it is independent of the choice of simplicial homeomorphism $T' \to X$ among the $(r + 1)!$ possibilities.

Finally suppose X is any simplicial space. By subdividing the triangulation of each simplex of X by the above method, we obtain a triangulation of X, again called the *barycentric subdivision* of the original triangulation.

Note that the barycentric subdivision of a (locally) finite simplicial space is (locally) finite, and if X is a polyhedron so is ba X.

It is clear that the set of vertices of ba X is precisely the set of centres of the simplexes of X, and that the centres of the simplexes $s_1, \ldots, s_m$ of X are vertices of a simplex of ba X iff $s_1, \ldots, s_m$ form an increasing sequence under an appropriate enumeration. This enables us to give a concise description of barycentric subdivision in the language of schemes: if (M, S) is the scheme of a simplicial space X, then the scheme of ba X is $(S, \text{ba } S)$, where ba S is the collection of all finite subsets of X that can be ordered by inclusion. This also gives a canonical ordering of ba X: a vertex $a \in \text{sk}_0 \, \text{ba} \, X$ precedes a vertex $b \in \text{sk}_0 \, \text{ba} \, X$ if the simplex (in x) with centre a is contained in the simplex with centre b.

In particular it follows from this description of the scheme of ba X that if X is a subspace of a space Y, then ba X is a complete subspace of ba Y; indeed the scheme of ba X is obviously a complete subscheme of the scheme of ba Y.

If $f : X \to Y$ is a simplicial map, the map $f : \text{ba} \, X \to \text{ba} \, Y$ is in general not simplicial (the simplest example is a surjective simplicial map $T^2 \to T^1$). However, the map from the scheme of X to the scheme of Y induced by f naturally induces a map of the scheme ba X to the scheme of ba Y, and hence a simplicial map ba $X \to \text{ba} \, Y$, which is clearly always monotonic.

It is easy to see that if X is a polyhedron, then the maximum diameter of the simplexes of ba X is not greater than $n/(n+1)$ times the maximum diameter of the simplexes of X, where $n = \dim X$. Hence for any $\varepsilon > 0$ there is a positive integer m such that the diameters of all the simplexes of $\text{ba}^m X$ are less than ε.

B. *The simplicial product.* The cellular product $X \times_c Y$ of simplicial spaces X, Y with $\dim X > 0$, $\dim Y > 0$ clearly does not have a triangulation in which the interiors of the simplexes are the products of the interiors of the simplexes of X and Y. We shall show, however, that it can be triangulated, and give a construction to do this canonically if X and Y are ordered. The simplicial space obtained from $X \times_c Y$ by this construction is called the *simplicial product* of X and Y, denoted by $X \times_s Y$.

First suppose that X is a Euclidean simplex in $\mathbb{R}^m$ with vertices $a_0, \ldots, a_q$ and Y a Euclidean simplex in $\mathbb{R}^n$ with vertices $b_0, \ldots, b_r$. Consider the sequences with $(q+r+1)$ elements of pairs (a_i, b_j) such that in them the pair (a_i, b_j) is followed either by (a_i, b_{j+1}) or by (a_{i+1}, b_j). It is easily verified that each such sequence is the sequence of vertices of an ordered $(q+r)$-dimensional Euclidean simplex in $X \times Y$, and that these simplexes together with their faces form a triangulation of $X \times Y$. This triangulation converts $X \times Y$ into $X \times_s Y$. Clearly if X and Y are the faces of ordered simplexes X', Y', then the inclusion $X \times_s Y \to X' \times_s Y'$ is a simplicial embedding. Figure 12 shows the cases $q = r = 1$, and $q = 2, r = 1$.

To construct the simplicial product of ordered topological simplexes X, Y, we construct by the previous method a triangulation of the product of the unit simplexes $T^q, T^r, q = \dim X, r = \dim Y$, and transfer it to $X \times Y$ by the product $T^q \times T^r \to X \times Y$ of the canonical simplicial homeomorphisms $T^q \to X, T^r \to Y$. The resulting triangulation converts $X \times Y$ into $X \times_s Y$.

Finally, suppose X and Y are any ordered simplicial spaces. An obvious verification shows that the covering of the space $X \times_c Y$ by the products $s \times t$, triangulated as above, where s and t are simplexes of X and Y respectively, forms a triangulation of $X \times_c Y$, which converts it into $X \times_s Y$.

Note that each cell of $X \times_c Y$ is the union of a finite number of cells of $X \times_s Y$ of the same or lower dimension. In particular, the map id $: X \times_c Y \to X \times_s Y$ is cellular.

The construction of the simplicial product obviously has the property that if $f : X \to X'$ and $g : Y \to Y'$ are monotonic simplicial maps, then $f \times g :$

$X \times Y \to X' \times Y'$ is a simplicial map. Also if X and Y are subspaces of ordered simplicial spaces X' and Y', then $X \times_s Y$ is a subspace of $X' \times_s Y'$.

We conclude by describing the simplicial product in the language of schemes. Let (M_1, S_1) and (M_2, S_2) be the simplicial schemes of spaces X_1 and X_2 respectively. Then the simplicial scheme of $X_1 \times_s X_2$ is $(M_1 \times M_2, S)$, where S is the collection of sets $A \subset M_1 \times M_2$ with the two properties: (i) $\mathrm{pr}_1(A) \in S_1$, $\mathrm{pr}_2(A) \in S_2$; (ii) if $(a_1, a_2) \in A$, $(a_1', a_2') \in A$ and $a_1 \le a_1'$, then $a_2 \le a_2'$.

C. *The join, cone and suspension.* If X, Y are topological simplexes, then the join $X * Y$ is a topological simplex in a natural sense: the characteristic homeomorphisms are maps of the form

$$T^{\dim X + \dim Y + 1} = T * S \xrightarrow{\phi * \psi} X * Y,$$

where T, S are opposite faces of the simplex $T^{\dim X + \dim Y + 1}$ with $\dim T = X$, $\dim S = \dim Y$, ϕ, ψ simplicial homeomorphisms, and the equality $T^{\dim X + \dim Y + 1} = T * S$ denotes the simplicial homeomorphism set up in 8.1.A. This enables us to endow the cellular join $X *_c Y$ of arbitrary simplicial spaces X, Y with a canonical triangulation: its simplexes are the images of the simplexes of X, Y under the inclusions $X \to X *_c Y$, $Y \to X *_c Y$ and the images of the simplexes $s * t$, where s is a simplex of X and t a simplex of Y, under the embedding $\mathrm{in} * \mathrm{in} : s * t \to X *_c Y$. The resulting simplicial space is called the *simplicial join* of X and Y, denoted by $X *_s Y$. As a cellular space it is the same as $X *_c Y$.

If (M_1, S_1) is the scheme of X, and (M_2, S_2) the scheme of Y, then the scheme of $X *_s Y$ is obviously $(M_1 \coprod M_2, S)$, where S is the collection of nonempty subsets of $M_1 \coprod M_2$, whose inverse image under $\mathrm{in}_1 : M_1 \to M_1 \coprod M_2$ belongs to S_1 or is empty, and whose inverse image under $\mathrm{in}_2 : M2 \to M_1 \coprod M_2$ belongs to S_2 or is empty.

Since for any topological space X the cone CX is canonically homeomorphic to $X * D^0$, and the suspension ΣX is canonically homeomorphic to $X * S^0$ (see 1.4.F), the construction above makes the cone and suspension of any simplicial space into simplicial spaces.

8.4. Stars, Links, Regular Neighbourhoods.

A. The *star* of a simplex s of a simplicial space X is the union of the simplexes of X that contain s; it is denoted by $\mathrm{St} s$ or $\mathrm{St}_X s$. It is clearly a subspace of X.

The *open star* of a simplex s is the union of the interiors of the simplexes that contain s; it is denoted by $\mathrm{st} s$ or $\mathrm{st}_X s$. It is clearly an open set, determined by the equations

$$b_{a_0}(x) > 0, \dots, b_{a_q}(x) > 0,$$

where $a_0, \dots, a_q$ are the vertices of s; also $\mathrm{Cl}\, \mathrm{st} s = \mathrm{St} s$.

The *link* of a simplex s is the union of the simplexes in $\mathrm{St} s$ that do not intersect s. It is denoted by $\mathrm{lk} s$ or $\mathrm{lk}_X s$, and is a subspace of X and $\mathrm{St} s$.

Clearly if s is a face of a simplex t, then $\mathrm{St} t \subset \mathrm{St} s$, $\mathrm{st} t \subset \mathrm{st} s$, $\mathrm{lk} t \subset \mathrm{lk} s$. If $a_0, \dots, a_q$ are vertices not lying in the same simplex then $\cap_{i=0}^{q} \mathrm{st}\, a_i$ is empty, and if $a_0, \ldots, a_q$ are vertices of a simplex s, then $\cap_{i=0}^{q} \operatorname{st} a_i = \operatorname{st} s$. If X is a subspace of Y, then for each simplex s in X

$$\operatorname{st}_X s = X \cap \operatorname{st}_Y s,$$

and if X is a complete subspace, then also

$$\operatorname{St}_X s = X \cap \operatorname{St}_Y s \quad \text{and} \quad \operatorname{lk}_X s = X \cap \operatorname{lk}_Y s.$$

Finally, if s' is a simplex in $\operatorname{lk}_X s$, and s'' is the smallest simplex containing s and s', then

$$\operatorname{lk}_{\operatorname{lk}_X s} s' = \operatorname{lk}_X s''.$$

Stars, open stars and links are defined also for the points of a simplicial space: the star $\operatorname{St} x = \operatorname{St}(x, X)$, open star $\operatorname{st} x = \operatorname{st}(x, X)$, and link $\operatorname{lk} x = \operatorname{lk}(x, X)$ of a point x are defined by

$$\operatorname{St} x = \operatorname{St} s, \quad \operatorname{st} x = \operatorname{st} s, \quad \operatorname{lk} x = \operatorname{St} s \setminus \operatorname{st} s,$$

where s is the smallest simplex containing x. Clearly $\operatorname{st} x$ is a neighbourhood of x, and $\operatorname{lk} x = \operatorname{Fr} \operatorname{St} x = \operatorname{Fr} \operatorname{st} x$. Moreover $\operatorname{St} x$ is homeomorphic to the cone over $\operatorname{lk} x$; there is even a canonical homeomorphism $C \operatorname{lk} x \to \operatorname{St} x$ defined by

$$\operatorname{pr}(y, t) \to \phi((1 - t)\phi^{-1}(x) + t\phi^{-1}(y)),$$

where $y \in \operatorname{lk} x$, $t \in I$, and ϕ is a characteristic homeomorphism of some simplex containing x and y.

Warning: The link of a point x is only the same as the link of the smallest simplex containing x when x is a vertex.

It is easy to see that for any simplex s of X there is a canonical simplicial homeomorphism between the join $s * \operatorname{lk}_X s$ and $\operatorname{St}_X s$. This means that the star $\operatorname{St}_X x$ of any point of X is canonically simplicially homeomorphic to $s * \operatorname{lk}_X s$, where s is the smallest simplex containing x, and this simplicial homeomorphism clearly maps the join of the boundary of s and $\operatorname{lk}_X s$ onto $\operatorname{lk}_X x$. Further, since the boundary of s is homeomorphic to the sphere $S^{\dim s - 1}$, and the join $S^{\dim s - 1} * \operatorname{lk}_X s$ is homeomorphic to the suspension $\Sigma^{\dim s} \operatorname{lk}_X s$, it follows that $\operatorname{lk}_X x$ is homeomorphic to $\Sigma^{\dim s} \operatorname{lk}_X s$.

B. *The homotopy invariance of the link of a point*. It is not hard to show that if T_1, T_2 are subspaces of a topological space X, each having a finite triangulation, and if $x_0 \in X$ is in the interior of T_1 and T_2, then the links $\operatorname{lk}_{T_1} x_0$ and $\operatorname{lk}_{T_2} x_0$ are homotopy equivalent.

C. The *regular neighbourhood* of a subspace A of a simplicial space X is the neighbourhood consisting of all open stars $\operatorname{st}_X a$ with $a \in A$, or, equivalently, of all open stars $\operatorname{st}_X a$ with $a \in \operatorname{sk}_0 A$.

If A is complete, then it is obviously a deformation retract of its regular neighbourhood. It follows from this in particular that every subspace of a simplicial space X is a deformation retract of its regular neighbourhood in ba X.

D. The *barycentric star* of a simplex s in the simplicial space X is the union of the simplexes of ba X whose first vertex is at the centre of s. It is denoted by bst s or bst$_X s$. Equivalently: bst s is the set of $x \in X$ for which $b_a(x) = b_b(x)$ if $a, b \in s \cap \mathrm{sk}_0 X$, and $b_a(x) \geq b_b(x)$ if $a \in s \cap \mathrm{sk}_0 X$, $b \in (X \setminus s) \cap \mathrm{sk}_0 X$.

The barycentric stars of the simplexes of X obviously cover X and are subspaces of ba X. Clearly bst $s \neq$ bst s' if $s \neq s'$, and bst $s \subset$ bst s' iff $s \supset s'$.

The union of the simplexes of the barycentric star bst s that do not contain the centre of s is called the *barycentric link* of s. The star bst s is clearly simplicially homeomorphic to the cone over the barycentric link, and the barycentric link is simplicially homeomorphic to ba lk s. Hence the pair consisting of bst s and the barycentric link of s is homeomorphic to the pair $(C \, \mathrm{lk} \, s, \mathrm{lk} \, s)$.

8.5. Simplicial Approximation of a Continuous Map.

A. A simplicial map g from a simplicial space X to a simplicial space Y is called a *simplicial approximation* to a continuous map $f : X \to Y$ if for each point $x \in X$, the point $g(x)$ belongs to the smallest simplex of Y containing $f(x)$.

A simplicial approximation g to $f : X \to Y$ is canonically homotopic to f; the obvious canonical homotopy is stationary on the set of $x \in X$ for which $g(x) = f(x)$.

It is easy to see that a simplicial map $g : X \to Y$ is a simplicial approximation to a continuous map $f : X \to Y$ iff $f(\mathrm{st} \, a) \subset \mathrm{st} \, g(a)$ for each vertex a of X. Hence a continuous map f from a simplicial space X to a simplicial space Y has a simplicial approximation iff for each vertex a of X there exists a vertex b of Y such that $f(\mathrm{st} \, a) \subset \mathrm{st} \, b$. The following theorem is an obvious consequence of this.

B. For any continuous map f of a finite simplicial space X into a simplicial space Y, there exists a positive integer m such that the map $f : \mathrm{ba}^m X \to Y$ has a simplicial approximation.

§9. Cellular Approximation of Maps and Spaces

9.1. Cellular Approximation of a Continuous Map.

A. Recall that a continuous map f from a cellular space X to a cellular space Y is *cellular* if $f(\mathrm{sk}_r X) \subset \mathrm{sk}_r Y$ for each r. From the point of view of homotopy theory this condition is not onerous. In fact the following fundamental theorem holds.

Cellular approximation theorem.

Every continuous map of a cellular space to a cellular space is homotopic to a cellular map. A continuous map of a cellular space X to a cellular space Y that is cellular on a subspace A of X is A-homotopic to a cellular map. Homotopic cellular maps are cellular homotopic. Further, if cellular maps $f, g : X \to Y$ are A-homotopic, where A is a subspace of X, then there is a cellular A-homotopy from f to g. $\square$

The usual proof of this theorem depends on several auxiliary results, which have an independent interest and are given below.

B. A cellular pair is a Borsuk pair. It follows in particular that if (X, A) is a cellular pair and the inclusion $A \to X$ is a homotopy equivalence, then A is a strong deformation retract of X (see 2.6), and also that if X is a cellular space and A a contractible subspace (contractible means that id : $A \to A$ is homotopic to a constant), then the projection $(X, A) \to (X/A, A/A)$ is a homotopy equivalence. It follows in turn from this that if (X, A) is any cellular pair, then the obvious map $X \cup_{\text{in}: A \to X} CA \to x/A$ is a homotopy equivalence.

C. Let k be a non-negative integer or ∞. If (X, A) is a cellular pair in which each cell of $X \setminus A$ has dimension not greater than k, and (Y, B) is any k-connected topological pair, then each continuous map $f : X \to Y$ with $f(A) \subset B$ is A-homotopic to a map taking X into a subset of B. In particular, each continuous map of a k-dimensional cellular space into a k-connected topological space is homotopic to a constant map.

Such homotopies may be constructed cell by cell, starting with the cells of lowest dimension. A homotopy can be extended from successive cells to the whole space since cellular pairs are Borsuk pairs.

If f is taken as the identity map, the above assertions become the following: if the cellular pair (X, A) is k-connected and each cell of $X \setminus A$ has dimension not greater than k, then A is a strong deformation retract of X. In particular, a k-connected k-dimensional cellular space is contractible.

D. Finally we state a basic technical lemma. This, together with the results in **B** and **C**, will give the results in **A**.

Lemma. *Let* $X = A \cup_\phi \coprod_{\mu \in M}(D_\mu = D^{k+1})$, *where A is a topological space, and* ϕ *a continuous map of* $\coprod_{\mu \in M}(S_\mu = S^k)$ *into A. Then the pair (X, A) is k-connected.* □

To prove this, by taking an arbitrary continuous map $(D^r, S^{r-1}) \to (X, A)$, $r \leq k$, and using for example the simplicial approximation theorem (8.5.**B**), we can free a point in each D_μ from the image of the map (by means of an arbitrarily small homotopy), and then take the composition of the resulting map with the obvious deformation retraction of the complement of these points onto A. □

9.2. Cellular k-connected Pairs.

A. Each connected cellular space contains a contractible one-dimensional subspace that contains all the 0-cells. To construct such a subspace, we may for example take an arbitrary 0-cell, and adjoin to it some of the closed 1-cells whose boundaries contain it but are not identical with it, taking care that all 0-cells which can be adjoined in this way are adjoined, but that each of them is joined to the initial cell by only one of the chosen 1-cells. Then adjoin the closed 1-cells whose boundaries intersect the subspace already constructed but are not wholly contained in it, taking care as before that all 0-cells which can be adjoined in this way are adjoined, but that each of them appears only once, and so on, repeating this process infinitely often if necessary.

B. If the subspace described in **A** is contracted to a point, the homotopy type of the space is unchanged. Hence every connected cellular space is homotopy equivalent to a cellular space of the same or lower dimension with a single 0-cell. In particular, every connected one-dimensional cellular space is homotopy equivalent to a bouquet of circles.

C. Assertion **A** has the following generalization. Every k-connected cellular pair $(0 \leq k \leq \infty)$ is homotopy equivalent to a cellular pair (Y, B) with $B \supset \mathrm{sk}_k Y$.

The simplest case is when the complement $X \setminus A$ in the original cellular pair (X, A) has only one cell e with $\dim e \leq k$. The construction giving (Y, B) in this case is as follows: $Y := X \cup_H D^{\dim e} \times I^2$, where $H : D^{\dim e} \times I \to X$ is a cellular $S^{\dim e - 1}$-homotopy between the characteristic map of the cell e, and a map whose image is contained in A, and $B := A \cup (D^{\dim e} \times (\partial I^2 \setminus \mathrm{Int}\, I))$. In the general case repeated application of this construction produces the desired effect.

D. It follows from assertion **C** in particular that every k-connected cellular space X $(0 \leq k \leq \infty)$ is homotopy equivalent to a cellular space Y whose k-skeleton consists of a single point. This generalizes assertion **B**. However, in contrast to the case $k = 0$ discussed in **B** where a space Y was obtained with $\dim Y \leq \dim X$, here our construction only guarantees that $\dim Y \leq \max(\dim X, k + 2)$.

E. Assertion **D** obviously implies the following homotopy properties of cellular constructions. If a cellular space X is k-connected, then the spaces ΣX and $\Sigma(X, x_0)$ where x_0 is a 0-cell, are $(k + 1)$-connected. If cellular spaces X, Y are k-connected and l-connected respectively, and $x_0 \in X$, $y_0 \in Y$ are 0-cells, then the tensor products $(X, x_0) \otimes (Y, y_0)$ and $(X, x_0) \otimes_c (Y, y_0)$ are $(k + l + 1)$-connected, and the joins $X * Y$, $X *_c Y$, and $(X, x_0) * (Y, y_0)$, $(X, x_0) *_c (Y, y_0)$ are $(k + l + 2)$-connected.

9.3. Simplicial Approximation of Cellular Spaces.

A. Theorem. *If X is a cellular space, there is a simplicial space of the same dimension that is homotopy equivalent to X, which is finite if X is finite, and countable if X is countable.* $\square$

B. The simplicial space whose existence is asserted in Theorem **A** is constructed as follows. If X is 0-dimensional, then X itself can be regarded as a simplicial space with all the required properties. If X has dimension r, and a simplicial space Y and homotopy equivalence $f' : \mathrm{sk}_{r-1} X \to Y$ have already been constructed for its $(r - 1)$-skeleton, then by representing X in the form $(\mathrm{sk}_{r-1} X) \cup_\phi \Delta$, where $\Delta = \bigsqcup_{e \in M_r} D_e^r$ and M_r is the set of r-cells of X and ϕ is a continuous map of $\Sigma = \cup_{e \in M_r} S_e^{r-1}$ into $\mathrm{sk}_{r-1} X$, and taking a simplicial approximation $g : \Sigma \to Y$ of $f' \circ \phi$ as a suitable triangulation of the pair ϕ, Σ, we can obtain the required simplicial space as the result of attaching Δ to Y by g. Applying this construction successively to all the skeletons of any cellular space X gives a sequence of simplicial spaces, whose union is the required cellular space.

C. Corollary to Theorem A. *If X is a finite cellular space, and Y a countable cellular space, then the set $\pi(X, Y)$ of homotopy classes of maps $X \to Y$ is countable.*

In fact, by Theorem **A** it is sufficient to consider the case when X and Y are simplicial, but in this case the cardinality of the set $\pi(X, Y)$ is not greater than the cardinality of the set of all simplicial maps $\mathrm{ba}^m X \to Y$ $(m = 0, 1, \ldots)$ by 8.5.**B**.

D. Theorem **A** can be generalized to cellular pairs in an obvious way.

9.4. Weak Homotopy Equivalence.

A. A continuous map f of a topological space X to a topological space Y is a *weak homotopy equivalence* if the homomorphism $f_* : \pi_r(X, x) \to \pi_r(Y, f(x))$ is an isomorphism for all $r \geq 0$ and $x \in X$. This terminology is justified by two facts: a homotopy equivalence is always a weak homotopy equivalence, but the converse is not true. The first fact has already been mentioned in 3.1; to establish the second it is sufficient to consider the map of the set of natural numbers defined by

$$f(x) = \begin{cases} 0, & \text{if } x = 1, \\ x^{-1}, & \text{if } x \neq 1, \end{cases}$$

onto its image regarded as a subspace of $\mathbb{R}$.

The composition of two weak homotopy equivalences is obviously a weak homotopy equivalence.

B. The following theorem is easily proved using the results of 9.1.

Let $f : X \to Y$ be a weak homotopy equivalence; then for any cellular pair (K, L) and any continuous maps $\phi : K \to Y$, $\psi : L \to X$ with $f \circ \psi = \phi|_L$, there exists a continuous map $\chi : K \to X$ such that $\chi|_L = \psi$ and $f \circ \chi$ is L-homotopic to ϕ. The converse is also true: what is more, if $f : X \to Y$ is continuous, and for any continuous maps $\phi : D^r \to Y$, $\psi : S^{r-1} \to X$ ($r \geq 0$) with $f \circ \psi = \phi|_{S^{r-1}}$, there exists a continuous map $\chi : D^r \to X$ such that $\chi|_{S^{r-1}} = \psi$ and $f \circ \chi$ is S^{r-1}-homotopic to ϕ, then f is a weak homotopy equivalence.

C. It obviously follows from this theorem that if $f : X \to Y$ is a weak homotopy equivalence, then for any cellular space M the map $\pi(\mathrm{id}, f) : \pi(M, X) \to \pi(M, Y)$ is invertible.

D. If X and Y are cellular spaces, then every weak homotopy equivalence $f : X \to Y$ is a homotopy equivalence. For by C, the map $\pi(\mathrm{id}, f) : \pi(Y, X) \to \pi(Y, Y)$ is invertible, and hence there is a continuous map $g : Y \to X$ whose class is mapped to the class of id_Y by $\pi(\mathrm{id}, f)$. The map g is easily seen to be a homotopy inverse of f.

E. Theorem **D** states that connected cellular spaces X, Y are homotopy equivalent if there is a continuous map $X \to Y$ that induces isomorphisms of their homotopy groups, but certainly does not state that they are homotopy equivalent if their homotopy groups are merely isomorphic. This is incorrect, as simple examples show such as $X = S^p \times \mathbb{R}P^q$ and $Y = \mathbb{R}P^p \neq S^q$ with $p \neq q$, or $X = S^2 \times \mathbb{C}P$ and $Y = S^3$.

F. A topological space is *homotopy valid* if it is homotopy equivalent to a cellular space. It follows from Theorem **D** that if X and Y are homotopy valid then every weak homotopy equivalence $X \to Y$ is a homotopy equivalence.

All CNRS spaces and all topological manifolds are homotopy valid spaces. The product of two homotopy valid spaces is a homotopy valid space. If Y is a homotopy valid space then $C(X, Y)$ is homotopy valid for any compact X. If Y is homotopy equivalent to a countable cellular space then $C(X, Y)$ is homotopy equivalent to a countable cellular space for any compact space X with a countable basis.

The simplest example of a space that is not homotopy valid is the subspace of the line consisting of the points 0 and $1/n$, $n = 1, 2, \ldots$. This space is not connected; an example of a connected (indeed ∞-connected) space that is not homotopy valid is the union of the graph of the function $x \mapsto \sin(1/x)$ on the interval $0 < x \le 1/\pi$ and the four-segment broken line with vertices $(1/\pi, 0)$, $(1/\pi, 2)$, $(-1, 2)$, $(-1, 0)$, $(0, 0)$; see Fig. 13.

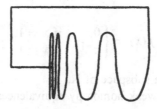

Fig. 13

G. *The relative case.* A map f of a topological pair (X, A) into a topological pair (Y, B) is a *weak homotopy equivalence* if the corresponding maps $X \to Y$ and $A \to B$ are weak homotopy equivalences.

Note that if $F : (X, A) \to (Y, B)$ is a weak homotopy equivalence, then the homomorphism $f_* : \pi_r(X, A, x) \to \pi_r(Y, B, f(x))$ is obviously an isomorphism for any $r \ge 1$ and $x \in A$. Conversely, if $f : (X, A) \to (Y, B)$ is a continuous map such that one of the corresponding maps $X \to Y$, $A \to B$ is a weak homotopy equivalence and all the homomorphisms $f_* : \pi_r(X, A, x) \to \pi_r(Y, B, f(x))$, $r \ge 1$, $f_* : \pi_0(X, x) \to \pi_0(Y, f(x))$, $f_* : \pi_0(A, x) \to \pi_0(B, f(x))$, $x \in A$ are isomorphisms, then f is a weak homotopy equivalence.

Theorem **C** has the following relative version. If $f : (X, A) \to (Y, B)$ is a weak homotopy equivalence, then for any cellular pair (M, N) the map of the set of homotopy classes of maps $(M, N) \to (X, A)$ into the set of homotopy classes of maps $(M, N) \to (Y, B)$ induced by f, is invertible. It follows from this (cf. the absolute case in **D**) that if the pairs (X, A), (Y, B) are cellular, then every weak homotopy equivalence $(X, A) \to (Y, B)$ is a homotopy equivalence.

H. *k-equivalence.* A continuous map f from a topological space X to a topological space Y is a *k-equivalence* if for any $x \in X$ the homomorphism $f_* : \pi_r(X, x) \to \pi_r(Y, f(x))$ is an isomorphism for $r < k$ and an epimorphism for $r = k$. Here k is a non-negative integer; a weak homotopy equivalence is sometimes called an ∞-equivalence. The composition of two k-equivalences is clearly a k-equivalence. Theorem **B** can be transferred to the case of k-equivalence as follows. Let $f : X \to Y$ be a k-equivalence; then for any cellular pair (K, L) with $K \setminus L \subset \mathrm{sk}_k K$ and any continuous maps $\phi : K \to Y$, $\psi : L \to X$ with $f \circ \psi = \phi|_L$, there exists a continuous map $\chi : K \to X$ such that $\chi|_L = \psi$ and $f \circ \chi$ is L-homotopic to ϕ. The converse is also true; what is more, if $f : X \to Y$ is continuous, and for any continuous maps $\phi : D^r \to Y$, $\psi : S^{r-1} \to X$ $(0 \le r \le k)$ with $f \circ \psi = \phi|_{S^{r-1}}$, there exists a continuous map $\chi : D^r \to X$ such that $\chi|_{S^{r-1}} = \psi$, and $f \circ \chi$ is S^{r-1}-homotopic to ϕ, then f is a k-equivalence.

This theorem has corollaries similar to those of Theorem **B**. First, if $f : X \to Y$ is a k-equivalence, then the map $\pi(\mathrm{id}, f) : \pi(M, X) \to \pi(M, Y)$ is invertible for any cellular space M with $\dim M < K$, and is onto $\pi(M, Y)$ when $\dim M = k$. Second, if X is a cellular space with $\dim X < k$ and Y a cellular space with $\dim Y \leq k$, then every k-equivalence $X \to Y$ is a homotopy equivalence.

I. Many of the operations described in 1.4 that can be performed on maps transform weak homotopy equivalences into weak homotopy equivalences. For example, it is clear that for any family of weak homotopy equivalences $\{f_\mu : X_\mu \to Y_\mu\}_{\mu \in M}$, the map $\coprod_\mu f_\mu : \coprod_\mu X_\mu \to \coprod_\mu Y_\mu$ is a weak homotopy equivalence, and if $f_1 : X_1 \to Y_1$, $f_2 : X_2 \to Y_2$ are weak homotopy equivalences then $f_1 \times f_2 : X_1 \times X_2 \to Y_1 \times Y_2$ is a weak homotopy equivalence.

In less obvious cases the proofs are based on the following property of weak homotopy equivalences. If f is a map from the triad (X, A, B) to the triad (Y, C, D) such that $f| : A \to C$, $f| : B \to D$, and $f| : A \cap B \to C \cap D$ are weak homotopy equivalences, and if $\mathrm{Int}\, A \cup \mathrm{Int}\, B = X$ and $\mathrm{Int}\, C \cup \mathrm{Int}\, D = Y$, then $f : X \to Y$ is also a weak homotopy equivalence. Using this result it is not hard to show that under certain conditions the result of gluing weak homotopy equivalences is a weak homotopy equivalence. More precisely, let (X, C), (X', C') be Borsuk pairs, Y, Y' topological spaces, $\phi : C \to Y$, $\phi' : C' \to Y'$ continuous maps, and $f : (X, C) \to (X', C')$, $g : Y \to Y'$ weak homotopy equivalences such that $g \circ \phi = \phi' \circ f|_C$; then the map $Y \cup_\phi X \to Y' \cup_{\phi'} X'$ determined by f and g is also a weak homotopy equivalence.

J. We now state some consequences of these assertions. Let f be a map of the triad (X, A, B) into the triad (X', A', B') such that $f| : A \to A'$, $f| : B \to B'$, and $f| : A \cap B \to A' \cap B'$ are weak homotopy equivalences. If $(A, A \cap B)$ and $(A', A' \cap B')$ are Borsuk pairs, then $f : X \to X'$ is a weak homotopy equivalence.

K. If (X, A), (Y, B) are Borsuk pairs, and $f : (X, A) \to (Y, B)$ is a weak homotopy equivalence, then $f/ : X/A \to Y/B$ is also a weak homotopy equivalence.

It immediately follows from this that if $f : X \to Y$ is a weak homotopy equivalence, then so is $\Sigma f : \Sigma X \to \Sigma Y$, and if spaces X_μ and Y_μ form Borsuk pairs with their base points x_μ and y_μ, and $f_\mu : (X_\mu, x_\mu) \to (Y_\mu, y_\mu)$ are weak homotopy equivalences, then $\vee_\mu f_\mu : \vee_\mu (X_\mu, x_\mu) \to \vee_\mu (Y_\mu, y_\mu)$ is a weak homotopy equivalence.

L. If $f : X \to X'$, $g : Y \to Y'$ are weak homotopy equivalences then so is $f * g : X * Y \to X' * Y'$.

M. If $f : X \to Y$ is a weak homotopy equivalence and K is a locally finite cellular space, then $C(\mathrm{id} f) : C(K, X) \to C(K, Y)$ is a weak homotopy equivalence.

9.5. Cellular Approximation to Topological Spaces.

A. A *cellular approximation to a topological space* X is any pair (K, ϕ), where K is a cellular space and $\phi : K \to X$ is a weak homotopy equivalence.

B. Example: if X is a Hausdorff space endowed with a cell decomposition such that every compact subset of X intersects only finitely many cells, then the cellular space obtained from X by cellular weakening of the topology together with id_X is a cellular approximation to X. In particular, if X and Y are cellular spaces, then

$(X \times_c Y, \mathrm{id})$ is a cellular approximation to $X \times Y$, and $(X *_c Y, \mathrm{id})$ is a cellular approximation to $X * Y$.

C. Theorem. *Every topological space has a cellular approximation.* $\square$

A cellular approximation to X can be constructed as the limit of an increasing sequence of cellular spaces $K_0 \subset K_1 \subset \cdots$ and maps $\phi_i : K_i \to X$, such that ϕ_i is an i-equivalence. The space K_0 and map ϕ_0 are constructed in an obvious way. The space K_r, $r > 0$, is constructed by gluing r-cells to K_{r-1}: in each component of K_{r-1}, one cell is glued to each generator of the r-dimensional homotopy group of that component, and one cell is glued to each generator of the kernel of map of $(r - 1)$-dimensional homotopy groups induced by ϕ_{r-1}. By an obvious method of gluing the map ϕ_{r-1} can be extended to a map ϕ_r that is an r-equivalence.

D. If (K, ϕ) and (L, ψ) are cellular approximations to topological spaces X and Y, then for any continuous map $f : X \to Y$ there exists a map $g : K \to L$ such that the diagram

$$
\begin{array}{ccc}
K & \xrightarrow{g} & L \\
\phi \downarrow & & \downarrow \psi \\
X & \xrightarrow{f} & Y
\end{array}
$$

is homotopy commutative (that is, the maps $f \circ \phi$, $\psi \circ g$ are homotopic). The homotopy class of g is determined uniquely by this condition. It is the inverse image of $f \circ \phi$ under the bijection $\pi(\mathrm{id}, \psi) : \pi(K, L) \to \pi(K, Y)$.

E. It follows from **D** that a cellular approximation to a topological space is unique up to homotopy equivalence: for any cellular approximations (K, ϕ), (L, ψ) to a given topological space, there is a homotopy equivalence $g : K \to L$ such that $\psi \circ g$ is homotopic to ϕ.

F. *Weak homotopy equivalence as a relation.* Two topological spaces are called *weak homotopy equivalent* if they have cellular approximations (K, ϕ), (L, ψ) with $K = L$. This relation is obviously an equivalence relation in the usual sense.

If there is a weak homotopy equivalence $f : X \to Y$ between spaces X, Y, then X and Y are weak homotopy equivalent. The converse is false, as simple examples show. The sets $\mathbb{Q}$ and $\mathbb{Z} \subset \mathbb{R}$ are weak homotopy equivalent, but there is no weak homotopy equivalence $\mathbb{Q} \to \mathbb{Z}$: a continuous map $\mathbb{Q} \to \mathbb{Z}$ is necessarily constant on entire intervals and hence the map $\pi_0(\mathbb{Q}) \to \pi_0(\mathbb{Z})$ that it induces cannot be one-one. The spaces $(\mathbb{Q} \times S^1) \coprod \mathbb{Z}$ and $(\mathbb{Z} \times S^1) \coprod \mathbb{Q}$ are also weak homotopy equivalent, but there is no weak homotopy equivalence $(\mathbb{Q} \times S^1) \coprod \mathbb{Z} \to (\mathbb{Z} \times S^1) \coprod \mathbb{Q}$, nor a weak homotopy equivalence $(\mathbb{Z} \times S^1) \coprod \mathbb{Q} \to (\mathbb{Q} \times S^1) \coprod \mathbb{Z}$.

If two spaces are homotopy equivalent then they are of course also weakly homotopy equivalent. The converse is false; for example, every topological space is weak homotopy equivalent to a cellular space, but not every topological space is homotopy equivalent to a cellular space (see 9.4.F). On the other hand, it follows from Theorem 9.4.**D** that in the cellular case weak homotopy equivalent spaces are homotopy equivalent.

G. *Cellular approximation of topological pairs.* A *cellular approximation of a topological pair* (X, A) is any pair $[(K, L), \phi]$ consisting of a cellular pair (K, L) and a weak homotopy equivalence $\phi : (K, L) \to (X, A)$. Jn the case when A

and L are single points, a cellular approximation $[(K, L), \phi]$ of (X, A) is called a *cellular approximation to the pointed space* X.

Relative version of Theorem **C**. Every topological pair (X, A) has a cellular approximation. Moreover, for each cellular approximation (L, ψ) of the subspace A there exists a cellular approximation $[(K, L), \phi]$ of (X, A) such that $\psi = \phi|$. Theorems **D** and **E** also go over to the case **G** in an obvious way.

H. *Cellular approximations and constructions.* If (K, ϕ), (L, ψ) are cellular approximations to spaces X, Y, then $(K \coprod L, \phi \coprod \psi)$, $(K \times_c L, \phi \times \psi)$, $(K *_c L, \phi * \psi)$ and $(\Sigma K, \Sigma \phi)$ are cellular approximations to $X \coprod Y$, $X \times Y$, $X * Y$, and ΣX. If $[(K_\mu, y_\mu), \phi_\mu]$ are cellular approximations to spaces X_μ with base points x_μ, such that (X_μ, x_μ) are Borsuk pairs, then $[\vee_\mu (K_\mu, y_\mu), \phi_\mu]$ is a cellular approximation to the bouquet $\vee_\mu/X_\mu, x_\mu)$. If $[(K, L), \phi]$ is a cellular approximation to a Borsuk pair (X, A), then $(K/L, \phi/ : K/L \to X/A)$ is a cellular approximation to the space X/A.

9.6. The Covering Homotopy Theorem.

A. If $\xi = (E, p, B)$ is a Serre bundle and (X, A) a cellular pair, then for any continuous map $f : X \to E$, any homotopy $F : X \times I \to B$ of $p \circ f$, and any homotopy $G : A \times I \to X$ of $f|_A$ covering $F|_{A \times I}$, there exists a homotopy of f that covers F and extends G. Such a homotopy can be constructed cell by cell for each cell in $X \setminus A$ in order of increasing dimension by using the Serre condition.

B. The absolute version of this theorem is the following: if $\xi = (E, p, B)$ is a Serre bundle and f is a continuous map of a cellular space X into E, then every homotopy of $p \circ f$ can be covered by a homotopy of f.

C. The following theorem, which generalizes the theorem on homotopies of covering paths (Theorem (ii) in 6.2), can easily be deduced from Theorem **A**. Let $\xi = (E, p, B)$ be a covering in the wide sense, X a connected cellular space with base point the 0-cell x_0, and $f, g : X \to E$ continuous maps. If $p \circ f$ and $p \circ g$ are x_0-homotopic and $f(x_0) = g(x_0)$, then f and g are x_0-homotopic.

D. *Weak homotopy equivalence of the fibres of Serre bundles.* Any two fibres of a Serre bundle with connected base are weak homotopy equivalent. In fact, given a cellular approximation of one fibre, we may obtain a cellular approximation of the other fibre by applying Theorem **B** to the composition of the first cellular approximation and the inclusion of the first fibre in the total space, and to the homotopy constructed by means of a path in the base joining the image of the first fibre to the image of the second.

Chapter 4
The Simplest Calculations

§10. The Homotopy Groups of Spheres
and Classical Manifolds

10.1. Suspension in the Homotopy Groups of Spheres.

A. Consider a q-dimensional spheroid $f : (S^r, (1, 0, \ldots, 0)) \to (X, x_0)$ of a topological space X with base point x_0. The map $\Sigma f : \Sigma S^r = S^{r+1} \to \Sigma X : \mathrm{pr}(y, t) \mapsto (f(y), t)$ is an $(r + 1)$-dimensional spheroid of X. If spheroids $f, g : S^r \to X$ are homotopic, then clearly so are $\Sigma f, \Sigma g : S^{r+1} \to \Sigma X$. It is easy to verify that $\Sigma(f + g)$ is homotopic to $\Sigma f + \Sigma g$. Thus the correspondence $f \to \Sigma f$ induces a homomorphism $\pi_r(X, x_0) \to \pi_{r+1}(\Sigma(X, x_0))$, called the *suspension homomorphism* and denoted by Σ. In particular for any q and n there is a homomorphism $\Sigma : \pi_r(S^n) \to \pi_{q+1}(S^{n+1})$.

B. If $f : (X, x_0) \to (Y, y_0)$ is a continuous map from one pointed space to another then the diagram

$$
\begin{array}{ccc}
\pi_r(X, x_0) & \overset{\Sigma}{\longrightarrow} & \pi_{r+1}(\Sigma(X, x_0)) \\
\downarrow f_* & & \downarrow \Sigma f_* \\
\pi_r(Y, y_0) & \overset{\Sigma}{\longrightarrow} & \pi_{r+1}(\Sigma(Y, y_0))
\end{array}
$$

(where Σf is understood in the sense of 1.5 and 1.4) is commutative for any $r \geq 0$.

C. Another description of the suspension homomorphism $\Sigma : \pi_r(X, x_0) \to \pi_{r+1}(\Sigma(X, x_0))$ is provided by the map $(X, x_0) \to \Omega(\Sigma(X, x_0))$ defined by $x \mapsto [t \mapsto \mathrm{pr}(x, t)]$. Namely, the composition of the homomorphism $\pi_r(X, x_0) \to \pi_r(\Omega(\Sigma(X, x_0)))$ induced by this map with the isomorphism $\pi_r(\Omega(\Sigma(X, x_0))) \to \pi_{r+1}(\Sigma(X, x_0))$ (see 3.1.F) is identical with Σ, as an automatic verification shows.

D. Yet another description of the homomorphism Σ arises from the interpretation of the suspension $\Sigma(X, x_0)$ as the quotient space of the cone $C(X, x_0)$ by its base (identified with X). Namely, $\Sigma : \pi_r(X, x_0) \to \pi_{r+1}(\Sigma(X, x_0))$ is the same as the composite homomorphism

$$
\pi_r(X, x_0) \overset{\partial^{-1}}{\longrightarrow} \pi_{r+1}(C(X, x_0), X, x_0) \overset{\mathrm{pr}_*}{\longrightarrow} \pi_{r+1}(\Sigma(X, x_0))
$$

(the boundary homomorphism $\partial : \pi_{r+1}(C(X, x_0), X, x_0) \to \pi_r(X, x_0)$ of the sequence of the pair $(C(X, x_0), X)$ is invertible because the cone is contractible).

E. The most well-known, the first historically, and the most elementary of the substantial theorems on the suspension homomorphism is the following:

Theorem. *The homomorphism $\Sigma : \pi_r(S^n) \to \pi_{r+1}(S^{n+1})$ is an isomorphism for $r \leq 2n - 2$ and an epimorphism for $r = 2n - 1$.*

We shall not discuss the proof of this theorem, which can be found in all the textbooks. For a generalization see 12.3 below.

F. The main content of Theorem E is the fact that each of the series

$$\cdots \longrightarrow \pi_r(S^n) \xrightarrow{\ \Sigma\ } \pi_{r+1}(S^{n+1}) \xrightarrow{\ \Sigma\ } \pi_{r+2}(S^{n+2}) \xrightarrow{\ \Sigma\ } \cdots$$

into which the suspension divides the homotopy groups of spheres, becomes stable. In more detail: in the kth series $\{\pi_{n+k}(S^n)\}$ the groups $\pi_{n+k}(S^n)$ with $n \geq k+2$ are related by the composite isomorphism established by the suspension. This canonical isomorphism enables the groups $\pi_{n+k}(S^n)$ with $n \geq k+2$ to be identified with a single group, which is called the kth stable group and denoted by Π_k.

10.2. The Simplest Homotopy Groups of Spheres.

A. The groups $\pi_r(S^n)$ with $r < n$ are trivial. In particular $\Pi_k = 0$ if $k < 0$. This follows from 9.1.A.

B. The groups $\pi_r(S^1)$ with $r > 1$ are trivial, and $\pi_1(S^1)$ is an infinite cyclic group generated by the homotopy class of the identity map.

The proof follows instantly by considering the homotopy sequence of the covering $\mathbb{R}^1 \to S^1 : t \to e^{2\pi i t}$ (see 5.2), although the homotopy groups of the circle can be calculated by far more elementary methods than the general theory just mentioned.

C. Corollary. The pair (D^2, S^1) is simple, the homotopy groups $\pi_r(D^2, S^1)$ are trivial for $r \neq 2$, and $\pi_2(D^2, S^1)$ is an infinite cyclic group generated by the homotopy class of the identity map.

D. For each $n \geq 1$, the suspension $\Sigma : \pi_n(S^n) \to \pi_{n+1}(S^{n+1})$ is an isomorphism, and maps the homotopy class of the identity map to the homotopy class of the identity map.

This follows from Theorem 10.1.E and the fact that $\Sigma : \pi_1(S^1) \to \pi_2(S^2)$ is not only an epimorphism (as stated in 10.1.E) but actually an isomorphism. This easily follows from consideration of the homotopy sequence of the Hopf bundle $S^3 \to S^2$.

E. From **B** and **D**, the group $\pi_n(S^n)$ is an infinite cyclic group for $n \geq 1$, generated by the homotopy class of the identity map. In particular, $\Pi_0 = \mathbb{Z}$.

F. Corollary. $\pi_n(D^n, S^{n-1}) = \mathbb{Z}$.

G. For $n \geq 1$, the theorem in **E** establishes a canonical isomorphism $\pi_n(S^n) \to \mathbb{Z}$, and in particular associates an integer with each continuous map $f : S^n \to S^n$. This is called the *degree of the map* f, denoted by $\deg f$. In exactly the same way for $n \geq 2$ the canonical isomorphism $\pi_n(D^n, S^{n-1}) \to \mathbb{Z}$ associates with each continuous map $f : (D^n, S^{n-1}) \to (D^n, S^{n-1})$ its degree $\deg f \in \mathbb{Z}$.

H. It follows from the exactness of the homotopy sequence of the Hopf bundle $S^3 \to S^2$ and theorem **B** that for $r \geq 3$, the homomorphism $\pi_r(S^3) \to \pi_r(S^2)$ induced by the projection of this bundle is an isomorphism. In particular, $\pi_3(S^2)$ is canonically isomorphic to $\mathbb{Z}$ and is generated by the homotopy class of the Hopf map.

I. It follows from theorem 5.3.F applied to the Hopf bundle $S^7 \to S^4$ that for any $r \geq 1$, the homomorphism $\pi_r(S^7) \to \pi_r(S^4)$ induced by the Hopf map $S^7 \to S^4$ maps $\pi_r(S^7)$ isomorphically onto a subgroup of $\pi_r(S^4)$ which has a direct complement isomorphic to $\pi_{r-1}(S^3)$. In particular, $\pi_7(S^4) \simeq \mathbb{Z} \oplus \pi_6(S^3)$.

J. In exactly the same way theorem 5.3.F applied to the Hopf bundle $S^{15} \to S^8$ gives the following result: for any $r \geq 1$, the homomorphism $\pi_r(S^{15}) \to \pi_r(S^8)$ induced by the Hopf map $S^{15} \to S^8$ maps $\pi_r(S^{15})$ isomorphically onto a subgroup of $\pi_r(S^8)$ which has a direct complement isomorphic to $\pi_{r-1}(S^7)$. In particular, $\pi_{15}(S^8) \simeq \mathbb{Z} \oplus \pi_{14}(S^7)$.

K. The composite maps $\pi_{r-1}(S^1) \xrightarrow{\Sigma} \pi_r(S^2) \xrightarrow{\Delta} \pi_{r-1}(S^1)$, $\pi_{r-1}(S^3) \xrightarrow{\Sigma}$ $\pi_r(S^4) \xrightarrow{\Delta} \pi_{r-1}(S^3)$, $\pi_{r-1}(S^7) \xrightarrow{\Sigma} \pi_r(S^8) \xrightarrow{\Delta} \pi_{r-1}(S^7)$, in which Δ is the boundary homomorphism of the homotopy sequences of Hopf bundles, are the identity for any $r \geq 1$, as is easily verified. It follows from this and 5.3.A that the homomorphisms $\Sigma : \pi_5(S^3) \to \pi_6(S^4)$ and $\Sigma : \pi_{13}(S^7) \to \pi_{14}(S^8)$ are isomorphisms, that is, in the series $\{\pi_{n+2}(S^n)\}$ and $\{\pi_{n+6}(S^n)\}$ (as in the series $\{\pi_n(S^n)\}$) stabilization takes place at least one step earlier than the suspension theorem 10.1.E guarantees.

10.3. The Composition Product.

A. Let X be a space with base point x_0. For any spheroids $\phi : S^p \to X$ and $\psi : S^q \to S^p$ the composition $\phi \circ \psi : S^q \to X$ is a spheroid whose homotopy class is determined by the classes of ϕ and ψ. Hence the *composition* $\alpha \circ \beta \in \pi_q(X, x_0)$ is defined for any $\alpha \in \pi_p(X, x_0)$ and $\beta \in \pi_q(S^p)$. An equivalent definition is $\alpha \circ \beta = \phi_*(\beta)$, where ϕ is a spheroid in the class α.

B. We state some obvious properties of composition. For any $\alpha \in \pi_p(X, x_0)$, its composition with the homotopy class of the identity map $S^p \to S^p$ is α. If $\alpha \in \pi_p(X, x_0)$, $\beta \in \pi_q(S^p)$, $\gamma \in \pi_r(S^q)$, then $\alpha \circ (\beta \circ \gamma) = (\alpha \circ \beta) \circ \gamma$. If $\alpha \in \pi_p(X, x_0)$, $\beta \in \pi_q(S^p)$, then $f_*(\alpha \circ \beta) = (f_*(\alpha)) \circ \beta$ for any continuous map f from X to another space. If $\alpha \in \pi_p(X, x_0)$, $\beta \in \pi_q(S^p)$, then $\Sigma(\alpha \circ \beta) = \Sigma\alpha \circ \Sigma\beta$. If $\alpha \in \pi_p(X, x_0)$ and $\beta_1, \beta_2 \in \pi_q(S^p)$, then $\alpha \circ (\beta_1 + \beta_2) = \alpha \circ \beta_1 + \alpha \circ \beta_2$ (in particular $\alpha \circ k[\mathrm{id}_{S^p}] = k\alpha$ for any $\alpha \in \pi_p(X, x_0)$, $k \in \mathbb{Z}$). This is known as *right* distributivity, in distinction to *left* distributivity, which says that if $\alpha_1, \alpha_2 \in \pi_p(X, x_0)$ and $\beta \in \pi_q(S^p)$, then $(\alpha_1 + \alpha_2) \circ \beta = \alpha_1 \circ \beta + \alpha_2 \circ \beta$; this is by no means always true.

C. For example, it follows from the commutativity of the diagram

$$
\begin{array}{ccc}
S^3 & \longrightarrow & S^3 \\
H \downarrow & & \downarrow H \\
S^2 & \longrightarrow & S^2
\end{array}
$$

in which the horizontal maps are defined by

$$(x_1, x_2, x_3, x_4) \to (x_1, -x_2, x_3, -x_4) \quad \text{and} \quad (x_1, x_2, x_3) \to (x_1, -x_2, x_3)$$

and H is the Hopf map, that

$$(-[\mathrm{id}_{S^2}]) \circ H_*[\mathrm{id}_{S^3}] = H_*[\mathrm{id}_{S^3}] \in \pi_3(S^3),$$

and hence

$$[\mathrm{id}_{S^2}]) \circ H_*[\mathrm{id}_{S^3}] + (-[\mathrm{id}_{S^2}]) \circ H_*[\mathrm{id}_{S^3}] = 2H_*[\mathrm{id}_{S^3}] \neq 0$$
$$= 0 \circ H_*[\mathrm{id}_{S^3}] = ([\mathrm{id}_{S^2}] + (-[\mathrm{id}_{S^2}])) \circ H_*[\mathrm{id}_{S^3}].$$

D. However, left distributivity holds for elements in the image of the suspension homomorphism: for any $\alpha_1, \alpha_2 \in \pi_p(X, x_0)$ and $\beta \in \pi_{q-1}(S^{p-1})$

$$(\alpha_1 + \alpha_2) \circ \Sigma\beta = \alpha_1 \circ \Sigma\beta + \alpha_2 \circ \Sigma\beta,$$

since the corresponding equality already holds at the spheroid level, as is easily verified.

E. *The ring Π_*.* We put $\Pi_* = \overset{\infty}{\underset{k=0}{\oplus}} \Pi_k$ and identify the stable homotopy groups Π_k of spheres with their images under the natural embeddings $\Pi_k \to \Pi_*$. The operation $\circ$ makes Π_* into a ring: if $\alpha \in \pi_{n+k}(S^n)$, $\beta \in \pi_{n+k+l}(S^{n+k})$, then $\Sigma(\alpha \circ \beta) = (\Sigma\alpha) \circ (\Sigma\beta)$ (as already mentioned in **B**), so that the operation $\circ$ is well-defined as a distributive multiplication of elements in Π_k by elements in Π_l with values in Π_{k+l} (see **B** and **D**) and can be extended by two-sided distributivity to a multiplication of elements of Π_* with values in Π_*. The ring Π_* is clearly associative with an identity (the natural generators of Π_n are the homotopy classes of the identity maps $S^n \to S^n$). A simple verification shows that it is skew-commutative in the sense that $\beta \circ \alpha = (-1)^{kl} \alpha \circ \beta$ if $\alpha \in \Pi_k$, $\beta \in \Pi_l$.

10.4. Homotopy Groups of Spheres.

A. For a long time the study and computation of the homotopy groups of spheres was a central preoccupation of topologists. They hoped that they would succeed in solving this problem, and that other more complicated problems of homotopy theory could be reduced to it to a significant extent. Deep results have been obtained in both directions, but the original hopes have not been realized. It gradually became clear that from a homotopy point of view, a sphere is not an elementary but an intricate and complex object. On the other hand, the information obtained about the homotopy groups of spheres has found unexpected applications, above all in differential topology.

B. It is doubtful whether on a first acquaintance with homotopy theory it is worth becoming involved with the problem of computing homotopy groups of spheres more seriously than has already been outlined in 10.1–10.3. However, it easily follows from what was said there that the group Π_1 contains at most two elements (this follows from 10.3.C and 10.3.D). Nevertheless we shall list a rather small part of the information known about homotopy groups of spheres.

C. The only groups $\pi_r(S^n)$ with $r > n$ that are infinite are the groups $\pi_{4m-1}(S^{2m})$, $m = 1, 2, \ldots$. Each of these infinite groups is isomorphic to the direct sum of $\mathbb{Z}$ and a finite group.

D. Among the groups $\pi_r(S^n)$ that have been computed are all groups $\pi_{n+k}(S^n)$ with $k \le 30$ and all groups Π_k with $k \le 37$. The groups $\pi_{n+k}(S^n)$ with $n \ge 2$ and $1 \le k \le 7$ are listed in the following table:

n/k	1	2	3	4	5	6	7
2	$\mathbb{Z}$	$\mathbb{Z}_2$	$\mathbb{Z}_2$	$\mathbb{Z}_{12}$	$\mathbb{Z}_2$	$\mathbb{Z}_2$	$\mathbb{Z}_3$
3	$\mathbb{Z}_2$	$\mathbb{Z}_2$	$\mathbb{Z}_{12}$	$\mathbb{Z}_2$	$\mathbb{Z}_2$	$\mathbb{Z}_3$	$\mathbb{Z}_{15}$
4		$\mathbb{Z}_2$	$\mathbb{Z}\otimes\mathbb{Z}_{12}$	$\mathbb{Z}_2\otimes\mathbb{Z}_2$	$\mathbb{Z}_2\otimes\mathbb{Z}_2$	$\mathbb{Z}_3\otimes\mathbb{Z}_{24}$	$\mathbb{Z}_{15}$
5			$\mathbb{Z}_{24}$	$\mathbb{Z}_2$	$\mathbb{Z}_2$	$\mathbb{Z}_2$	$\mathbb{Z}_{30}$
6				0	$\mathbb{Z}$	$\mathbb{Z}_2$	$\mathbb{Z}_{60}$
7					0	$\mathbb{Z}_2$	$\mathbb{Z}_{120}$
8						$\mathbb{Z}_2$	$\mathbb{Z}\otimes\mathbb{Z}_{120}$
9							$\mathbb{Z}_{240}$

The table below, in which H always denotes one of the Hopf maps $S^3 \to S^2$, $S^7 \to S^4$, $S^{15} \to S^8$, gives the generators of the groups Π_k with $k = 1, \ldots, 7$.

Group	Generator
$\Pi_1 = \pi_4(S^3) \simeq \mathbb{Z}_2$	$\Sigma H_*[\mathrm{id}_{S^3}]$
$\Pi_2 = \pi_6(S^4) \simeq \mathbb{Z}_2$	$\Sigma^2 H_*[\mathrm{id}_{S^3}] \circ \Sigma^3 H_*[\mathrm{id}_{S^3}]$
$\Pi_3 = \pi_8(S^5) \simeq \mathbb{Z}_{24}$	$\Sigma H_*[\mathrm{id}_{S^3}]$
$\Pi_4 = \Pi_5 = 0$	
$\Pi_6 = \pi_{14}(S^8) \simeq \mathbb{Z}_2$	$\Sigma^4 H_*[\mathrm{id}_{S^4}] \circ \Sigma^7 H_*[\mathrm{id}_{S^7}]$
$\Pi_7 = \pi_{16}(S^2) \simeq \mathbb{Z}_{240}$	$\Sigma H_*[\mathrm{id}_{S^{15}}]$

In addition,

$$\Sigma^3 H_*[\mathrm{id}_{S^3}] \circ \Sigma^4 H_*[\mathrm{id}_{S^3}] \circ \Sigma^5 H_*[\mathrm{id}_{S^3}] = 12 H_*[\mathrm{id}_{S^7}],$$
$$\Sigma^7 H_*[\mathrm{id}_{S^3}] \circ \Sigma^6 H_*[\mathrm{id}_{S^7}] \circ \Sigma^9 H_*[\mathrm{id}_{S^7}] = 120 H_*[\mathrm{id}_{S^{15}}].$$

Together with the previous tables these relations give a complete description of the part $\bigoplus_{k=1}^{7} \Pi_k$ of the ring Π_*. The groups Π_k with $k = 8, \ldots, 15$ are given in the table:

k	Π_k	k	Π_k
8	$\mathbb{Z}_2 \oplus \mathbb{Z}_2$	12	0
9	$\mathbb{Z}_2 \oplus \mathbb{Z}_2 \oplus \mathbb{Z}_2$	13	$\mathbb{Z}_3$
10	$\mathbb{Z}_2 \oplus \mathbb{Z}_3$	14	$\mathbb{Z}_2 \oplus \mathbb{Z}_2$
11	$\mathbb{Z}_{504}$	15	$\mathbb{Z}_{480} \oplus \mathbb{Z}_2$

For an odd prime p, the order of the group $\Pi_{2m(p-1)-1}$ with $m = 1, 2, \ldots, p-1$ is divisible by p, but not divisible by p^2; the order of the group Π_k with $k > 2p(p-1)-1$ is not divisible by p if $k \neq -1 \bmod 2(p-1)$.

10.5. The Homotopy Groups of Projective Spaces and Lens Spaces.

A. For $2 \le n \le \infty$, the group $\pi_1(\mathbb{R}P^n)$ consists of two elements and is generated by the class of the loop $I \to \mathbb{R}P^n : t \mapsto (\cos \pi t : \sin \pi t : 0 : 0 : \ldots)$, and $\pi_r(\mathbb{R}P^n)$ with $r \ne 1$ is isomorphic to $\pi_r(S^n)$ (in particular it is trivial if $n = \infty$), where the isomorphism is induced by the projection $S^n \to \mathbb{R}P^n$ (this follows from 5.3.H). The space $\mathbb{R}P^n$ is simple for odd n, and is not n-simple for even n. This is related to the fact that the non-trivial automorphism of the covering $S^n \to \mathbb{R}P^n$ for odd n is obviously homotopic to the identity map of S^n, while for even n it induces a non-trivial automorphism of $\pi_n(S^n)$.

B. For $1 \le n \le \infty$ $\pi_2(\mathbb{C}P^n)$ is isomorphic to $\mathbb{Z}$ and is generated by the class of the spheroid in : $S^2 = \mathbb{C}P^1 \to \mathbb{C}P^n$ (since the pair $(\mathbb{C}P^n, \mathbb{C}P^1)$ is 3-connected). The group $\pi_r(\mathbb{C}P^n)$ with $r \ne 2$ is isomorphic to $\pi_r(S^{2n+1})$ (in particular it is trivial for $n = \infty$), where the isomorphism is induced by the projection $S^{2n+1} \to \mathbb{C}P^n$; this result is obtained from the homotopy sequence of the bundle $S^{2n+1} \to \mathbb{C}P^n$.

C. For $1 \le n \le \infty$ and any $r \ge 1$, the homomorphism induced by the projection $S^{4n+3} \to \mathbb{H}P^n$ maps $\pi_r(S^{4n+3})$ isomorphically onto a subgroup of $\pi_r(\mathbb{H}P^n)$ having a direct complement isomorphic to $\pi_{r-1}(S^3)$. In particular, for $r \ge 1$ the group $\pi_r(\mathbb{H}P^\infty)$ is isomorphic to $\pi_{r-1}(S^3)$. To prove this it is sufficient to apply theorem 5.3.F to the bundle $S^{4n+3} \to \mathbb{H}P^n$.

D. *Lens spaces.* Let m be a positive integer, and $l_1, \ldots, l_n$ be integers relatively prime to m. The complex formula

$$(k, (z_1, \ldots, z_n)) \mapsto (z_1 e^{2\pi i k l_1/m}, \ldots, z_n e^{2\pi i k l_n/m}) ,$$

where k is an integer and $(z_1, \ldots, z_n)$ is a point of S^{2n-1}, defines an action $\mathbb{Z} \times S^{2n-1} \to S^{2n-1}$ with kernel of non-effectiveness $m\mathbb{Z}$; after factoring by this kernel we obtain a free action of $\mathbb{Z}_m = \mathbb{Z}/m\mathbb{Z}$ on S^{2n-1}. The quotient space $S^{2n-1}/\mathbb{Z}_m$ is denoted by $L(m; l_1, \ldots, l_n)$ and called a *lens space*. There are also *infinite lens spaces* $L(m; l_1, l_2, \ldots)$ with $l_1, l_2, \ldots$ relatively prime to m. The lens $L(m; l_1, l_2, \ldots)$ is defined as the quotient space $S^\infty/\mathbb{Z}_m$ under the free action induced by making the action

$$(k, (z_1, z_2, \ldots)) \mapsto (z_1 e^{2\pi i k l_1/m}, z_2 e^{2\pi i k l_2/m}, \ldots)$$

of $\mathbb{Z}$ on S^∞ effective. An equivalent description is

$$L(m; l_1, l_2, \ldots) = \lim(L(m; l_1, \ldots l_n), \text{ in : } L(m; l_1, \ldots, l_n) \to L(m; l_1, \ldots, l_{n+1})).$$

The infinite lens $L(m; 1, 1, \ldots)$ is denoted more briefly by $L(m)$. The natural projections $S^{2n-1} \to L(m; l_1, \ldots, l_n)$ and $S^\infty \to L(m; l_1, l_2, \ldots)$ are coverings.

E. Since the automorphisms of these coverings are homotopic to the identity maps, $L(m; l_1, \ldots, l_n)$ and $L(m; l_1, l_2, \ldots)$ are simple. The groups $\pi_1(L(m; l_1, \ldots, l_n)$ and $\pi_1(L(m; l_1, l_2, \ldots))$ are isomorphic to $\mathbb{Z}_m$. The groups $\pi_r(L(m; l_1, \ldots, l_n))$ with $r \ge 2$ are isomorphic to $\pi_r(S^{2n-1})$, where the isomorphisms are induced by the projections $S^{2n-1} \to L(m; l_1, \ldots, l_n)$. The groups $\pi_r(L(m; l_1, l_2, \ldots))$ with $r \ge 2$ are trivial.

10.6. Homotopy Groups of the Classical Groups.

A. The inclusion homomorphism $\pi_r(SO(n)) \to \pi_r(SO(n+1))$ is an isomorphism for $r \leq n-2$ and an epimorphism for $r = n-1$. This follows by considering the homotopy sequence of the bundle $SO(n+1) \to S^n$: $A \mapsto (0, \ldots, 0, 1)A$ with base point id $\in SO(n+1)$. The group $\pi_1(SO(n))$ is isomorphic to $\mathbb{Z}$ for $n = 2$ (since $SO(2) = S^1$), and to $\mathbb{Z}_2$ for $n \geq 3$ (since $SO(3) = RP^3$ and in virtue of the stabilization mentioned above) and is generated by the class of the inclusion $S^1 = SO(2) \to SO(n)$. The group $\pi_2(SO(n))$ is trivial for any n (since $SO(2) = S^1$, $SO(3) = \mathbb{R}P^3$, $SO(4) = \mathbb{R}P^3 \times S^3$, and from the stabilization). The group $\pi_3(SO(n))$ is trivial for $n \leq 2$, isomorphic to $\mathbb{Z}$ for $n = 3$, isomorphic to $\mathbb{Z} \oplus \mathbb{Z}$ for $n = 4$, and isomorphic to the quotient of $\mathbb{Z} \oplus \mathbb{Z}$ by a cyclic subgroup for $n \geq 5$ (for similar reasons).

B. The inclusion homomorphism $\pi_r(U(n)) \to \pi_r(U(n+1))$ is an isomorphism for $r \leq 2n-1$ and an epimorphism for $r = 2n$. For $n \geq 1$ $\pi_1(U(n))$ is isomorphic to $\mathbb{Z}$, generated by the class of the inclusion $S^1 = U(1) \to U(n)$. The group $\pi_2(U(n))$ is trivial for any n. The group $\pi_3(U(1))$ is trivial, and $\pi_3(U(n))$ is isomorphic to $\mathbb{Z}$ for $n \geq 2$. The inclusion homomorphism $\pi_1(U(n)) \to \pi_1(SO(2n))$ is an epimorphism for any n.

These results all follow from the equalities [in : $U(1) \to SO(2)$] = id, $U(2) = S^1 \times S^3$, and the homotopy sequence of the bundle $U(n+1) \to S^{2n+1}$: $A \mapsto (0, 0, \ldots, 0, 1)A$ with base point id $\in U(n+1)$.

C. The inclusion homomorphism $\pi_r(Sp(n)) \to \pi_r(Sp(n+1))$ is an isomorphism for $r \leq 4n + 1$ and an epimorphism for $r = 4n + 2$. In particular, $\pi_r(Sp(n))$ is isomorphic to $\pi_r(Sp(1)) = \pi_r(S^3)$ for $r \leq 5$ and any $n \geq 1$. This is clear from the homotopy sequence of the bundle $Sp(n+1) \to S^{4n+3}$: $A \mapsto (0, 0, \ldots, 0, 1)A$ with base point id $\in Sp(n+1)$.

D. *Stabilizations.* As has been shown in **A–C**, for any r each of the series of groups

$$\pi_r(SO(1)) \to \pi_r(SO(2)) \to \pi_r(SO(3)) \to \cdots$$

$$\pi(U(1)) \to \pi_r(U(2)) \to \pi_r(U(3)) \to \cdots$$

$$\pi_r(Sp(1)) \to \pi_r(Sp(2)) \to \pi_r(Sp(3)) \to \cdots$$

stabilizes: the first starting from $\pi_r(SO(r+2))$, the second from $\pi_r(U([(r+2)/2]))$, and the third from $\pi_r(Sp([(r+2)/4]))$. The groups $\pi_r(SO(n))$ with $n \geq r + 2$, $\pi_r(U(n))$ with $n \geq [(r+2)/2]$, and $\pi_r(Sp(n))$ with $n \geq [(r+2)/4]$ are called *stable*, and denoted by $\pi_r(SO)$, $\pi_r(U)$, $\pi_r(Sp)$. According to **A**, $\pi_1(SO) \simeq \mathbb{Z}$, $\pi_2(SO) = 0$ and $\pi_3(SO)$ is isomorphic to the quotient of $\mathbb{Z} \oplus \mathbb{Z}$ by a cyclic subgroup. According to **B**, $\pi_1(U) \simeq \mathbb{Z}$, $\pi_2(U) = 0$, $\pi_3(U) \simeq \mathbb{Z}$. According to **C**, $\pi_1(Sp) = 0$, $\pi_2(Sp) = 0$, $\pi_3(Sp) \simeq \mathbb{Z}$. The notations $\pi_r(SO)$, $\pi_r(U)$, $\pi_r(Sp)$ also have their primary meaning: they are the usual r-dimensional homotopy groups of the limit spaces $SO = \lim SO(n)$, $U = \lim U(n)$, $Sp = \lim Sp(n)$.

E. These are all the results on homotopy groups of classical groups that can be obtained using only the methods set out above. We add some important results whose proofs can be found in Milnor (1963) for example.

The groups $\pi_r(SO)$, $\pi_r(U)$, $\pi_r(Sp)$ have been completely calculated. In fact for $r \geq 1$, there are canonical isomorphisms $\pi_r(SO) \to \pi_{r+8}(SO)$, $\pi_r(Sp) \to \pi_{r+8}(Sp)$, $\pi_r(U) \to \pi_{r+2}(U)$. The initial eight groups of SO and Sp, and the initial two groups of U are given in the following tables:

r	1	2
$\pi_r(U)$	$\mathbb{Z}$	0

r	1	2	3	4	5	6	7	8
$\pi_r(SO)$	$\mathbb{Z}_2$	0	$\mathbb{Z}$	0	0	0	$\mathbb{Z}$	$\mathbb{Z}_2$
$\pi_r(Sp)$	0	0	$\mathbb{Z}$	$\mathbb{Z}_2$	$\mathbb{Z}_2$	0	$\mathbb{Z}$	0

Many unstable homotopy groups of the manifolds $SO(n)$, $U(n)$, $Sp(n)$ are also known; for example, $\pi_{2n}(U(n)) \simeq \mathbb{Z}_{n!}$, $\pi_{4n+2}(Sp(n)) \simeq \mathbb{Z}_{(2n+1)!}$ for even n, and $\pi_{4n+2}(Sp(n)) \simeq \mathbb{Z}_{2[(2n+1)!]}$ for odd n. For details and references see Fuchs (1971).

10.7. Homotopy Groups of Stiefel Manifolds and Spaces.

A. For $0 \leq k \leq n$ let $V(n,k)$ denote the manifold of linear isometric maps $R^k \to R^n$. Such a map is determined by the images of the basis vectors $(1, 0, \ldots, 0), \ldots, (0, \ldots, 0, 1) \in \mathbb{R}^k$, that is, by an orthonormal k-frame in $\mathbb{R}^n$. The coordinates of the vectors of this frame form the $n \times k$ matrix of the map. Thus $V(n,k)$ may be interpreted as the set of orthonormal k-frames in $\mathbb{R}^n$, or as the set of $n \times k$ matrices $\|v_{si}\|$ such that $\sum_{s=1}^n v_{si}v_{sj} = \delta_{ij}$ for $1 \leq i \leq j \leq k$. Hence it is clearly a smooth submanifold of $\mathbb{R}^{nk}$, called a *Stiefel manifold*.

For $0 \leq k \leq n$, $\mathbb{C}V(n,k)$ denotes the set of linear isometric maps of $\mathbb{C}^k$ into $\mathbb{C}^n$, or, equivalently, the set of orthonormal k-frames in $\mathbb{C}^n$, or, equivalently, the set of complex $n \times k$ matrices $\|v_{si}\|$ such that $\sum_{s=1}^n v_{si}\bar{v}_{sj} = \delta_{ij}$ $(1 \leq i \leq j \leq k)$. Clearly from the last description $\mathbb{C}V(n,k)$ is a closed $(2nk - k^2)$-dimensional analytic submanifold of $\mathbb{R}^{2nk}$, called a *complex Stiefel manifold*. Warning: $\mathbb{C}V(n,k)$ is not a complex manifold in the generally accepted sense (in fact for odd k it is odd-dimensional).

The same can be repeated almost word for word, replacing the complex field by the skew-field of quaternions. Here $\mathbb{H}^n$ is regarded as a left vector space, a linear map $\mathbb{H}^k \to \mathbb{H}^n$ is understood to be a map that is left-linear, and the scalar product of vectors $(u_1, \ldots, u_n)$, $(v_1, \ldots, v_n)$ is defined as $\sum_{i=1}^n u_i \bar{v}_i$. The resulting closed real-analytic manifold $\mathbb{H}V(n,k)$, $0 \leq k \leq n$, of dimension $4nk - (2k^2 - k)$ without boundary, is called a *quaternion Stiefel manifold*.

B. If $k > n$, then $V(n,k)$ is simple; the inclusion homomorphism $\pi_r(V(n,k)) \to \pi_r(V(n+1, k+1))$ is an isomorphism for $r < n - 1$, and an epimorphism for $r = n - 1$; it is an isomorphism for $r = n - 1$ when n is odd and $k = 1$.

All the manifolds $\mathbb{C}V(n,k)$ and $\mathbb{H}V(n,k)$ are simple. The inclusion homomorphism $\pi_r(\mathbb{C}V(n,k)) \to \pi_r(\mathbb{C}V(n+1, k+1))$ is an isomorphism for $r < 2n$ and an epimorphism for $r = 2n$. The inclusion homomorphism $\pi_r(\mathbb{H}V(n,k)) \to$

$\pi_r(\mathbb{H}V(n+1, k+1))$ is an isomorphism for $r < 4n+2$ and an epimorphism for $r = 4n+2$.

The fact that these manifolds are simple follows from the relations $V(n, k) = SO(n)/SO(n-k)$, $\mathbb{C}V(n, k) = U(n)/U(n-k)$, $\mathbb{H}V(n, k) = Sp(n)/Sp(n-k)$ (see Rokhlin and Fuchs (1977) and 3.4.F). The other results are obtained from the homotopy sequences of the bundles $(V(n+1, k+1), \text{pr}, S^n)$, $(\mathbb{C}V(n+1, k+1), \text{pr}, S^{2n+1})$, $(\mathbb{H}V(n+1, k+1), \text{pr}, S^{4k+3})$. In the real case we need an additional fact, that for odd n the bundle $(V(n+1, 2), \text{pr}, S^n)$ is the bundle of unit tangent vectors of S^n, and so has a section; thanks to this, for odd n and $k = 1$, the first of the above homotopy sequences splits at the term $\pi_r(V(n+1, k+1))$; see 5.3.D.

C. For $k > n$, $V(n, k)$ is $(n - k - 1)$-connected. The group $\pi_{n-k}(V(n, k))$ with $0 < k < n$ is cyclic, generated by the class of the inclusion $S^{n-k} = V(n - k + 1, 1) \to V(n, k)$; it is infinite if $n - k$ is odd, and also if $k = 1$.

This follows from **B**: if $r < n - k$, then $\pi_r(V(n, k)) \simeq \pi_r(V(n - 1, k - 1)) \simeq \cdots \simeq \pi_r(V(n - k + 1, 1)) = \pi_r(S^{n-k}) = 0$; in the sequence $\pi_{n-k}(S^{n-k}) = \pi_{n-k}(V(n - k + 1, 1)) \to \pi_{n-k}(V(n - k + 2, 2)) \to \cdots \to \pi_{n-k}(V(n, k))$, all the arrows except the first are isomorphisms, and the first is an isomorphism for even $n - k$ and an epimorphism for odd $n - k$.

D. The manifold $\mathbb{C}V(n, k)$ is $2(n-k)$-connected. The group $\pi_{2n-2k+1}(\mathbb{C}V(n, k))$ is isomorphic to $\mathbb{Z}$, generated by the class of the inclusion $S^{2n-2k+1} = \mathbb{C}V(n - k + 1, 1) \to \mathbb{C}V(n, k)$. This also follows from **B**: for $r \leq 2n - 2k + 1$, all the arrows in the sequence $\pi_r(S^{2n-2k+1}) = \pi_r(\mathbb{C}V(n - k + 1, 1)) \to \pi_r(\mathbb{C}V(n - k + 2, 2)) \to \cdots \to \pi_r(\mathbb{C}V(n, k))$ are isomorphisms.

E. The manifold $\mathbb{H}V(n, k)$ is $(4n - 4k + 2)$-connected. The group $\pi_{4n-4k+3}(\mathbb{H}V(n, k))$ is isomorphic to $\mathbb{Z}$, generated by the class of the inclusion $S^{4n-4k+3} = \mathbb{H}V(n - k + 1, 1) \to \mathbb{H}V(n, k)$. This also follows from **B**: for $r \leq 4n - 4k + 3$ (even for $r \leq 4n - 4k + 5$) all the arrows are isomorphisms in the sequence $\pi_r(S^{4n-4k+3}) = \pi_r(\mathbb{H}V(n - k + 1, 1)) \to \pi_r(\mathbb{H}V(n - k + 2, 2)) \to \cdots \to \pi_r(\mathbb{H}V(n, k))$.

F. The spaces $V(\infty, k) = \lim(V(n, k), \text{in}: V(n, k) \to V(n+1, k))$, $\mathbb{C}V(\infty, k) = \lim(\mathbb{C}V(n, k), \text{in}: \mathbb{C}V(n, k) \to \mathbb{C}V(n+1, k))$, $\mathbb{H}V(\infty, k) = \lim(\mathbb{H}V(n, k), \text{in}: \mathbb{H}V(n, k) \to \mathbb{H}V(n+1, k))$, are ∞-connected. This follows from **C**, **D**, **E** and 3.1.**J**.

10.8. Homotopy Groups of Grassmann Manifolds and Spaces.

A. In this section the computation of the most important homotopy groups of the Grassmann manifolds $G(n, k)$, $G_+(n, k)$, $\mathbb{C}G(n, k)$, $\mathbb{H}G(n, k)$ and the Grassmann spaces $G(\infty, k)$, $G_+(\infty, k)$ and $\mathbb{H}G(\infty, k)$ is reduced to the computation of the homotopy groups of the corresponding classical groups. Grassmann manifolds and spaces are considered together, so that n may take the value ∞.

B. If $k > 0$ and $0 < r < n-k$, then $\pi_r(G_+(n, k))$ is isomorphic to $\pi_{r-1}(SO(n))$, and the inclusion homomorphism $\pi_r(G_+(n, k)) \to \pi_r(G_+(m, k))$ is an isomorphism for any $m > n$. The first result follows from the theorems of 10.7.**C** and 10.7.**F** and the homotopy sequence of the bundle $(V(n, k), \text{pr}, G_+(n, k))$, where pr maps each k-frame in R^n to the oriented k-dimensional subspace that it determines.

The second result follows from the commutativity of the diagram

$$
\begin{array}{ccc}
\pi_r(G_+(n,k)) & \xrightarrow{\Delta} & \pi_{r-1}(SO(k)) \\
\downarrow{\scriptstyle in_*} & & \downarrow{\scriptstyle in_*=id} \\
\pi_r(G_+(m,k)) & \xrightarrow{\Delta} & \pi_{r-1}(SO(k))
\end{array}
$$

see 5.2.**D**.

C. The group $\pi_r(G(n,k))$ with $0 < k < n$ and $r \geq 2$ is isomorphic to $\pi_r(G_+(n,k))$. The group $\pi_1(G(n,k))$ is isomorphic to $\mathbb{Z}$ if $n = 2$ and $k = 1$, and to $\mathbb{Z}_2$ if $0 < k < n$ and $n \geq 3$. The isomorphism $\pi_1(G(2,1)) \simeq \mathbb{Z}$ follows because $G(2,1)$ is homeomorphic to a circle. The other results follow from the theorem in 5.3.**H** applied to the canonical two-sheeted covering $G_+(n,k) \to G(n,k)$.

D. The following two theorems are the analogues of the theorem in **B**, and are proved in the same way with obvious modifications.

If $0 < r < 2n - 2k + 1$, then $\pi_1(\mathbb{C}G(n,k))$ is isomorphic to $\pi_{r-1}(U(k))$ and the inclusion homomorphism $\pi_r(\mathbb{C}G(n,k)) \to \pi_r(\mathbb{C}G(m,k))$ is an isomorphism for any $m > n$.

If $0 < p < 4n - 4k + 3$, then $\pi_r(\mathbb{H}G(n,k))$ is isomorphic to $\pi_{r-1}(Sp(k))$ and the inclusion homomorphism $\pi_r(\mathbb{H}G(n,k)) \to \pi_r(\mathbb{H}G(m,k))$ is an isomorphism for any $m > n$.

§11. Application of Cellular Techniques

11.1. Homotopy Groups of a 1-dimensional Cellular Space.

A. The fundamental group of a connected 1-dimensional cellular space is a free group, and its other homotopy groups are trivial. It is sufficient to prove this for a bouquet of circles, since every connected 1-dimensional cellular space is homotopy equivalent to a bouquet of circles (see 9.2.**B**).

A universal covering with a contractible covering space can be constructed in an explicit way for a bouquet of circles. We shall not describe the construction of this covering in the general case, but confine ourselves to a figure showing a fragment of this covering in the case of a bouquet of two circles. The covering space is glued together from countably many crosses, each of which is mapped surjectively by the covering projection, while its interior is mapped injectively.

The fundamental group of a bouquet of circles is freely generated by the homotopy classes of the natural embeddings of the circles, so that its rank is equal to the number of circles. That it is generated by these classes follows from the cellular approximation theorem 9.1.**A**, and that there are no relations is proved by constructing paths covering the loops, and belonging to the various products of the basic classes (cf. 6.2 (ii)).

11.2. The Effect of Attaching Balls.

A. Let $X = A \cup_\phi [\coprod_{\mu \in M}(D_\mu = D^{k+1})]$, where A is a connected topological space and ϕ a continuous map of the space $\coprod_{\mu \in M}(S_\mu = S^k)$ into A. Further, let

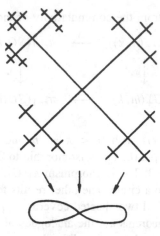

Fig. 14

x_0 be a point of A. In this section we shall determine a system of generators of $\pi_{k+1}(X, A, x_0)$.

Note that the groups $\pi_r(X, A)$ with $r \leq k$ are trivial (see 9.1.**D**), and the groups $\pi_r(X, A)$ with $r > k + 1$ already present a considerably more difficult problem: in the simplest case when A is a single point, and the family $\{D_\mu\}$ consists of a single ball, then $\pi_r(X, A)$ is $\pi_r(S^{k+1})$.

Let f_μ denote the map $D^{k+1} \to X$ defined as the composition of the inclusion of the μth ball $D^{k+1} \to \coprod_\nu D_\nu$ and the projection $\coprod_\nu D_\nu \to X$, and let α_μ denote the element of $\pi_{k+1}(X, A, f_\mu(1, 0, \ldots, 0))$ defined by the spheroid f_μ : $(D^{k+1}, S^k, (1, 0, \ldots, 0)) \to (X, A, f_\mu(1, 0, \ldots, 0))$.

B. Theorem. *Let $w_\mu : I \to A$ be any paths joining the points $f_\mu(1, 0, \ldots, 0)$ to x_0. If $k \geq 1$, the group $\pi_{k+1}(X, A, x_0)$ is generated over $\pi_1(A, x_0)$ by the classes $\beta_\mu = T_w \alpha_\mu$ (that is, is generated in the usual sense by the classes $T_w \beta_\mu$ with $w \in \pi_1(A, x_0)$, and T_w is the translation along the path w).* □

To prove this theorem we may for example take an arbitrary spheroid $(D^{k+1}, S^k) \to (X, A)$, apply the simplicial approximation theorem to its restriction to the inverse images of certain $(k + 1)$-simplexes of $X \setminus A$ (one for each attached ball) and deform the resulting spheroid in such a way that it becomes represented in the desired form.

C. Corollary. *Under the conditions of theorem **B** the inclusion homomorphism $\pi_r(A, x_0) \to \pi_r(X, x_0)$ is an isomorphism for $r \leq k - 1$, and an epimorphism for $r = k$. The kernel of this epimorphism is generated over $\pi_1(A, x_0)$ by the classes $\partial \beta_\mu = T_{w_\mu}(\partial \alpha_\mu)$ (that is, by the classes of the attaching spheroids ∂f_μ translated to the point x_0).* □

D. Let (X, A) be a cellular pair with base point $x_0 \in A$. If A is connected and contains the k-skeleton $sk_k X$ with $k \geq 1$, then the groups $\pi_r(X, A)$ with $r \leq k$ are trivial, and $\pi_{k+1}(X, A, x_0)$ is generated over $\pi_1(A, x_0)$ by the classes of the characteristic maps corresponding to the $(k + 1)$-cells of $X \setminus A$ (considered as spheroids) translated in any way to the point x_0; the inclusion homomorphism $\pi_r(A, x_0) \to \pi_r(X, x_0)$ is an isomorphism for $r \leq k - 1$, and an epimorphism for $r = k$; the kernel of the last epimorphism is generated over $\pi_1(A, x_0)$ by the classes of the attaching spheroids corresponding to the $(k + 1)$-cells of $X \setminus A$, translated in any way to the point x_0.

11.3. The Fundamental Group of a Cellular Space.

A. Let X be a cellular space with a single 0-cell x_0. This cell is also the only 0-cell in the 1-skeleton, so that $\mathrm{sk}_1 X$ is cellularly homeomorphic to a bouquet of spheres. Hence $\pi_1(\mathrm{sk}_1 X, x_0)$ is a free group, generated by the homotopy classes of the characteristic maps corresponding to the 1-cells.

According to 11.2.C, the homomorphism $\mathrm{in}_* : \pi_1(\mathrm{sk}_1 X, x_0) \to \pi_1(X, x_0)$ is an epimorphism and its kernel is generated over $\pi_1(\mathrm{sk}_1 X, x_0)$ by the homotopy classes of the attaching maps of the 2-cells, translated in any way to x_0. In our case $\pi_1(\mathrm{sk}_1 X, x_0)$ acts as the group of inner automorphisms. This means that $\mathrm{Ker\,in}_*$ is the smallest normal subgroup of $\pi_1(X, x_0)$ containing the homotopy classes specified above, and the fundamental group $\pi_1(X, x_0)$ is canonically isomorphic to the quotient group of $\pi_1(\mathrm{sk}_1 X, x_0)$ by this normal subgroup.

B. From what has been said it follows that to compute $\pi_1(X, x_0)$ it is sufficient to know the 1-skeleton of X and the attaching maps for the 2-cells. In terms of this data we can specify a system of generators and relations for $\pi_1(X, x_0)$: for each 1-cell there is a generator, namely the class of the corresponding characteristic loop, and for each 2-cell a relation, namely the word which represents the class of the corresponding attaching map (translated to x_0) in terms of the generators is put equal to the identity.

Note that, unlike the system of generators, the system of relations is not completely canonical: it depends on the choice of translation paths, so that the left-hand sides of the relations are determined only up to conjugacy.

C. Corollary. *The fundamental group of a finite connected cellular space has a presentation with a finite system of generators and relations.* $\square$

D. *The addition theorem (the van Kampen-Seifert theorem).* If A, B are subspaces of a topological space X with inclusion maps

$$
\begin{array}{ccc}
& A & \\
i \nearrow & & \searrow i' \\
A \cap B & & X \\
i \searrow & & \nearrow j' \\
& B &
\end{array}
$$

and $x_0 \in A \cap B$, then the formula $\alpha * \beta \to i'_*(\alpha) j'_*(\beta)$ defines a homomorphism $\pi_1(A, x_0) * \pi_1(B, x_0) \to \pi_1(X, x_0)$ (where $*$ denotes the free product) whose kernel contains all elements of the form $i_*(\delta) j_*(\delta^{-1})$, where $\delta \in \pi_1(A \cap B, x_0)$. Hence the above formula also defines an induced homomorphism

$$[\pi_1(A, x_0) * \pi_1(B, x_0)]/\mathrm{Vk}(X, A, B, x_0) \to \pi_1(X, x_0) \, ,$$

where $\mathrm{Vk}(X, A, B, x_0)$ is the smallest normal subgroup of $\pi_1(A, x_0) * \pi_1(B, x_0)$ containing the elements described above (it is called the *van Kampen subgroup*). This last homomorphism is natural in the sense that the diagram

$$
\begin{array}{ccc}
[\pi_1(A, x_0) * \pi_1(B, x_0)]/\mathrm{Vk}(X, A, B, x_0) & \longrightarrow & \pi_1(X, x_0) \\
\downarrow & & \downarrow \\
[\pi_1(A', x_0') * \pi_1(B', x_0')]/\mathrm{Vk}(X', A', B', x_0') & \longrightarrow & \pi_1(X', x_0')
\end{array}
$$

induced by a continuous map of (X, A, B, x_0) into (X', A', B', x_0') is always commutative.

Theorem. *Let (X, A, B) be a cellular triad (that is, X is a cellular space with subspaces $A, B,$ and $A \cup B = X$) and $x_0 \in A \cap B$. If $A, B,$ and $A \cap B$ are connected, then the homomorphism*

$$[\pi_1(A, x_0) * \pi_1(B, x_0)]/\mathrm{Vk}(X, A, B, x_0) \to \pi_1(X, x_0)$$

is an isomorphism. $\square$

This theorem is a direct corollary of what has been said in 11.3.B.

E. Corollary. *If A, B are cellular spaces with 0-cells a, b as base points, then $\pi_1((A, a) \vee (B, b)) \simeq \pi_1(A, a) * \pi_1(B, b)$.*

11.4. Homotopy Groups of Compact Surfaces.

A. Recall that a sphere with g handles (or with h crosscaps) and l holes ($l > 1$) is homotopy equivalent to a bouquet of $2g + l - 1$ (or $h + l - 1$) circles. Hence the fundamental group of such a space is a free group, with $2g + l - 1$ or $h + l - 1$ generators respectively, and the higher homotopy groups are trivial.

B. *The fundamental groups of closed surfaces.* The cell decomposition of a sphere with g handles, constructed in 7.2.G, contains one 0-cell e_0, $2g$ 1-cells $a_1, b_1, \ldots, a_g, b_g$, and one 2-cell, whose attaching map takes $(1, 0, \ldots, 0)$ to e_0. The homotopy class of this map (regarded as a loop) is represented in terms of the generators $\alpha_1, \beta_1, \ldots, \alpha_g, \beta_g$ of the fundamental group of the 1-skeleton of the surface associated with the cells $a_1, b_1, \ldots, a_g, b_g$, by the word

$$\alpha_1 \beta_1 \alpha_1^{-1} \beta_1^{-1} \ldots \alpha_g \beta_g \alpha_g^{-1} \beta_g^{-1} \, .$$

Thus the fundamental group of the surface at e_0 can be described as the group with generators $a_1, b_1, \ldots, a_g, b_g$ and the relation

$$a_1 b_1 a_1^{-1} b_1^{-1} \ldots a_g b_g a_g^{-1} b_g^{-1} = 1 \, .$$

The cell decomposition of a sphere with h crosscaps, constructed in 7.2.G contains one 0-cell e_0, h 1-cells $c_1, \ldots, c_h$, and one 2-cell. The fundamental group of the surface at e_0 can be described as the group with generators $c_1, \ldots, c_h$ and the relation

$$c_1 c_1 \ldots c_h c_h = 1 \, .$$

C. No two of the groups computed in **B** are isomorphic. (To prove this, it is sufficient to make the groups commutative: the fundamental group of a sphere with g handles becomes a free Abelian group of rank $2g$, while the fundamental group of the sphere with h crosscaps becomes the direct sum of a free Abelian group of rank $h - 1$ and a group of order 2.) Hence the model closed surfaces are pairwise non-homeomorphic.

It is easy to deduce from this that all the model compact surfaces are pairwise non-homeomorphic: it is sufficient to close the holes by gluing on discs. The number of holes is a topological invariant because it is equal to the number of components of the boundary.

Higher homotopy groups.

D. Let P be a sphere with g handles. If $g \geq 1$ and $r \geq 2$, then $\pi_r(P) = 0$. This may be proved by, for example, constructing a covering of P by a non-compact surface (see for example Fig. 15) and noting that any element of a homotopy group of the latter comes from an appropriate compact subset of this surface.

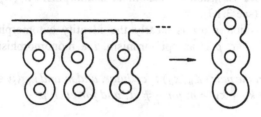

Fig. 15

E. Let P be a sphere with h crosscaps. If $h \geq 2$ and $r \geq 2$, then $\pi_r(P) = 0$.

This follows from **D**, since the sphere with $h - 1$ handles is a double covering of the sphere with h crosscaps.

11.5. Homotopy Groups of Bouquets.

A. Suppose we are given a family $\{(X_\mu, x_\mu)\}_{\mu \in M}$ of pointed T_1-spaces. We consider the bouquet $B = \vee_{\mu \in M}(X_\mu, x_\mu)$. For $r \geq 2$ the inclusions $\mathrm{in}_\mu : X_\mu \to B$ define a homomorphism

$$\bigoplus_{\mu \subset M} \pi_r(X_\mu, x_\mu) \to \pi_r(B)$$

which we shall denote by i. This homomorphism is natural in the sense that if $B' = \vee_{\mu' \in M'}(X'_{\mu'}, x'_{\mu'})$ is another bouquet, $\alpha : M' \to M$ any map, and $f_{\mu'} : (X'_{\mu'}, x'_{\mu'}) \to (X'_{\alpha(\mu')}, x'_{\alpha(\mu')})$ a continuous map, then the diagram

$$\begin{array}{ccc} \displaystyle\bigoplus_{\mu' \in M'} \pi_r(X'_{\mu'}, x'_{\mu'}) & \xrightarrow{\ i\ } & \pi_r(B') \\ {\scriptstyle (\oplus_{\mu' \in M'} f_{\mu'})_*}\downarrow & & \downarrow{\scriptstyle (\vee_{\mu' \in M'} f_{\mu'})_*} \\ \displaystyle\bigoplus_{\mu \in M} \pi_r(X_\mu, x_\mu) & \xrightarrow{\ i\ } & \pi_r(B) \end{array}$$

is commutative.

B. Lemma. *For any* $\alpha \in \pi_r(B)$ *with* $r \geq 1$, *there is a finite set* $M' \subset M$ *such that for* $\mu \in M \setminus M'$ *the homomorphism* $\pi_r(B) \to \pi_r(X_\mu, x_\mu)$ *induced by the natural projection takes* α *to* 0, *and in addition is in the image of the homomorphism* $\pi_r(B') \to \pi_r(B)$ *induced by the natural inclusion of* $B' = \vee_{\mu \in M'}(X_\mu, x_\mu)$. $\square$

To prove this it is enough to note that for any spheroid $\phi \in \mathrm{Sph}_r(B)$, the image $\phi(I^r)$ is covered by a finite number of the sets $\mathrm{in}_\mu(X_\mu)$. This follows from the fact that X_μ is a T_1-space and $\phi(I^r)$ is compact: if we choose a point in each nonempty intersection $\phi(I^r) \cap \mathrm{in}_\mu(X_\mu \setminus x_\mu)$, we obtain a set which is both compact and discrete, and hence finite.

C. For each $r \geq 2$, we define $p : \pi_r(B) \to \oplus_{\mu \in M} \pi_r(X_\mu, x_\mu)$ as the map that takes a class α to the sequence of its images under the homomorphisms $\pi_r(B) \to \pi_r(X_\mu, x_\mu)$ induced by the natural projections (this is a valid definition because of lemma **B**). Clearly p is a homomorphism and is natural in an obvious sense (cf. the naturality of the homomorphism i in **A**). If M is finite, then obviously p is equal to the composition $\pi_r(B) \to \pi_r(\times_{\mu \in M}(X_\mu, x_\mu))$, induced by the inclusion $B \to \times_{\mu \in M} X_\mu$ (see 1.5), and the canonical isomorphism $\pi_r(\times_{\mu \in M}(X_\mu, x_\mu)) \to \oplus_{\mu \in M} \pi_r(X_\mu, x_\mu)$ (see 3.1.I).

D. The composition $p \circ i$ is clearly the identity automorphism of $\oplus_{\mu \in M} \pi_r(X_\mu, x_\mu)$. In particular, p is an epimorphism, i a monomorphism, and $\pi_r(B) = \mathrm{Ker}\, p \oplus \mathrm{Im}\, i$.

E. *Suppose that the pair* (X_μ, x_μ) *is cellular, and* m *and* k_μ ($\mu \in M$) *are positive integers such that* $k_\mu + k\nu \geq m$ *for* $\nu \neq \mu$, *and for each* μ

$$\pi_s(X_\mu, x_\mu) = 0 \text{ for } 1 \leq s \leq k_\mu .$$

If $2 \leq r \leq m$, *then the homomorphisms*

$$i : \oplus_{\mu \in M} \pi_r(X_\mu, x_\mu) \to \pi_r(B) ,$$
$$p : \pi_r(B) \to \oplus_{\mu \in M} \pi_r(X_\mu, x_\mu)$$

are isomorphisms. $\square$

Proof. By theorem 9.2.**D** and the naturality of i, we may asssume that the skeletons $\mathrm{sk}_{k_\mu} X$ are reduced to single points x_μ; in this case B clearly contains the skeleton $\mathrm{sk}_{m+1} X$ of the cellular product X of the X_μ. By theorem 9.1.**A**, it follows that $\mathrm{in}_* : \pi_r \to (X)$ is an isomorphism for $r \leq m$. If M is finite, the composition $\mathrm{in}_* \circ i : \oplus_{\mu \in M} \pi_r(X_\mu, x_\mu) \to \pi_r(X)$ is an isomorphism for any r (see 3.1.I). Consequently when M is finite, i is an isomorphism for $r \leq m$.

Now that this has been proved, it is easy to deduce statements **B** and **D**, using the naturality of the homomorphisms i and p.

F. Corollary. *Let* B *be a bouquet of* n-spheres. *If* $n \geq 2$, *then* $\pi_r(B)$ *is trivial for* $r < n$, *and* $\pi_r(B)$ *is a free Abelian group, with the homotopy classes of the inclusions of the spheres in* B *as free generators.* $\square$

11.6. Homotopy Groups of a k-connected Cellular Pair.

A. Theorem. *Let* (X, A) *be a cellular pair with base point* $x_0 \in A$. *If* A *is connected and* $\pi_r(X, A) = 0$ *for* $r \leq k$, *then* $\mathrm{pr}_* : \pi_{k+1}(X, A, x_0) \to$

$\pi_{k+1}(X/A, \mathrm{pr}(x_0))$ *is an epimorphism, and for* $k \geq 1$, *the kernel* $\mathrm{Ker}\,\mathrm{pr}_*$ *is the smallest subgroup of* $\pi_{k+1}(X, A, x_0)$ *containing all the differences* $(T_\sigma \alpha) - \alpha$ *with* $\alpha \in \pi_{k+1}(X, A, x_0)$, $\sigma \in \pi_1(A, x_0)$. *For* $k = 0$ *the situation is described by the commutative diagram*

$$\pi_1(A, x_0) \xrightarrow{\mathrm{in}_*} \pi_1(X, x_0) \xrightarrow{\mathrm{rel}_*} \pi_1(X, A, x_0)$$

$$\searrow \mathrm{pr}_* \qquad\qquad \downarrow \mathrm{pr}_*$$

$$\pi_1(X/A, \mathrm{pr}(x_0))$$

in which rel_* *and* $\mathrm{pr}_* : \pi_1(X, x_0) \to (X/A, \mathrm{pr}(x_0))$ *are also epimorphisms and* $\mathrm{Ker}(\mathrm{pr}_*)$ *is the smallest normal subgroup of* $\pi_1(X, x_0)$ *containing* $\mathrm{Ker}\,\mathrm{rel}_* = \mathrm{Im}\,\mathrm{in}_*$.
□

The proof for the case $k \geq 1$ uses theorems 11.2.D and 11.5.F and is based on the fact that we can make the pair (X, A) k-connected by discarding the components of X that do not contain x_0, and then replace it by a homotopy equivalent pair (X', A') with $\mathrm{sk}_k X' \subset A'$. For $k = 0$ the proof follows in an obvious way from the results of 11.3.B in the case when X has a single 0-cell; in the general case we may replace the triple (X, A, x_0) by a homotopy equivalent triple with a single 0-cell.

B. The quotient space of a cellular space X by a simply connected subspace A is k-connected ($0 \leq k \leq \infty$) iff (X, A) is k-connected. If this condition is satisfied for some $k < \infty$, then the homomorphism $\mathrm{pr}_* : \pi_{k+1}(X, A, x_0) \to \pi_{k+1}(X/A, \mathrm{pr}(x_0))$ is an isomorphism.

The second part of this theorem is an obvious consequence of theorems **A** and 3.4.E. The first part follows from the second by induction on k.

C. If a cellular space X with base point the 0-cell x_0 is k-connected, then the homomorphism $\Sigma : \pi_{k+1}(X, x_0) \to \pi_{k+2}(\Sigma(X, x_0))$ is an isomorphism. (This is a corollary of theorem **B**.)

11.7. Spaces with Given Homotopy Groups.
    A. Lemma. *Let* π *be a group and* n *a positive integer. If* π *is Abelian or* $n = 1$, *then there exists a connected cellular space* X *for which* $\pi_r(X)$ *is trivial for* $r \neq n$, *and* $\pi_n(X) \simeq \pi$. □

The proof is based on the inductive construction of connected cellular spaces $X_0, X_1, \ldots$ with base points $x_0, x_1, \ldots$ and cellular embeddings $\phi_0 : X_0 \to X_1$, $\phi_1 : X_1 \to X_2, \ldots$, taking base points to base points, such that
    (i) the groups $\pi_r(X_k, x_k)$ with $r < n$ and $n < r \leq n + k$ are trivial,
    (ii) $\pi_n(X_k, x_k) \simeq \pi$,
    (iii) $\phi_{k*} : \pi_n(X_k, x_k) \to \pi_n(X_{k+1}, x_{k+1})$ is an isomorphism.
The space $X = \lim(X_k, \phi_k)$ will then have the desired properties (see 3.1.J). Let us choose a system of generators $\{g_\alpha\}_{\alpha \in A}$ and relations $\{r_\beta\}_{\beta \in B}$ for π (if $n > 1$, the relations that make π Abelian are understood, that is, the relations expressing commutativity are omitted), and put $Y = \vee_{\alpha \in A}(S_\alpha = S^n)$; then $\pi_i(Y) = 0$ for $i < n$, and $\pi_n(Y) = \oplus_{\alpha \in A} \mathbb{Z}$ if $n > 1$, and $\pi_n(Y) = *_{\alpha \in A} \mathbb{Z}$ if $n = 1$. The generators of $\pi_n(Y)$ are in one-one correspondence with the g_α, and we denote them by γ_α. On

the left-hand side of each relation r_β we replace g_α by γ_α, to obtain elements ρ_β of $\pi_n(Y)$. We fix spheroids $S^{n+1} \to Y$ representing these elements, and attach $(n+1)$-cells to Y by means of these spheroids. The resulting space is X_0. We then take any system of generators of $\pi_{n+1}(X_0)$, represent them by spheroids $S^{n+1} \to X_0$ and attach $(n+2)$-cells to X_0 by these spheroids. The resulting space is X_1. We attach cells to X_1 using π_{n+2}, and so on.

B. *Remark*. This construction is wholly non-effective: we do not know what the groups $\pi_{n+1}(X_0)$, $\pi_{n+2}(X_1)$ are, nor do we have any means of calculating them. We can of course attach cells to X_0 by all conceivable spheroids $S^{n+1} \to X_0$. and then do the same to X_1, and so on. This frees the construction from arbitrariness, but makes it very unwieldy.

C. The spaces whose existence is asserted by lemma **A** are called *Eilenberg-MacLane spaces*, or *spaces of type* $K(\pi, n)$; we often say for short that "X is a $K(\pi, n)$". In view of **B**, the not very numerous explicit constructions of such spaces are of interest.

D. *Explicit constructions*.

(1) The space $\mathbb{C}P^\infty$ is of type $K(\mathbb{Z}, 2)$; it is the single visible space of type $K(\pi, n)$ with $n > 1$.

(2) S^1 is a $K(\mathbb{Z}, 1)$.

(3) $\mathbb{R}P^\infty$ is a $K(\mathbb{Z}_2, 1)$.

(4) The infinite lens space $L(m; 1, 1, \ldots)$ is a $K(\mathbb{Z}_m, 1)$.

(5) Since the product of a $K(\pi_1, n)$ space and a $K(\pi_2, n)$ space is obviously a $K(\pi_1 \times \pi_2, n)$, constructions (2)–(4) allow us to construct $K(\pi, 1)$ spaces for any finitely generated Abelian group π. Very many $K(\pi, 1)$ spaces are also known with π non-Abelian. They include for example all closed connected surfaces except S^2, $\mathbb{R}P^2$, and $S^1 \times S^1$ (the last is a $K(\mathbb{Z} \times \mathbb{Z}, 1)$ but $\mathbb{Z} \times \mathbb{Z}$ is Abelian). As well as those listed above, we add three more classes of spaces of type $K(\pi, 1)$: all complete Riemannian manifolds of non-positive curvature; the complement of any closed connected curve (knot) in S^3, and, finally, the space of all non-ordered sets of n distinct points of the plane is a space of type $K(B_n, 1)$, where B_n is the Artin braid group of n threads, $B_n = \{\sigma_1, \ldots, \sigma_{n-1} | \sigma_i \sigma_j = \sigma_j \sigma_i$ for $|i - j| \geq 2$ and $\sigma_i \sigma_{i+1} \sigma_i = \sigma_{i+1} \sigma_i \sigma_{i+1}\}$.

E. If X is a $K(\pi, n)$, then clearly ΩX is a $K(\pi, n+1)$.

F. *Uniqueness*. Any two spaces of type $K(\pi, n)$ are weakly homotopy equivalent (since any two cellular spaces of type $K(\pi, n)$ are homotopy equivalent). To prove this, given any space Z of type $K(\pi, n)$ we may construct a weak homotopy equivalence $f : X \to Z$, where X is the $K(\pi, n)$ space constructed in **A**. Recall that the n-skeleton Y of X is a bouquet $\vee_{\alpha \in A} S^n_\alpha$ of n-spheres corresponding to the generators of π. Since $\pi_n(\mathbb{Z}) = \pi$, we can identify these generators with elements of $\pi_n(\mathbb{Z})$. We map $\vee_{\alpha \in A} S^n_\alpha$ to $\mathbb{Z}$ by spheroids representing these elements. All we now have to do is to extend this map somehow to a continuous map $X \to Z$; any such extension will have the desired properties. The extension can be constructed separately for each cell of X, and each time it turns out to be possible because the composition of the attaching map of the cell with the map already constructed is a spheroid homotopic to a constant. For $(n + 1)$-cells this follows because they correspond to generators of π, and for cells of higher dimension because $\pi_i(Y) = 0$ for $i > n$.

G. Theorem. *For any group π_1, and any Abelian groups $\pi_2, \pi_3, \ldots$, there exists a connected cellular space X with $\pi_r(X) \simeq \pi_r$, $r = 1, 2, \ldots$.* $\square$

To prove this we first construct spaces $X_1, X_2, \ldots$, of types $K(\pi_1, 1)$, $K(\pi_2, 2)$, $\ldots$ (see **A**), and then define spaces $Y_0, Y_1, \ldots$ inductively by $Y_0 = D^0$, $Y_{k+1} = Y_k \times X_{k+1}$, and cellular maps $\psi_k : Y_k \to Y_{k+1}$ (as fibres). The space $X = \lim(Y_k, \psi_k)$ then has the desired properties, as follows from 3.1.**I** and **J**.

Note that in contrast to the case of $K(\pi, n)$ spaces, the weak homotopy type of a space with several non-trivial homotopy groups is not determined by them (see 9.4.**E**).

§12. Appendix

12.1. The Whitehead Product.

A. In this section we define and study in part an operation on the elements of homotopy groups that generalizes the action of the fundamental group on homotopy groups in a certain sense. This operation is a map

$$\pi_m(X, x_0) \times \pi_n(X, x_0) \to \pi_{m+n-1}(X, x_0) \,,$$

defined for any topological space X with base point x_0. We begin with a very concrete preliminary construction.

The product $S^m \times S^n$ of two spheres can be decomposed into four cells of dimensions 0, m, n, $m+n$. The union of the first three is the bouquet $S^m \vee S^n$. The characteristic map $D^{m+n} \to S^m \times S^n$ of the fourth cell maps S^{m+n-1} to $S^m \vee S^n$; the resulting map $S^{m+n-1} \to S^m \vee S^n$ is called the *Whitehead map*. This definition needs to be made more precise (the characteristic map is not determined by the cell decomposition): the Whitehead map is the map

$$w : S^{m+n-1} \to S^m \vee S^n$$

defined as follows. The sphere S^{m+n-1} can be split into the union of closed regions

$$U = \{(x_1, \ldots, x_{m+n}) \in S^{m+n-1} | x_1^2 + \cdots + x_m^2 \le \tfrac{1}{2}\} \,,$$
$$V = \{(x_1, \ldots, x_{m+n}) \in S^{m+n-1} | x_1^2 + \cdots + x_m^2 \ge \tfrac{1}{2}\} \,.$$

Clearly $U \simeq D^m \times S^{n-1}$, $V \simeq S^{m-1} \times D^n$, $U \cap V \simeq S^{m-1} \times S^{n-1}$ (the case $m = 2$, $n = 1$ is illustrated in Fig. 16; an interesting and important case is the decomposition of S^3 by the torus $S^1 \times S^1$ into the union of two full tori). Alternatively the decomposition $S^{m+n-1} = U \cup V$ can be constructed thus: $S^{m+n-1} = \partial D^{m+n} \simeq \partial(D^m \times D^n) = (D^m \times \partial D^n) \cup (\partial D^m \times D^n) = U \cup V$.

The map w is defined on U and V as the composite projection

$$U = D^m \times S^{n-1} \to D^m \to D^m/S^{m-1} = S^m \hookrightarrow S^m \vee S^n \,,$$
$$V = S^{m-1} \times D^n \to D^n \to D^n/S^{n-1} = S^n \hookrightarrow S^m \vee S^n \,,$$

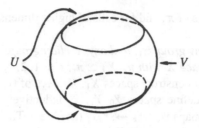

Fig. 16

and takes the dividing surface $U \cap V = S^{m-1} \times S^{n-1}$ to a point.

Now let $f : S^m \to X$, $g : S^n \to X$ be two spheroids in a pointed space X. Together they define a map $S^m \vee S^n \to X$, and the composition of this map with $w : S^{m+n-1} \to S^m \vee S^n$ is a certain spheroid $h : S^{m+n-1} \to X$. The homotopy class of this spheroid depends only on the classes of f and g, and is called their *Whitehead product*. It is denoted by square brackets, so that if $\alpha \in \pi_m(X, x_0)$, $\beta \in \pi_n(X, x_0)$ then $[\alpha, \beta] \in \pi_{m+n-1}(X, x_0)$.

Note that from this definition the homotopy class of w itself is the Whitehead product of the classes of the natural embeddings $S^m \to S^m \vee S^n$ and $S^n \to S^m \vee S^n$.

B. Clearly $f_*([\alpha, \beta]) = [f_*(\alpha), f_*(\beta)]$ for any $\alpha \in \pi_m(X, x_0)$, $\beta \in \pi_n(X, x_0)$ and any continuous map $f : X \to Y$. Also $T_s([\alpha, \beta]) = [T_s\alpha, T_s\beta]$ for any $\alpha \in \pi_m(X, x_0)$, $\beta \in \pi_n(X, x_0)$ and any path $s : I \to X$ starting from x_0.

C. If $\alpha \in \pi_m(X, x_0)$, $\beta \in \pi_n(X, x_0)$ then $[\beta, \alpha] = (-1)^{mn}[\alpha, \beta]$. This follows in an obvious way from the following immediately verifiable property of the map w: the isomorphism $\pi_{m+n-1}(S^m \vee S^n) \to \pi_{m+n-1}(S^n \vee S^m)$ induced by interchanging the spheres S^m, S^n, maps the homotopy class of $w : S^{m+n-1} \to S^m \vee S^n$ into $(-1)^{mn} \times$ (the homotopy class of $w : S^{m+n-1} \to S^n \vee S^m$)

D. If $\alpha \in \pi_m(X, x_0)$ and $\beta_1, \beta_2 \in \pi_n(X, x_0)$ with $n > 1$, then $[\alpha, \beta_1 + \beta_2] = [\alpha, \beta_1] + [\alpha, \beta_2]$. If $\alpha_1, \alpha_2 \in \pi_m(X, x_0)$ with $m > 1$, $\beta \in \pi_n(X, x_0)$, then $[\alpha_1 + \alpha_2, \beta] = [\alpha_1, \beta] + [\alpha_2\beta]$.

By **C** it is sufficient to prove the first assertion. Just as in the proof of **C**, the required assertion can be reduced to an assertion about the map w which can easily be verified.

E. If $\alpha \in \pi_1(X, x_0)$, $\beta \in \pi_n(X, x_0)$ for any $n \geq 1$, then $[\alpha, \beta] = \beta(T_\alpha\beta)^{-1}$ [or, if the group operation in $\pi_n(X, x_0)$ is written additively, as is usual for $n \geq 2$, then $[\alpha, \beta] = \beta - T_\alpha\beta \in \pi_n(X, x_0)$. In particular, if $\alpha, \beta \in \pi_1(X, x_0)$ then $[\alpha, \beta] = \beta\alpha\beta^{-1}\alpha^{-1}$. This too is an obvious consequence of the definition of the Whitehead product, and follows from an immediately verifiable property of the map $w : S^{m+n-1} \to S^m \vee S^n$ for $m = 1$.

F. Let ι_n denote the canonical generator of $\pi_n(S^n)$ (that is the class of the identity map) and η_2 the canonical generator of $\pi_3(S^2)$ (the class of the Hopf map $S^3 \to S^2$). It is not hard to show that $[\iota_2, \iota_2] = 2\eta_2$.

G. The Whitehead product is in general not associative. This is already evident from **E**: for example, if α, β, γ denote the natural generators of $\pi_1(S^1 \vee S^1 \vee$

S^1) (they are its free generators; see 11.1), then it turns out that $[[\alpha, \beta], \gamma] = \gamma\beta\alpha\beta^{-1}\alpha^{-1}\gamma^{-1}\alpha\beta\alpha^{-1}\beta^{-1}$, but $[\alpha, [\beta, \gamma]] = \gamma\beta\gamma^{-1}\alpha\beta\gamma\beta^{-1}\gamma^{-1}\alpha^{-1}$.

H. It can be shown that the Whitehead product satisfies an identity like the Jacobi identity but with differing signs. If $\alpha \in \pi_m(X, x_0)$, $\beta \in \pi_n(X, x_0)$, $\gamma \in \pi_p(X, x_0)$ and $m > 1$, $n > 1$, $p > 1$, then $(-1)^{pm}[[\alpha, \beta], \gamma] + (-1)^{mn}[[\beta, \gamma], \alpha] + (-1)^{np}[[\gamma, \alpha], \beta] = 0$.

I. The product $[\alpha, \beta]$ of $\alpha \in \pi_m(X, x_0)$, $\beta \in \pi_n(X, x_0)$ is always in the kernel of the suspension homomorphism

$$\Sigma : \pi_{m+n-1}(X, x_0) \to \pi_{m+n}(\Sigma(X, x_0)) .$$

This follows from the corresponding immediately verifiable fact about $S^m \vee S^n$, namely, that the spaces $\Sigma(S^m \times S^n)$ and $S^m \vee S^n \vee S^{m+n+1}$ are homotopy equivalent.

J. For any $\alpha \in \pi_m(X, x_0)$, $\beta \in \pi_n(Y, y_0)$, the Whitehead product of their images under the homomorphisms induced by the natural inclusions $X, Y \hookrightarrow X \vee Y$ are in the kernel of each of the following three homomorphisms:

$$\text{pr}_{1*} : \pi_{m+n-1}(X \vee Y) \to \pi_{m+n-1}(X) ,$$

$$\text{pr}_{2*} : \pi_{m+n-1}(X \vee Y) \to \pi_{m+n-1}(Y) ,$$

$$\text{pr}_{3*} : \pi_{m+n-1}(X \vee Y) \to \pi_{m+n-1}(X \times Y) .$$

K. If X is an H-space, then $[\alpha, \beta] = 0$ for any $x_0 \in X$ and $\alpha \in \pi_m(X, x_0)$, $\beta \in \pi_n(X, x_0)$. This generalizes the theorem in 3.3.E. The proof consists in showing that the map $S^m \vee S^n \to X$, through which the map defining $[\alpha, \beta]$ is factored, can be extended to a map $S^m \times S^n \to X$ by using the H-space structure of X; but in $S^m \times S^n$ the Whitehead product of the classes of the inclusions $S^m, S^n \hookrightarrow S^m \times S^n$ is obviously zero.

L. *The hard part of the suspension theorem* (supplement to the theorem of 10.1.E). The kernel of the homomorphism $\Sigma : \pi_{2k+1}(S^{k+1}) \to \pi_{2k+2}(S^{k+2})$ is a cyclic group generated by the Whitehead square $[\iota'_{k+1}\iota_{k+1}]$ of the natural generator ι_{k+1} of $\pi_{k+1}(S^{k+1})$. The proof of this theorem is an immediate generalization of the proof of the suspension theorem 10.1.E. For any odd k, the class $[\iota_{k+1}\iota_{k+1}]$ has infinite order in $\pi_{2k+1}(S^{k+1})$. In particular, for any $n \geq 1$, $\pi_{4n-1}(S^{2n})$ and the kernel of $\Sigma : \pi_{4n-1}(S^{2n}) \to \pi_{4n}(S^{2n+1})$ are infinite (cf. C, 10.1, 10.4).

12.2. The Homotopy Sequence of a Triad.

A. A *triad* is a triple (X, A, B) where X is a topological space and A and B are subsets with $A \cup B = X$. Let $(X; A, B)$ be a triad with $x_0 \in A \cap B$. The homomorphism

$$\text{in}_* : \pi_r(A, A \cap B, x_0) \to \pi_r(X, B, x_0)$$

is called the *excision homomorphism*. The explanation of this name is that the pair $(A, A \cap B)$ is obtained by removing the same set $X \setminus A = B \setminus A$ from both X and B. The analogous homomorphism in homology theory is an isomorphism under very broad assumptions. It is one of the most important properties of homology groups, and is elevated to the status of an axiom, the "excision axiom", in the axiomatic development of homology theory. In fact thanks to this property the computation of homology groups is a relatively easy task. The homotopy excision homomorphism is not an isomorphism even in very simple examples. Let $S_{\pm}^2 = \{(x_1, x_2, x_3) \in S^2 | \pm x \geq 0\}$; then $\pi_3(S^2, S_-^2, (1, 0, 0)) \simeq \pi_3(S^2, (1, 0, 0))$ $(\simeq z)$, where the isomorphism is induced by the inclusion, which is a homotopy equivalence, $\pi_3(S_+^2, S^1, (1, 0, 0)) \overset{\partial}{=} \pi_2(S^1, (1, 0, 0)) = 0$, and the excision homomorphism $\text{in}_* : \pi_3(S_+^2, S^1) \rightarrow \pi_3(S^2, S_-^2)$ is not an isomorphism. In this section we shall include the homotopy excision homomorphism in an exact sequence, and describe how far it fails to be an isomorphism.

B. Let $(X; A, B)$ be a triad, $x_0 \in A \cap B$ and $r \geq 2$. Let $\pi_r(X; A, B, x_0)$ denote the set of homotopy classes of maps $D^r \rightarrow X$ taking the hemisphere $S_+^{r-1} = \{(x_1, \ldots, x_r) \in S^r | x_r \geq 0\}$ into A, the hemisphere $S_-^{r-1} = \{(x_1, \ldots, x_r) \in S^r | x_r \leq 0\}$ into B, and the base point $(1, 0, \ldots, 0)$ of the equator S^{r-2} to x_0. For $r \geq 3$ the set $\pi_r(X; A, B, x_0)$ is endowed with a natural group structure, which is commutative for $r \geq 4$. For $r \geq 2$, the set $\pi_r(X; A, B, x_0)$ is called the *rth homotopy group of the triad* $(X; A, B)$. We give two other descriptions of it. Since there is a continuous map

$$(D^{r-1} \times I, S^{r-2} \times I, D^{r-1} \times 1, D^{r-1} \times 0 \cup (1, 0, \ldots, 0) \times I)$$
$$\rightarrow (D^r, S_+^{r-1}, S_-^{r-1}, (1, 0, \ldots, 0)),$$

inducing a homeomorphism

$$(D^{r-1} \times I/D^{r-1} \times 0 \cup (1, 0, \ldots, 0) \times I,$$
$$S^{r-2} \times I/S^{r-2} \times 0 \cup (1, 0, \ldots, 0) \times I), D^{r-1} \times 1)$$
$$\rightarrow (D^r, S_+^{r-1}, S_-^{r-1}),$$

the group $\pi_r(X; A, B, x_0)$ is

$$C(D^{r-1} \times I, S^{r-2} \times I, D^{r-1} \times 1, D^{r-1} \times 0 \cup (1, 0, \ldots, 0) \times I; X, A, B, x_0),$$

and, by the exponential law, this is the same as

$$\pi_{r-1}(C(I, 0, 1; X, x_0, B), C(I, 0, 1; A, x_0, A \cap B), I \rightarrow x_0 \hookrightarrow X).$$

This last description gives a superfluous description of the group operation in $\pi_r(X; A, B, x_0)$ and other structures.

C. For any triad $(X; A, B)$ and $x_0 \in A \cap B$, there is a π-sequence

$$\cdots \longrightarrow \pi_{r+1}(X; A, B, x_0) \overset{\alpha}{\longrightarrow} \pi_r(A, A \cap B, x_0) \overset{\text{in}_*}{\longrightarrow} \pi_r(X, B, x_0)$$
$$\overset{\beta}{\longrightarrow} \pi_r(X; A, B, x_0) \overset{\alpha}{\longrightarrow} \cdots \longrightarrow \pi_2(X; A, B, x_0) \overset{\alpha}{\longrightarrow} \pi_1(A, A \cap B, x_0)$$
$$\overset{\text{in}_*}{\longrightarrow} \pi_1(X, B, x_0).$$

It is called the *homotopy sequence of the triad* $(X; A, B)$ at x_0. It is obtained from the homotopy sequence of the pair $(C(I, 0, 1; X, x_0, B), C(I, 0, 1; A, x_0, A \cap B))$ by means of the isomorphisms

$$\pi_r(X, B, x_0) = \pi_{r-1}(C(I, 0, 1; X, B, x_0), I \to x_0 \hookrightarrow X),$$
$$\pi_r(A, A \cap B, x_0) = \pi_{r-1}(C(I, 0, 1; A, A \cap B, x_0), I \to x_0 \hookrightarrow A).$$

In particular it is exact. Its homomorphisms can be described in the language of spheroids $(D^r; S_+^{r-1}, S_-^{r-1}, (1, 0, \ldots, 0)) \to (X; A, B, x_0)$ as follows: the homomorphism α takes the class of a spheroid $(D^{r+1}; S_+^r, S_-^r, (1, 0, \ldots, 0)) \xrightarrow{\phi} (X; A, B, x_0)$ to the class of the composition of the homeomorphism $(D^r, S^{r-1}) \to (S_+^r, S^{r-1})$ and the restriction of ϕ; the homomorphism β takes the class of the spheroid $\psi : (D^r, S^{r-1}, (1, 0, \ldots, 0)) \to (X, B, x_0)$ to the class of the composition of the map $(D^r, S_+^{r-1}, S_-^{r-1}, (1, 0, \ldots, 0)) \to (D^r, S^{r-1}, (1, 0, \ldots, 0), (1, 0, \ldots, 0))$, that maps $\mathrm{Int}\, D^r \to \mathrm{Int}\, D^r$ and $\mathrm{Int}\, S_+^{r-1} \to S^{r-1} \setminus (1, 0, \ldots, 0)$ homeomorphically, with the spheroid ψ.

12.3. Homotopy Excision, Quotient and Suspension Theorems.

A. Lemma. *Let C be a topological space, A the space obtained from C by attaching an n-cell e^n (that is, $A = C \cup_f D^n$, where $f : S^{n-1} \to C$ is continuous and $e^n = A \setminus C$), B the space obtained from C by attaching an m-cell e^m, and $X = A \cup B$. Then for all r with $2 \le r \le n + m - 2$, the group $\pi_r(X; A, B, x_0)$ is trivial.* $\square$

This lemma is proved with the aid of the simplicial approximation theorem. For an elementary proof see Chapter 6 of Switzer (1975).

B. The following homotopy excision theorem can be deduced from lemma **A** by standard arguments. *Let $(X; A, B)$ be a triad in which the pair $(A, A \cap B)$ is n-connected with $n \ge 1$, and $(B, A \cap B)$ is m-connected; then the excision homomorphism $\mathrm{in}_* : \pi_r(A, A \cap B) \to \pi_r(X, B)$ is an isomorphism for $1 \le r < n + m$, and an epimorphism for $r = n + m$.* $\square$

C. Corollary. Quotient theorem. *Let (X, A) be an n-connected Borsuk pair $(n \ge 1)$ with an m-connected subspace A. Then the projection $(X, A) \to (X/A, A/A)$ induces a homomorphism $\pi_r(X, A) \to \pi_r(X/A)$ which is an isomorphism for $2 \le r \le m + n$, and an epimorphism for $r = m + n + 1$.* $\square$

D. Generalized suspension theorem (cf. 10.1.E). *For any n-connected space X $(n \ge 0)$, the suspension homomorphism $\Sigma : \pi_r(X) \to \pi_{r-1}(\Sigma X)$ is an isomorphism for $1 \le r \le 2n$, and an epimorphism for $r = 2n + 1$.* $\square$

This follows from **C** and the description of the suspension homomorphism given in 10.1.**D**.

II. Homology and Cohomology

O.Ya. Viro, D.B. Fuchs

Translated from the Russian
by C.J. Shaddock

Contents

Chapter 1
Additive Theory

§1. Algebraic Preparation

1.1. Complexes and Their Homology.

A. A *complex* is a sequence of Abelian groups and homomorphisms of the form

$$\ldots \longrightarrow C_3 \xrightarrow{\partial_3} C_2 \xrightarrow{\partial_2} C_1 \xrightarrow{\partial_1} C_0 \xrightarrow{\partial_0} C_{-1} \longrightarrow \ldots,$$

in which $\partial_q \circ \partial_{q+1} = 0$ for each q. A complex is *positive* if $C_q = 0$ for $q < 0$, *free* if all the C_q are free Abelian groups, and a *complex of finite type* if the sum $\bigoplus_q C_q$ is finitely generated. (In dealing with positive complexes, the sequence $\{C_q, \partial_q\}$ is often cut short, and we speak of the sequence $\ldots \longrightarrow C_1 \xrightarrow{\partial_1} C_0$.) The homomorphisms ∂_q are called the *differentials* or *boundary operators*; elements of C_q are called the *q-chains* of the complex; chains that are annihilated by the differential are called *cycles*, and cycles whose difference lies in the image of the differential are said to be *homologous*.

It will be convenient to regard an Abelian group as a special case of a complex: an Abelian group G is identified with the complex in which $C_0 = G$, $C_q = 0$ for $q \neq 0$, and $\partial_q = 0$ for all q.

B. The equality $\partial_{q+1} \circ \partial_q = 0$ is equivalent to the inclusion $\operatorname{Im} \partial_{q+1} \subset \operatorname{Ker} \partial_q$; the factor group $\operatorname{Ker} \partial_q / \operatorname{Im} \partial_{q+1}$ is called the *q*th *homology group* of the complex, and is denoted by $H_q(C)$ or H_q. (Thus the elements of the homology group are classes of homologous cycles.)

Note that if $\partial_q = \partial_{q+1} = 0$, then $H_q = C_q$; in particular, if $\partial_q = 0$ for all q, then $H_q = C_q$ for all q. For example, $H_0(G) = G$, $H_q(G) = 0$ for $q \neq 0$.

C. Given a complex $C = \{C_q, \partial_q\}$ and an Abelian group G, we may form the complex $C \otimes G$

$$\ldots \longrightarrow C_1 \otimes G \xrightarrow{\partial_1 \otimes G} C_0 \otimes G \xrightarrow{\partial_0 \otimes G} C_{-1} \otimes G \longrightarrow \ldots$$

and the complex $\operatorname{Hom}(C, G)$

$$\ldots \longrightarrow \operatorname{Hom}(C_{-1}, G) \xrightarrow{\operatorname{Hom}(\partial_0, G)} \operatorname{Hom}(C_0, G) \xrightarrow{\operatorname{Hom}(\partial_1, G)} \operatorname{Hom}(C_1, G) \longrightarrow \ldots$$

(the groups in these complexes are numbered so that $C_0 \otimes G$ and $\operatorname{Hom}(C_0, G)$ are numbered 0). The homology of the complex $C \otimes G$ is called the *homology of the complex C with coefficients in G*, denoted by $H_q(C; G)$; the homology of $\operatorname{Hom}(C, G)$ is called the *cohomology of C with coefficients in G* (or *with values in G*), denoted by $H^q(C; G)$; thus

$$H_q(C; G) = \operatorname{Ker}(\partial_q \otimes G)/\operatorname{Im}(\partial_{q+1} \otimes G),$$

$$H^q(C; G) = \operatorname{Ker} \operatorname{Hom}(\partial_{q+1}, G)/\operatorname{Im} \operatorname{Hom}(\partial_q, G).$$

The chains, cycles and boundary operators of $\text{Hom}(C, G)$ are called the *cochains*, *cocycles*, and *coboundary operators* of C.

1.2. Maps and Homotopies.
A. A *map* (*chain map, homomorphism*) of the complex $C' = \{C'_q, \partial'_q\}$ into the complex $C'' = \{C''_q, \partial''_q\}$ is a sequence of homomorphisms $f_q : C'_q \to C''_q$ such that the diagram

$$\cdots \longrightarrow \; C'_1 \; \xrightarrow{\partial'_1} \; C'_0 \; \xrightarrow{\partial'_0} \; C'_{-1} \; \longrightarrow \cdots$$

$$\downarrow{f_1} \qquad\quad \downarrow{f_0} \qquad\quad \downarrow{f_{-1}}$$

$$\cdots \longrightarrow \; C''_1 \; \xrightarrow{\partial''_1} \; C''_0 \; \xrightarrow{\partial''_0} \; C''_{-1} \; \longrightarrow \cdots$$

is commutative. Since each map f_q takes cycles to cycles and homologous cycles to homologous cycles, the map $f = \{f_q\} : C' \to C''$ induces a sequence of homomorphisms

$$f_* = f_{*q} : H_q(C') \to H_q(C'');$$

moreover, for any Abelian group G, f induces maps $f \otimes G : C' \otimes G \to C'' \otimes G$, $\text{Hom}(f, G) : \text{Hom}(C'', G) \to \text{Hom}(C', G)$, and hence homomorphisms

$$f_* = f_{*q} : H_q(C'; G) \to H_q(C''; G),$$

$$f^* = f^{*q} : H^q(C'; G) \to H^q(C''; G).$$

Clearly $(f \circ g)_* = f_* \circ g_*$, $(f \circ g)_* = g^* \circ f^*$.

B. A *homotopy* (*chain homotopy*) between maps $f = \{f_q\}$, $g = \{g_q\} : C' \to C''$ is a sequence of maps $F_q : C_q \to C_{q+1}$ such that for each q

$$\partial_{q+1} \circ F_q + F_{q-1} \circ \partial_q = g_q - f_q.$$

If there is a homotopy between maps f and g, they are said to be *homotopic*, written $f \sim g$. (The relation between these homotopies and those considered in the previous part will be explained in §2.)

Proposition 1. *Homotopic maps induce the same homology homomorphisms.* ☐

In fact, if c' is a q-cycle of C', then $g_q(c') - f_q(c') = \partial_{q+1}(F_q(c')) \in \text{Im} \, \partial_{q+1}$, that is, the cycles $f_q(c')$ and $g_q(c')$ in C'' are homologous.

In addition, if we apply the operations $\otimes G$ and $\text{Hom}(\, , G)$ to a homotopy between $f, g : C' \to C''$, we obtain a homotopy between $f \otimes G, g \otimes G : C' \otimes G \to C'' \otimes G$, and a homotopy between $\text{Hom}(f, G), \text{Hom}(g, G) : \text{Hom}(C'', G) \to \text{Hom}(C', G)$. Thus we have the following stronger version of the preceding proposition.

Proposition 2. *If the maps $f, g : C' \to C''$ are homotopic, then for any q and G, the maps*

$$f_{*q}, g_{*q} : H_q(C'; G) \to H_q(C''; G)$$

are the same, and so are the maps

$$f^*_q, g^*_q : H^q(C''; G) \to H^q(C'; G). \; \square$$

Example. Consider the complexes

$$
\begin{array}{ccccccccc}
C': & \cdots \to & 0 & \to & \mathbb{Z} & \xrightarrow{.2} & \mathbb{Z} & \to & 0 & \to \cdots \\
C'': & \cdots \to & 0 & \to & \mathbb{Z}_2 & \to & \mathbb{Z} & \to & 0 & \to \cdots \\
& & 2 & & 1 & & 0 & & -1 &
\end{array}
$$

(the figures underneath indicate the number of the group) and the two maps f, g : $C' \to C''$, $f \neq 0$, $g = 0$ (the nonzero map $C' \to C''$ is unique). Clearly

$$
H_q(C') = \begin{cases} \mathbb{Z}_2 & \text{for } q = 0, \\ 0 & \text{for } q \neq 0, \end{cases} \qquad H_q(C'') = \begin{cases} \mathbb{Z}_2 & \text{for } q = 1, \\ 0 & \text{for } q \neq 1, \end{cases}
$$

and hence $f_* = g_* = 0$. At the same time

$$
H_q(C'; \mathbb{Z}_2) = \begin{cases} \mathbb{Z}_2 & \text{for } q = 0, 1, \\ 0 & \text{for } q \neq 0, 1, \end{cases}
$$

and $C'' \otimes \mathbb{Z}_2 = C''$. It is also clear that

$$
f_* : H_1(C'; \mathbb{Z}_2) \to H_1(C''; \mathbb{Z}_2)
$$

is the isomorphism $\mathbb{Z}_2 \to \mathbb{Z}_2$. In view of this, proposition 2 shows that f and g are not homotopic (which of course is also obvious directly); we see that the converse of proposition 1 is false.

C. Let C', C'' be complexes. A map $f : C' \to C''$ is called a *(chain) homotopy equivalence* if there is a map $g : C'' \to C'$ such that $f \circ g \sim \text{id}$ and $g \circ f \sim \text{id}$. In this situation the maps f and g are said to be *homotopy inverses* of each other, and the complexes C' and C'' are said to be *homotopy equivalent*.

Clearly if f is a homotopy equivalence, then $f \otimes G$ and $\text{Hom}(f, G)$ are homotopy equivalences for any Abelian group G. (Further, if f and g are homotopy inverse, then $f \otimes G$ and $g \otimes G$ are homotopy inverse, as are $\text{Hom}(f, G)$ and $\text{Hom}(g, G)$.) It is also obvious that if f is a homotopy equivalence then all the homomorphisms f_* and f^* are isomorphisms. Less obvious (though also not particularly difficult) is the following important assertion.

Theorem. *If the complexes C' and C'' are positive and free, and the map f : $C' \to C''$ induces an isomorphism $f_* : H_q(C') \to H_q(C'')$ for all q, then f is a homotopy equivalence.* □

We recommend the reader to find examples showing that all four assumptions of this theorem (positivity and freedom of C, and positivity and freedom of C'') are essential. (The example in **B** may be taken as a model, although it is not suitable literally as it stands.)

1.3. Homology sequences.

A. Let $C' = \{C'_q, \partial'_q\}$ be a subcomplex of the complex $C = \{C_q, \partial_q\}$ (this means that $C'_q \subset C_q$ and $\partial'c' = \partial c'$ for all q and $c' \in C'_q$) and let $C'' = \{C''_q, \partial''_q\}$ be the factor complex C/C' (that is, $C''_q = C_q/C'_q$ and $\partial''_q : C''_q \to C''_{q-1}$ is induced by $\partial_q : C_q \to C_{q-1}$). This situation is sometimes described by saying that we are given a short exact sequence of complexes

$$0 \to C' \to C \to C'' \to 0. \tag{1}$$

It turns out that as well as the homomorphisms

$$H_q(C') \to H_q(C), \quad H_q(C) \to H_q(C''),$$

there are *connecting homomorphisms*

$$\partial_* : H_q(C'') \to H_{q-1}(C'),$$

defined as follows. Let $\alpha \in H_q(C'')$ and let $c'' \in C''_q = C_q/C'_q$ be a cycle representing α. We pick an arbitrary representative $c \in C_q$ of the coset c''. Since $\partial'' c'' = 0$, ∂c belongs to the zero coset of C'_q in C_q, that is, $\partial c \in C'_q$. It is a cycle ($\partial' \partial c = \partial \partial c = 0$), and we define $\partial_* \alpha$ to be its homology class in $H_{q-1}(C')$. A routine verification shows that this is a valid definition (that is, ∂_* does not depend on the choices of c'' in α and c in c''), and that the resulting map ∂_* is a homomorphism.

B. The connecting homomorphisms together with the induced homomorphisms form a sequence infinite in both directions

$$\cdots \longrightarrow H_q(C') \longrightarrow H_q(C) \longrightarrow H_q(C'') \xrightarrow{\partial_*} H_{q-1}(C) \longrightarrow \cdots, \tag{2}$$

called the *homology sequence of the pair* (C, C') or the *(long) homology sequence associated with the short exact sequence* (1). The following proposition is exceptionally important for us.

Theorem. *The sequence* (2) *is exact.*   ☐

The proof can be found in any textbook on homological algebra or algebraic topology, but we recommend the beginner to construct it for himself.

The exactness of the homology sequence has a number of standard corollaries; in particular, if C'' is *acyclic* (a complex is acyclic if all its homology groups are trivial, that is, the complex itself is an exact sequence), then the inclusion $C' \to C$ induces isomorphisms of all the homology groups.

1.4. The Euler characteristic and the Lefschetz number.

A. Let C be a complex. Let h_q denote the rank of the group $H_q(C)$, if it is finite. If all the h_q are defined and only finitely many of them are nonzero, we put

$$\text{Eu}(C) = \sum (-1)^q h_q,$$

and call $\text{Eu}(C)$ the *Euler characteristic* of C.

Theorem. *If C is a complex of finite type, then*

$$\text{Eu}(C) = \sum (-1)^q c_q,$$

where c_q is the rank of the group C_q.   ☐

This follows from the formula $c_q = r_{q+1} + h_q + r_q$, where r_i is the rank of the image of ∂_i.

The theorem shows, in particular, that the Euler characteristic of a complex of finite type depends only on the groups C_q, and not on the differentials ∂_q.

B. Let G be a finitely generated Abelian group; then Tors G denotes the subgroup of elements of finite order in G. The factor group $G/\text{Tors}\,G$ is called the *free part* of G; it is a free Abelian group whose rank is equal to the rank of G. An endomorphism $\phi : G \to G$ maps Tors G to Tors G, and so induces an endomorphism of the free part of G; this endomorphism has a trace, which is called the *trace* of the endomorphism ϕ, denoted $\text{tr}\,\phi$.

Now suppose that C is a complex with finitely generated homology groups $H_q(C)$, of which only finitely many are nonzero, and let $f = \{f_q\} : C \to C$ be a map of C to itself. Let t_q denote the trace of the endomorphism $f_* : H_q(C) \to H_q(C)$, and put

$$\text{Le}(f) = \sum(-1)^q t_q;$$

$\text{Le}(f)$ is called the *Lefschetz number* of f. Note that

$$\text{Le}(\text{id}_C) = \text{Eu}(C).$$

Theorem. *If C is a complex of finite type, then*

$$\text{Le}(f) = \sum(-1)^q \text{tr}\, f_q. \quad \square$$

In virtue of the remarks made above, this theorem is a generalization of the previous one.

C. The theory set out in **A** and **B** above has a "vector space version" that arises when all the groups C_q in a complex C are vector spaces over a field k, and all the differentials ∂_q are linear operators over k. In this situation the homology groups H_q are also vector spaces over k. Typical example of vector complexes are the complexes $C \otimes V$ and $\text{Hom}(C, V)$, where C is any complex, and V a vector space over k. Thus for any vector space V over a field k and any complex C, the groups $H_q(C; V)$ and $H^q(C; V)$ have the structure of a vector space over k; in particular this is true for $H_q(C; k)$ and $H^q(C; k)$.

If C is a vector complex over k, there is an analogue of the Euler characteristic defined by

$$\text{Eu}_k(C) = \sum(-1)^q \dim_k H_q(C),$$

and we have the analogue of theorem **A**:

Theorem.

$$\text{Eu}_k(C) = \sum(-1)^q \dim_k C_q,$$

when the left-hand side is defined. $\square$

Note that there are two cases in which rank $G = \dim_k(G \otimes k)$ for a finitely generated Abelian group G and a field k: when G is free and when k has characteristic zero. This allows us to make the following deduction from the last theorem and theorem **A**:

Corollary. *Let C be a complex of finite type and k a field. If C is free or if k has characteristic zero, then*

$$\mathrm{Eu}(C) = \mathrm{Eu}_k(C \otimes k). \quad \square$$

In other words, in the cases mentioned the Euler characteristic of a complex can be calculated using its homology (or its cohomology, as the results of the following section will show) with coefficients in k; this result is particularly striking when we consider that in the case when k has nonzero characteristic (as we again shall see in the next section) the relation rank $H_q(C) = \dim_k(C \otimes k)$ is not in general true even if C is free.

To a certain extent what has been said can also be transferred to the Lefschetz numbers; in particular, the analogue of the previous theorem holds for them. However, $\mathrm{Le}_k(f)$ is an element of k, and hence the last equation will take the form

$$\mathrm{Le}_k(f \otimes k) = \mathrm{Le}(f)1,$$

where 1 is the identity of k. Hence $\mathrm{Le}(f \otimes k)$ determines $\mathrm{Le}(f)$ only modulo the characteristic of the field. But if the field has characteristic 0, this proviso loses its force, and we may write $\mathrm{Le}_k(f \otimes k) = \mathrm{Le}(f)$ (assuming that $\mathbb{Z} \subset k$). In other words, the Lefschetz number of an endomorphism of a complex of finite type can be calculated using its homology (or, in view of the results of the following section, its cohomology) with coefficients in any field of characteristic 0.

1.5. Change of coefficients.

A. Let

$$0 \to G' \to G \to G'' \to 0 \tag{3}$$

be a short exact sequence of Abelian groups, and C a free complex. In view of known properties of the operations $\otimes$ and Hom, there arise short exact sequences

$$0 \to C \otimes G' \to C \otimes G \to C \otimes G'' \to 0,$$
$$0 \to \mathrm{Hom}(C, G') \to \mathrm{Hom}(C, G) \to \mathrm{Hom}(C, G'') \to 0,$$

which induce exact homology sequences in accordance with the general constructions in 1.3. In our notations these have the form

$$\cdots \to H_q(C; G') \to H_q(C; G) \to H_q(C; G'') \to H_{q-1}(C; G') \to \cdots$$
$$\cdots \to H^q(C; G') \to H^q(C; G) \to H^q(C; G'') \to H^{q+1}(C; G') \to \cdots$$

called the *coefficient sequences* associated with the sequence (3). They establish a very simple relation between homology and cohomology with different coefficients.

B. To describe the deeper relation between them, we need two operations Tor and Ext, which are less elementary than $\otimes$ and Hom.

Let A and B be Abelian groups. Suppose in addition that $B = F_1/F_2$, where F_1 is a free Abelian group, and F_2 a subgroup, which will automatically be free also (such a presentation exists for any Abelian group). Taking the tensor product of the exact sequence

$$0 \to F_2 \to F_1 \to B \to 0$$

by A (on the left), we obtain the exact sequence

$$A \otimes F_2 \to A \otimes F_1 \to A \otimes B \to 0,$$

in which the first map is in general not a monomorphism. It turns out that the kernel of this map does not depend on the choice of presentation $B = F_1/F_2$ and is defined, to within a canonical isomorphism, by the groups A and B. It is called the *periodic product* of A and B, denoted by $A * B$, or (in a more modern way) by $\mathrm{Tor}(A, B)$.

Properties of the operation Tor. 1°. Tor is natural (covariant) in each argument. 2°. There is a natural isomorphism $\mathrm{Tor}(A, B) \cong \mathrm{Tor}(B, A)$ (A and B do not appear symmetrically in the definition of Tor). 3°. If A is a free Abelian group or $A = \mathbb{Q}$, $\mathbb{R}$, or $\mathbb{C}$, then $\mathrm{Tor}(A, B) = 0$. 4°. $\mathrm{Tor}(\mathbb{Z}_m, \mathbb{Z}_n) \cong \mathbb{Z}_m \otimes \mathbb{Z}_n$. Hence for finitely generated Abelian groups there is a (non-canonical) isomorphism

$$\mathrm{Tor}(A, B) \cong \mathrm{Tors}\, A \otimes \mathrm{Tors}\, B$$

(in the infinitely generated case the two sides are in general not isomorphic).

The definition of the operation Ext is "dual": we present A in the form F_1/F_2, where F_1 and F_2 are free Abelian groups, and write out the exact sequence

$$0 \to \mathrm{Hom}(A, B) \to \mathrm{Hom}(F_1, B) \to \mathrm{Hom}(F_2, B).$$

The cokernel of the last map is then $\mathrm{Ext}(A, B)$.

Properties of the operation Ext. 1°. Ext is contravariant in the first argument, and covariant in the second. 2°. $\mathrm{Ext}(\mathbb{Z}, B) = 0$ for any group B, $\mathrm{Ext}(\mathbb{Z}_m, \mathbb{Z}_n) \cong \mathbb{Z}_m \otimes \mathbb{Z}_n$, $\mathrm{Ext}(\mathbb{Z}_m, \mathbb{Z}) \cong \mathbb{Z}_m$. 3°. If one of the groups A, B is $\mathbb{Q}$, $\mathbb{R}$, or $\mathbb{C}$, then $\mathrm{Ext}(A, B) = 0$. 4°. There exists a natural one-one correspondence between the set $\mathrm{Ext}(A, B)$ and the set of equivalence classes of exact sequences $0 \to B \to C \to A \to 0$ with C an Abelian group (equivalence is defined in the obvious way).

C. Theorem (*universal coefficient formulae*). *For any free complex C, any Abelian group G and any q, there are natural exact sequences*

$$0 \to H_q(C) \otimes G \to H_q(C; G) \to \mathrm{Tor}(H_{q-1}(C), G) \to 0,$$

$$0 \to H^q(C; \mathbb{Z}) \otimes G \to H^q(C; G) \to \mathrm{Tor}(H^{q+1}(C; \mathbb{Z}), G) \to 0,$$

$$0 \leftarrow \mathrm{Hom}(H_q(C), G) \leftarrow H^q(C; G) \leftarrow \mathrm{Ext}(H_{q-1}(C), G) \leftarrow 0.$$

All these sequences split, so that there are (non-canonical) isomorphisms

$$H_q(C; G) \cong [H_q(C) \otimes G] \oplus \mathrm{Tor}(H_{q-1}(C), G),$$

$$H^q(C; G) \cong [H^q(C; \mathbb{Z}) \otimes G] \oplus \mathrm{Tor}(H^{q+1}(C; \mathbb{Z}), G),$$

$$H^q(C; G) \cong \mathrm{Hom}(H_q(C), G) \oplus \mathrm{Ext}(H_{q-1}(C), G). \quad \square$$

We shall give two methods of obtaining the first sequence. The cohomology analogue of the first method gives the second sequence, and the cohomology analogue of the second method gives the third sequence.

First method. Let $G = F_1/F_2$ where F_1 and F_2 are free Abelian groups. It is completely obvious that $H_q(C; F_i) = H_q(C) \otimes F_i$. ($F_i = \mathbb{Z} \oplus \mathbb{Z} \oplus \mathbb{Z} \cdots$, $C \otimes F_i = C \oplus C \oplus \cdots$, etc.). Hence from the homology coefficient sequence associated with the short sequence $0 \to F_2 \to F_1 \to G \to 0$, we obtain an exact sequence

$$H_q(C) \otimes F_2 \to H_q(C) \otimes F_1 \to H_q(C; G) \to$$
$$\to H_{q-1}(C) \otimes F_2 \to H_{q-1}(C) \otimes F_1,$$

that is, an exact sequence

$$0 \to \mathrm{Coker}[H_q(C) \otimes F_2 \to H_q(C) \otimes F_1] \to H_q(C; G) \to$$
$$\to \mathrm{Ker}[H_{q-1}(C) \otimes F_2 \to H_{q-1}(C) \otimes F_1] \to 0,$$

that is, an exact sequence

$$0 \to H_q(C) \otimes G \to H_q(C; G) \to \mathrm{Tor}(H_{q-1}(C), G) \to 0$$

(the splitting is not obvious from what has been said, but we leave the proof to the reader).

Second method. Since C_{q-1} is free, the image of $\partial_q : C_q \to C_{q-1}$ is also free. Hence there is a reverse homomorphism $e_{q-1} : \mathrm{Im}\,\partial_q \to C_q$, $\partial_q \circ e_{q-1} = \mathrm{Id}\,\mathrm{Im}\,\partial_q$. Clearly $C_q = \mathrm{Ker}\,\partial_q \oplus \mathrm{Im}\,e_{q-1} = \mathrm{Ker}\,\partial_q \oplus \mathrm{Im}\,e_q$, and the differential ∂_q equals 0 on $\mathrm{Ker}\,\partial_q$, and on $\mathrm{Im}\,\partial_q$ is the same as the inclusion $\mathrm{Im}\,\partial_q \to \mathrm{Ker}\,\partial_{q-1}$. This shows that the complex C decomposes into the sum of short complexes

$$\cdots \longrightarrow 0 \longrightarrow \underset{(q)}{\mathrm{Im}\,\partial_q} \overset{\subset}{\longrightarrow} \underset{(q-1)}{\mathrm{Ker}\,\partial_{q-1}} \longrightarrow 0 \longrightarrow 0 \cdots$$

The complex $C \otimes G$ is the direct sum of the complexes

$$\cdots 0 \longrightarrow 0 \longrightarrow (\mathrm{Im}\,\partial_q) \otimes G \longrightarrow (\mathrm{Ker}\,\partial_{q-1}) \otimes G \longrightarrow 0 \longrightarrow 0 \cdots.$$

Since $\mathrm{Im}\,\partial_q$ and $\mathrm{Ker}\,\partial_{q-1}$ are free, and $H_{q-1} = \mathrm{Ker}\,\partial_{q-1}/\mathrm{Im}\,\partial_q$, the homologies of the above summand are:

in dimension q, $\mathrm{Tor}(H_{q-1}, G)$,

in dimension $q - 1$, $H_{q-1} \otimes G$.

The homology of the sum of complexes is the sum of their homologies, and we obtain the promised formula for $H_q(C \otimes G)$. (We leave it to the reader to investigate whether the above constructions are natural or not.)

Finally we give direct descriptions of the three most important homomorphisms in the sequences constructed above. The homomorphism $H_q(C) \otimes G \to H_q(C; G)$ sends the tensor product of the homology class of a cycle c and $g \in G$ to the homology class of the cycle $c \otimes g$. The homomorphism $H^q(C; \mathbb{Z}) \otimes G \to H^q(C; G)$ sends the tensor product of the cohomology class of a cocycle $u : C_q \to \mathbb{Z}$ and $g \in G$ to the cohomology class of the cocycle $c \to u(c)g$. The homomorphism $H^q(C; G) \to \mathrm{Hom}(H_q(C), G)$ sends the cohomology class of a cocycle $u : C_q \to$

G to the homomorphism $H_q(C) \to G$ that maps the homology class of a cycle $c \in C_q$ to $u(c)$.

D. We list some consequences of the universal coefficient formulae and some related assertions.

$1°$. If $k = \mathbb{Q}, \mathbb{R}$, or $\mathbb{C}$, then $H_q(C; k) = H_q(C) \otimes k$ and $H^q(C; k) = \mathrm{Hom}(H_q(C), k)$.

$2°$. If the homology of the complex C is finitely generated, then the free parts of $H_q(C)$ and $H^q(C; \mathbb{Z})$ are the same, and

$$\mathrm{Tors}\, H^q(C; \mathbb{Z}) \cong \mathrm{Tors}\, H_q(C).$$

$3°$. For any field k

$$H^q(C; k) = \mathrm{Hom}_k(H_q(C; k), k).$$

1.6. Tensor products of complexes and the Künneth formula.

A. Let $C' = \{C'_q, \partial'_q\}$ and $C'' = \{C''_q, \partial''_q\}$ be complexes. The groups $C_q = \bigoplus_{r+s=q}(C'_r \otimes C''_s)$ and homomorphisms $\partial_q : C_q \to C_{q-1}$ defined by

$$\partial_q(c' \otimes c'') = \partial'_r(c') \otimes c'' + (-1)^r c' \otimes \partial''_s(c''), \quad \text{for } c' \in C'_r, \ c'' \in C''_s,$$

form a complex, as is easily verified. This is called the *tensor product of the complexes C' and C''*, denoted by $C' \otimes C''$. The tensor product of free complexes is obviously free, that of positive complexes is positive, and that of complexes of finite type is of finite type.

B. Theorem. (the *Künneth formula*). *For any free complexes C' and C'' and any q there is a natural exact sequence*

$$0 \to \bigoplus_{r+s=q} [H_r(C') \otimes H_s(C'')] \to$$

$$\to H_q(C' \otimes C'') \to \bigoplus_{r+s=q-1} \mathrm{Tor}(H_r(C'), H_s(C'')) \to 0.$$

This sequence splits, so that there is a (non-canonical) isomorphism

$$H_q(C' \otimes C'') \cong \bigoplus_{r+s=q} [H_r(C') \otimes H_s(C'')] \oplus \bigoplus_{r+s=q-1} \mathrm{Tor}(H_r(C'), H_s(C'')). \quad \square$$

The proof is similar to the proof of the universal coefficient formula. The homomorphism

$$\bigoplus_{r+s=q} [H_r(C') \otimes H_s(C'')] \to H_q(C' \otimes C'')$$

takes the product $\gamma' \otimes \gamma'' \in H_r(C') \otimes H_s(C'')$ to the homology class of the cycle $c' \otimes c''$, where c', c'' are cycles in the classes γ', γ''.

In addition, for any field k and any free complexes C', C'', we have the formula

$$H_q(C' \otimes C''; k) = \bigoplus_{r+s=q} [H_r(C'; k) \otimes_k H_s(C''; k)],$$

and if in addition the complexes C', C'' are positive, then

$$H^q(C' \otimes C''; k) = \bigoplus_{r+s=q} [H^r(C'; k) \otimes H^s(C''; k)].$$

§2. General singular homology theory

2.1. Basic definitions.

A. A *q-dimensional singular simplex* of a topological space X is any continuous map of the standard (unit) q-simplex T^q into X (see Part I, 8.1.B). A singular q-simplex $s : T^q \to X$ has $q+1$ faces $\Delta_0 s, \Delta_1 s, \ldots, \Delta_q s$; the ith face $\Delta_i s$ is the singular $(q-1)$-simplex of X defined as the the composition $s_0 \Gamma_i : T^{q-1} \to X$, where $\Gamma_i : T^{q-1} \to T^q$ is the affine embedding $(t_1, \ldots, t_q) \mapsto (t_1, \ldots, t_i, 0, t_{i+1}, \ldots, t_q)$.

The group $C_q(x)$ of *singular q-chains* of X is by definition the free Abelian group spanned by the set $\text{sing}_q X$ of all singular q-simplexes of X; thus, a singular q-chain of X is a (finite) formal linear combination $\sum a_j s_j$ of singular q-simplexes of X with integer coefficients (in particular, the singular simplexes are themselves singular chains). The *boundary* $\partial c = \partial_q c$ of the singular q-chain $c = \sum a_j s_j$ is defined by the formulae

$$\partial \sum a_j s_j = \sum a_j s_j, \quad \partial s_j = \sum_{i=0}^{q} (-1)^i \Delta_i s_j;$$

the map $c \to \partial c$ defines a homomorphism $\partial = \partial_q : C_q(X) \to C_{q-1}(X)$ for each q, called the *differential* or *boundary operator*. The case $q < 0$ is not excluded from these definitions: $C_q = 0$ for $q < 0$, and $\partial_q = 0$ for $q \leq 0$.

Fundamental lemma. *The composition*

$$C_q(X) \xrightarrow{\partial} C_{q-1}(X) \xrightarrow{\partial} C_{q-2}(X)$$

is equal to 0 *for any* q *and* X; *in other words,* $\partial \partial c = 0$ *for any singular chain c.*

Thus the sequence of groups and homomorphisms

$$\cdots \longrightarrow C_q(X) \xrightarrow{\partial} C_{q-1}(X) \longrightarrow \cdots$$

is a complex, called the *singular complex of X*, denoted by $S(X)$. This complex is free and positive, which will be important later.

B. The homology group $H_q(S(X))$ is denoted simply by $H_q(X)$ and called the *qth (singular) homology group of X*. Thus

$$H_q(X) = \frac{\text{Ker}[\partial_q : C_q(X) \to C_{q-1}(X)]}{\text{Im}[\partial_{q+1} : C_{q+1}(X) \to C_q(X)]}$$

We also use the notation $H_* = \bigoplus_q H_q(X)$. The elements of $\text{Ker} \, \partial_q$ are called *(singular) q-cycles* of X, and the elements $\text{Im} \, \partial_{q+1}$ are sometimes called *q-boundaries*, or q-cycles *homologous to* 0. Two q-cycles whose difference is homologous to 0 are called *homologous to each other*.

The ranks of the homology groups of X are called its *Betti numbers*. If they are finite, then the power series (or polynomial) $\sum_{q=0}^{\infty} \operatorname{rank} H_q(X)t^q$ is defined, and called the *Poincaré series* (or *polynomial*) of X.

The number $\operatorname{Eu}(X) = \sum_{q=0}^{\infty}(-1)^q \operatorname{rank} H_q(X)$ (which should be called the Euler characteristic of $S(X)$ according to the general definition in §1) will be called the *Euler characteristic of the space X*.

The homology and cohomology groups of $S(X)$ with coefficients in an Abelian group G are denoted by $H_q(X; G)$ and $H^q(X; G)$. (Thus $H_q(X) = H_q(X; \mathbb{Z})$). The groups forming the complexes $S(X) \otimes G$ and $\operatorname{Hom}(S(X), G)$ are denoted similarly; $C_q(X) \otimes G$ is denoted by $C_q(X; G)$ and $\operatorname{Hom}(C_q(X), G)$ by $C^q(X; G)$. The elements of these groups are called respectively (singular) *chains with coefficients in* C, and *cochains with coefficients* or with *values in* G. The former may be described explicitly as (finite) formal linear combinations $\sum g_i s_i$ of singular q-simplexes with coefficients in G, and the latter as arbitrary maps $\operatorname{sing}_q X \to G$. The differential of the complex $\operatorname{Hom}(S(X), G)$ will be denoted by $\delta = \delta^q : C^q(X; G) \to C^{q+1}(X; G)$; the elements of the kernel of δ are called *cocycles*. The term *cohomologous cocycles* and the notations $H_*(X; G)$ and $H^*(X; G)$ are also used.

C. A continuous map $f : X \to Y$ maps a singular q-simplex of X into a singular q-simplex of Y for each q ($s \mapsto f \circ s$), and hence defines homomorphisms

$$f_\# : C_q(X) \to C_q(Y).$$

These homomorphisms clearly commute with ∂ ($f_\# \partial_q c = \partial_q f_\# c$ for any chain $c \in C_q(X)$), so that they form a chain map of $S(X)$ into $S(Y)$ (also denoted by $f_\#$). The homomorphisms $f_\#$ induce maps

$$f_\# : C_q(X; G) \to C_q(Y; G), \quad f^\# : C^q(Y; G) \to C^q(X; G);$$
$$f_* : H_q(X) \to H_q(Y), \quad f_* : H_q(X; G) \to H_q(Y; G),$$
$$f^* : H^q(Y; G) \to H^q(X; G).$$

A homotopy $F : X \times I \to Y$ between $f : X \to Y$ and a continuous map $g : X \to Y$ induces a chain homotopy $F_\# : S(X) \to S(Y)$ between $f_\#$ and $g_\#$, constructed as follows. Let $v_0, \ldots, v_q$ be the canonically ordered vertices of T^q and $w_0, \ldots, w_{q+1}$ the canonically ordered vertices of T^{q+1}. Also let $Z_i : T^{q+1} \to T^q \times I$ ($i = 0, \ldots, q$) be the affine embedding taking the vertex w_j to the point $(v_j, 0)$ for $j \leq i$, and to the point $(v_{j-1}, 1)$ for $j > i$. (The simplexes $Z_i(T^{q+1})$ and their faces constitute the standard triangulation of the prism $T^q \times I = T^q \times T^1$; see Part I, 8.3.B). If s is a singular simplex of X, we put

$$F_\# s = \sum_{i=0}^{q}(-1)^i[(s \times \operatorname{id}_I) \circ Z_i],$$

and extend $F_\#$ by linearity to a map $F_\# : C_q(X) \to C_{q+1}(Y)$. It can be verified that $F_\#$ is a chain homotopy between $f_\#$ and $g_\#$.

This construction provides the proof of the following important facts.

Theorem. *(a) If the maps $f, g : X \to Y$ are homotopic, then for any q and G*

$$f_* = g_* : H_q(X; G) \to H_q(Y; G),$$
$$f^* = g^* : H^q(Y; G) \to H^q(X; G).$$

(b) If spaces X and Y are homotopy equivalent, then the complexes $S(X)$ and $S(Y)$ are also homotopy equivalent; in particular, in this case for any q and G

$$H_q(X; G) \cong H_q(Y; G), \quad H^q(X; G) \cong H^q(Y; G). \quad \square$$

D. If Y is a subspace of X, then the complex $S(Y)$ is a subcomplex of $S(X)$ (the inclusion in : $Y \to X$ induces the inclusion $in_* : S(Y) \to S(X)$). In this case the factor complex $S(X)/S(Y)$ is called the *singular complex of the pair* (X, Y) or the *relative singular complex of X modulo Y*, and is denoted by $S(X, Y)$. The notations $H_q(X, Y)$, $H_q(X, Y; G)$, $H^q(X, Y; G)$; $H_*(X, Y)$, $H_*(X, Y; G)$, $H^*(X, Y; G)$ are used for the homology and cohomology of this complex. The terms *relative cycle*, *relative cocycle*, etc. are also used.

What was said in **C** carries over to the relative case more or less word for word. In particular, a continuous map $X_1 \to X_2$ that takes $Y_1 \subset X_1$ into $Y_2 \subset X_2$ induces a chain map $S(X_1, Y_1) \to S(X_2, Y_2)$, and a similar statement holds for homotopies; homotopy equivalent pairs have singular complexes that are homotopy equivalent, and their homologies and cohomologies are the same.

The theorem of 1.3 can be applied to the pair $(S(X), S(Y))$. In particular there are homomorphisms

$$\partial_* : H_q(X, Y; G) \to H_{q-1}(Y; G), \quad \delta^* : H^{q-1}(Y; G) \to H^q(X, Y; G)$$

and sequences of groups and homomorphisms

$$\cdots \longrightarrow H_q(Y; G) \longrightarrow H_q(X; G) \longrightarrow H_q(X, Y; G) \overset{\partial_*}{\longrightarrow} H_{q-1}(Y; G) \longrightarrow \cdots \quad (4)$$

$$\cdots \longrightarrow H^{q-1}(Y; G) \overset{\delta^*}{\longrightarrow} H^q(X, Y; G) \longrightarrow H^q(X; G) \longrightarrow H^q(Y; G) \longrightarrow \cdots \quad (5)$$

which are called respectively the *homology* and *cohomology sequence of the pair* (X, Y) *with coefficients in G*.

For ease of reference we repeat the statement of the basic result of 1.3 in the context of singular complexes.

Theorem. *The sequences (4) and (5) are exact.* $\square$

(In view of the exceptional importance of these sequences we shall give a description of the homomorphisms ∂_* and δ^* that does not depend on §1, and we recommend the reader to satisfy himself as an exercise that the sequences (4) and (5) are exact. The class $\gamma \in H_q(X, Y; G)$ is represented by a relative cycle; this cycle is a chain $c \in C(X; G)$ given to within addition of a chain in $C_q(Y; G)$, and having the property that ∂c belongs to $C_{q-1}(Y; G)$. Clearly ∂c is a cycle, which need not be a boundary in $S(Y)$; the homology class of this cycle is precisely $\partial_* \gamma \in H_{q-1}(Y; G)$. Again, let $a : sing_{q-1} Y \to G$ be a cocycle representing a class

$\alpha \in H^{q-1}(Y; G)$. We pick any extension $a : \mathrm{sing}_{q-1} X \to G$ of a; the restriction of the cocycle $\delta a : \mathrm{sing}_q \to G$ to $\mathrm{sing}_q Y$ is equal to 0 (because a is a cocycle). Hence δa is a relative cocycle; the class of this cocycle is precisely $\delta^* \alpha \in H^q(X, Y; G)$.)

As generalizations of the sequences (4) and (5) we have the *homology* and *cohomology sequences of a triple*:

$$\cdots \to H_q(Y, Z; G) \to H_q(X, Z; G) \to H_q(X, Y; G) \to H_{q-1}(Y, Z; G) \to \cdots$$
$$\tag{6}$$
$$\cdots \to H^{q-1}(Y, Z; G) \to H^q(X, Y; G) \to H^q(X, Z; G) \to H^q(Y, Z; G) \to \cdots$$
$$\tag{7}$$

These sequences are defined for any triple (X, Y, Z), $X \supset Y \supset Z$, and are simply the homology and cohomology sequences of the pair of complexes $(S(X)/S(Z), S(Y)/S(Z))$. In view of the results of 1.3, we have

Theorem. *The sequences* (6) *and* (7) *are exact.*   □

E. The *reduced singular complex* $\tilde{S}(X)$ is obtained from $S(X)$ by making two changes: the group $C_{-1}(X)$ is replaced by $\tilde{C}_{-1}(X) = \mathbb{Z}$, and the differential $\partial_0 : C_0(X) \to C_{-1}(X)$ is replaced by the map $\tilde{\partial}_0 : C_0(X) \to \mathbb{Z}$ defined by $\tilde{\partial}_0(\sum a_i s_i) = \sum a_i$ (where the s_i are singular 0-simplexes). The remaining groups C_i and differentials ∂_i are unchanged, but the notation is changed to $\tilde{C}_i$ and $\tilde{\partial}_i$. The homology and cohomology of the reduced singular complex is called the *reduced singular homology* and *cohomology* of X, and is denoted by $\tilde{H}_q(X)$, $\tilde{H}_q(X; G)$ and $\tilde{H}^q(X; G)$. Of course these groups hardly differ from $H_q(X)$, $H_q(X; G)$ and $H^q(X; G)$, but we postpone a detailed investigation of these differences until 2.2.

The number $\sum a_i$ appearing in the definition of ∂_0 is called the *index* of the chain $\sum a_i s_i$.

All the previous theorems carry over to the case of reduced homology and cohomology without change. In particular, there are exact *reduced homology* and *cohomology sequences of a pair*:

$$\cdots \to \tilde{H}_q(Y; G) \to \tilde{H}_q(X; G) \to \tilde{H}_q(X, Y; G) \to \tilde{H}_{q-1}(Y; G) \to \cdots,$$
$$\cdots \to \tilde{H}^{q-1}(Y; G) \to \tilde{H}^q(X, Y; G) \to \tilde{H}^q(X; G) \to \tilde{H}^q(Y; G) \to \cdots.$$

2.2. The simplest calculations.

A. If X is a one-point space, then

$$H_q(X; G) \cong H^q(X; G) \cong \begin{cases} G & \text{for } q = 0, \\ 0 & \text{for } q \neq 0. \end{cases} \tag{8}$$

(In fact, for $q \geq 0$ there is a unique singular simplex $s^q : T^q \to X$; in particular, $\Gamma_i s^q = s^{q-1}$ for any $q \geq 1$ and i. Hence

$$\partial_q s^q = \sum_{i=0}^q (-1)^i \Gamma_i s^q = \left(\sum_{i=0}^q (-1)^i\right) s^{q-1} = \begin{cases} s^{q-1} & \text{for even } q > 1, \\ 0 & \text{for odd } q \geq 1, \end{cases}$$

and the complex $S(X)$ has the form

$$\cdots \xrightarrow{0} \mathbb{Z} \xrightarrow{\mathrm{id}} \mathbb{Z} \xrightarrow{0} \mathbb{Z} \xrightarrow{\mathrm{id}} \mathbb{Z} \xrightarrow{0} \mathbb{Z} \longrightarrow 0 \longrightarrow 0 \longrightarrow 0 \longrightarrow \cdots$$

from which our assertion follows.)

Corollary. *If X is contractible its homology and cohomology are given by formula* (8). □

B. Let X be a path-connected topological space. Then

$$H_0(X; G) \cong H^0(X; G) \cong G.$$

(In fact 0-chains $\sum g_i s_i$ and $\sum g'_j s'_j$ are always cycles, and are homologous iff $\sum g_i = \sum g'_j$, which establishes the isomorphism $H_0(X; G) \cong G$; similarly for cohomology.)

If A is a nonempty subset of a path-connected space X, then $H_0(X, A; G) \cong H^0(X, A; G) = 0$.

C. Let X_α ($\alpha \in A$) be all the path components of X. Then for any q and g

$$H_q(X; G) = \bigoplus_{\alpha \in A} H_q(X; G), \quad H^q(X; G) = \prod_{\alpha \in A} H^q(X; G).$$

In particular, (a) $H_0(X; G) = \bigoplus_{\alpha \in A} G$, $H^0(X; G) = \prod_{\alpha \in A} G$; (b) if the space X is discrete, then

$$H_q(X; G) = \begin{cases} \bigoplus_{x \in X} G & \text{for } q = 0, \\ 0 & \text{for } q \neq 0; \end{cases}$$

$$H^q(X; G) = \begin{cases} \prod_{x \in X} G & \text{for } q = 0, \\ 0 & \text{for } q \neq 0. \end{cases}$$

Also for any pair (X, A)

$$H_q(X, A; G) = \bigoplus_{\alpha \in A} H_q(X_\alpha, X_\alpha \cap A; G);$$

$$H^q(X, A; G) = \prod_{\alpha \in A} H^q(X_\alpha, X_\alpha \cap A; G).$$

D. If X is nonempty, then for any $q \neq 0$ and any G

$$\tilde{H}_q(X; G) = H_q(X; G), \quad \tilde{H}^q(X; G) = H^q(X; G);$$

the groups $\tilde{H}_0(X; G)$ and $\tilde{H}^0(X; G)$ are respectively the direct sum and direct product of groups isomorphic to G with the number of summands (factors) one less than in $H_0(X; G)$ and $H^0(X; G)$ (see C). In particular, if X is path-connected, then

$$\tilde{H}_0(X; G) = \tilde{H}^0(X; G) = 0.$$

A more precise statement is: for any point $x_0 \in X$, the composite projection

$$\tilde{S}(X) \longrightarrow S(X) \longrightarrow S(x_0)$$

is a homotopy equivalence.

If X is empty, then

$$\tilde{H}_q(X; G) \cong \tilde{H}^q(X; G) \cong \begin{cases} 0 & \text{for } q \neq -1, \\ G & \text{for } q = -1. \end{cases}$$

2.3. Natural transformations; refinement and approximation.

A. We shall say that we are given a *natural transformation* τ if for each space X there is given a chain map $\tau_X : s(X) \to s(X)$, such that:

(1) if $c \in C_0(X)$, then the chains c and $\tau_X(c)$ have the same index (see 2.1.E);

(2) the map τ_X is natural with respect to X; this means that for any continuous map $f : X \to Y$ the diagram

$$\begin{array}{ccc} S(X) & \xrightarrow{\tau_X} & S(X) \\ {\scriptstyle f_\#}\downarrow & & \downarrow{\scriptstyle f_\#} \\ S(Y) & \xrightarrow{\tau_Y} & S(Y) \end{array}$$

is commutative.

Transformation lemma. *If $\tau = \{\tau_X\}$ is a natural transformation, then the chain map τ_X is homotopic to the identity for any space X (and in particular induces the identity map in homology and cohomology).* $\square$

B. Recall (see Part I, 8.3.A) that the barycentric subdivision of the standard simplex T^q divides it into $(q + 1)!$ smaller q-simplexes. These simplexes are numbered by the permutations of the numbers $0, 1, \ldots, q$; the simplex $t^q(i_0, \ldots, i_q)$ corresponding to the permutation $(i_0, \ldots, i_q)$ is defined as the set of points of T^q whose barycentric coordinates satisfy the inequalities $x_{i_0} \leq \ldots \leq x_{i_q}$; the vertices of this simplex have a natural ordering (by increasing dimension of the open faces of T^q containing them), and hence there is an affine map $f(i_0, \ldots, i_q) : T^q \to T^q$, mapping T^q homeomorphically onto $t^q(i_0, \ldots, i_q)$. Now let $c = \sum a_i s_i$ be any singular q-chain of a space X. The *barycentric refinement* bc of the chain c is defined by

$$bc = \sum a_i bs_i, \quad bs_i = \sum_{(i_0, \ldots, i_q)} \text{sgn}(i_0, \ldots, i_q) s_i \circ f(i_0, \ldots, i_q).$$

An obvious verification shows that the resulting maps $b : C_q(X) \to C_q(X)$ constitute a natural transformation. From the transformation lemma follows the

Barycentric refinement lemma. *The maps b form a map $S(X) \to S(X)$ that is homotopic to the identity. In particular, if c is a cycle, then bc is a cycle homologous to c.* $\square$

Combined with the fact that under repeated barycentric subdivision the diameters of the simplexes become arbitrarily small (see Part I, 8.3.A), this lemma enables us to prove the following important result.

Let $\mathcal{U}$ be any open covering of X. Let $S_{\mathcal{U}}(X)$ denote the subcomplex of $S(X)$ generated by the singular simplexes $f : T^q \to X$ for which $f(T^q)$ is contained in some set of the covering $\mathcal{U}$.

Refinement theorem. *The embedding $S_{\mathcal{U}}(X) \to S(X)$ is a homotopy equivalence.*

C. For purposes of illustration (and for use in Chapter 2) we give another example of a natural transformation. Let $\omega : T^q \to T^q$ be an affine homeomorphism that reverses the order of the vertices. For a singular q-chain $c = \sum a_i s_i$ of X we define a chain $\operatorname{inv} c$ by

$$\operatorname{inv} c = \sum (-1)^{q(q-1)/2} a_i (\omega \circ s_i).$$

An immediate verification shows that inv is a natural transformation; in particular, if c is a cycle, then $\operatorname{inv} c$ is a cycle homologous to c.

D. In conclusion we give an interesting generalization of the refinement theorem (the proof of which also uses the barycentric refinement lemma). Let $\mathscr{A}$ be a subset of the set of all singular simplexes of X. A continuous map f of a simplicial space Y into the space X is called an $\mathscr{A}$-map if the restriction of f to any simplex of Y belongs to $\mathscr{A}$ when regarded as a singular simplex of X. The set $\mathscr{A}$ is called *sufficient* if it satisfies the following conditions:

(1) If a singular simplex belongs to $\mathscr{A}$ then so do all its faces.

(2) Any continuous map of an arbitrary simplicial space Y into X is homotopic to a map that is an $\mathscr{A}$-map with respect to a refinement Y' of Y; in addition, if the given map $f : Y \to X$ is already an $\mathscr{A}$-map on a refinement Z' of some simplicial subspace Z of Y, then there exists a refinement Y' of Y, coinciding with Z' on Z, and an $\mathscr{A}$-map $g : Y' \to X$ coinciding with f on Z' and homotopic to f by a homotopy that is fixed on Z.

Let $S_{\mathscr{A}}(X)$ denote the subcomplex of $S(X)$ generated by the simplexes of a sufficient set $\mathscr{A}$ (it is a subcomplex by condition (1)).

Theorem. *The inclusion $S_{\mathscr{A}}(X) \to S(X)$ is a homotopy equivalence.* $\square$

Hence the homology and cohomology of the space X may be calculated using only the simplexes in the set $\mathscr{A}$.

We give some examples of sufficient sets of singular simplexes (in each case the verification of sufficiency presents no difficulty).

$1°$. X any space, $\mathscr{A}$ the set of singular simplexes whose images are contained in the sets of a given covering. (In this case the preceding theorem reduces to the refinement theorem in **B**.)

$2°$. X a smooth manifold, $\mathscr{A}$ the set of smooth singular simplexes (the smoothness class can vary from C^1 to real analytic).

$3°$. X a simplicial space, $\mathscr{A}$ the set of affine maps of standard simplexes onto the simplexes of X.

$4°$. X a domain in Euclidean space, $\mathscr{A}$ the set of affine maps of standard simplexes into X.

2.4. Excision, factorization, suspension.

A. The excision theorem. *Let A and B be subsets of a topological space X such that B is closed, A is open, and $B \subset A$. Then the homomorphism $S(X \setminus B, A \setminus B) \to S(X, A)$ induced by the inclusion $X \setminus B \to X$ is a homotopy equivalence. In particular,*

$$H_q(X \setminus B, A \setminus B; G) = H_q(X, A; G),$$

$$H^q(X \setminus B, A \setminus B; G) = H^q(X, A; G)$$

for any q and any G.   □

For let $\mathcal{U}$ denote the two-element covering of X consisting of the sets A and $X \setminus B$. Clearly $S_{\mathcal{U}}(X, A) = S(X \setminus B, A \setminus B)$. The homotopy equivalence between $S_{\mathcal{U}}(X, A)$ and $S(X, A)$ follows from the relative version of the refinement theorem.

The conditions on **A** and **B** in the theorem above may be replaced by more convenient ones; see **C**.

B. Factorization theorem. *Let (X, A) be a Borsuk pair (see Part I, 2.9). Let $x_0 \in X/A$ be the the image of A under the projection $X \to X/A$. Then the map $S(X, A) \to S(X/A, x_0)$ induced by the projection is a homotopy equivalence.*   □

Let (X, A) be a Borsuk pair; then this theorem, combined with what was said in 2.1, establishes a homotopy equivalence

$$S(X, A) \sim \tilde{S}(X, A),$$

and isomorphisms

$$H_q(X, A; G) \cong \tilde{H}_q(X/A; G), \quad H^q(X, A; G) \cong \tilde{H}^q(X/A; G).$$

For the proof we adjoin to X the cone CA over A, and denote by C_1 and C_2 the open subsets of CA shown in Fig. 1. The homotopy equivalence $S(X, A) \sim S(X/A, x_0)$ is composed of the homotopy equivalences

$$S(X, A) \sim S(X \cup CA \setminus C_1, C_2 \setminus C_1) \sim S(X \cup CA, C_2) \sim S(X/A, x_0),$$

where the first and third equivalences come from the homotopy invariance of the relative singular complex (see 2.1.C), and the second from the excision theorem.

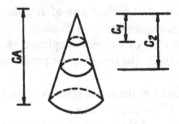

Fig. 1

C. The factorization theorem enables us to extend the range of application of the excision theorem. In fact, this theorem holds when (X, A) and $(X \setminus B, A \setminus B)$ are Borsuk pairs and the natural map $(X \setminus B)/(A \setminus B) \to X/A$ is a homeomorphism. An example of this situation: X is a smooth manifold, B an open domain with smooth boundary, $A = \text{Cl } B$ (the original formulation of the excision theorem does not apply to this situation).

Another way to extend the range of application of the excision theorem is to combine it with the theorem on the homotopy invariance of the singular complex. For example, the conclusion of the theorem is valid if $X \supset A \supset B$ and there exists

$B' \subset B$ such that $\mathrm{Cl}\, B' \subset \mathrm{Int}\, A$, and the inclusion $(X \setminus B, A \setminus B) \to (X \setminus B', A \setminus B')$ is a homotopy equivalence (the preceding example also illustrates this possibility).

D. Let X be a topological space, CX the cone over X, and ΣX the suspension of X. Since CX is contractible, for any q and G we have canonical isomorphisms

$$H_{q+1}(CX, X; G) = \tilde{H}_q(X; G), \quad H^{q+1}(CX, X; G) = \tilde{H}^q(X; G)$$

(see 1.3.**B** and 2.2.**E**), and from the factorization theorem we have

$$H_{q+1}(CX, X; G) = \tilde{H}_{q+1}(\Sigma X; G),$$
$$H^{q+1}(CX, X; G) = \tilde{H}^{q+1}(\Sigma X; G).$$

Combining these isomorphisms, we obtain canonical isomorphisms

$$\tilde{H}_q(X; G) = \tilde{H}^{q+1}(\Sigma X; G), \quad \tilde{H}^q(X; G) = \tilde{H}^{q+1}(\Sigma X; G),$$

called the *suspension isomorphisms* and denoted by Σ.

Corollary. *For $n \geq C$*

$$\tilde{H}_q(S^n; G) \cong \tilde{H}^q(S^n; G) \cong \begin{cases} G & \text{for } q = n, \\ 0 & \text{for } q \neq n \end{cases}$$

(for $n = 0$ we already know this, as S^0 consists of two points, and the case $n > 0$ follows by means of the suspension isomorphism); thus for $n > 0$,

$$H_q(S^n; G) \cong H^q(S^n; G) \cong \begin{cases} G & \text{for } q = 0, n, \\ 0 & \text{for } q \neq 0, n. \end{cases} \quad \square$$

2.5. Addition theorems.

A. Let $(X; A, B)$ be a *topological triad*, that is, X is a topological space and A and B are subsets with $A \cup B = X$. The fundamental theorem of this section expresses the homology and cohomology of X in terms of the homology and cohomology of the spaces A, B, and $A \cap B$. We begin by describing the class of triads for which such an expression is possible.

A triad $(X; A, B)$ is called *homologically proper* if it satisfies any of the following equivalent conditions:

(1) The inclusion of the subcomplex $S(A) \to S(B)$ in $S(X)$ is a homotopy equivalence.

(2) The inclusion $S(A, A \cap B) \to S(X, B)$ is a homotopy equivalence.

(3) The inclusion $S(B, A \cap B) \to S(X, A)$ is a homotopy equivalence.

Proposition. *Each of the following conditions is sufficient for $(X; A, B)$ to be homologically proper:*

(4) $\mathrm{Int}\, A \cup \mathrm{Int}\, B = X$.

(5) (X, B) *and* $(A, A \cap B)$ *are Borsuk pairs.*

(6) (X, A) *and* $(B, A \cap B)$ *are Borsuk pairs.*

(7) A *and* B *are closed and* $(A, A \cap B)$ *is a Borsuk pair.*

(8) A *and* B *are closed and* $(B, A \cap B)$ *is a Borsuk pair.* $\qquad \square$

This proposition and the equivalence of conditions (1)–(3) can be proved without great difficulty from previous results.

B. If $(X; A, B)$ is a homologically proper triad, then for any Abelian group G there are exact sequences

$$\cdots \to H_q(A \cap B; G) \to H_q(A; G) \oplus H_q(B; G) \to$$
$$\to H_q(X; G) \to H_{q-1}(A \cap B; G) \to \cdots,$$
$$\cdots \to H^{q-1}(A \cap B; G) \to H^q(X; G) \to$$
$$\to H^q(A; G) \oplus H^q(B; G) \to H^q(A \cap B; G) \to \cdots.$$

These sequences are called the *addition sequences* or *Mayer-Vietoris sequences*. They are obtained from the homology and cohomology sequences associated with the short exact sequence of complexes

$$0 \to S(A \cap B) \to S(A) \oplus S(B) \to S(A) \to S(B) \to 0,$$

in which the second and third arrows are defined by $c \mapsto (c, -c)$ and $(c_1, c_2) \mapsto c_1 + c_2$.

For the reader's pleasure we give a beautiful diagram which includes the Mayer-Vietoris sequence of the triad $(X; A, B)$ and the (undulating) homology sequences of the pairs (X, A), (X, B), $(A, A \cap B)$, $(B, A \cap B)$ (to see the last two, the homology of the pairs (X, A) and (X, B) must be replaced by the isomorphic homology of the pairs $(B, A \cap B)$ and $(A, A \cap B)$):

$$
\begin{array}{ccccccc}
H_{r+1}(X, A) & \longrightarrow & H_r(A) & \longrightarrow & H_r(X, B) & \longrightarrow & H_{r-1}(B) \\
& \searrow \quad \nearrow & & \searrow \quad \nearrow & & \searrow \quad \nearrow & \\
\cdots \to H_r(A \cap B) & \to * \to & H_r(X) & \to H_{r-1}(A \cap B) & \to ** \to & H_{r-1}(X) \\
& \nearrow \quad \searrow & & \nearrow \quad \searrow & & \nearrow \quad \searrow & \\
H_{r+1}(X, B) & \longrightarrow & H_r(B) & \longrightarrow & H_r(X, A) & \longrightarrow & H_{r-1}(A)
\end{array}
$$

$$* \quad H_r(A) \oplus H_r(B)$$
$$** \quad H_{r-1}(A) \oplus H_{r-1}(B)$$

These sequences have an obvious analogue for reduced homology. They have two relative analogues, which arise in the situation when we are given a homologically proper triad $(X; A, B)$ and a space Y, which is assumed to satisfy either $Y \supset X$ or $Y \subset A \cap B$. The resulting homology sequences have the form

$$\cdots \to H_q(Y, A \cap B); G) \to H_q(Y, A; G) \oplus H_q(Y, B; G) \to$$
$$\to H_q(Y, X; G) \to H_{q-1}(YA \cap B; G) \to \cdots,$$

or

$$\cdots \to H_q(A \cap B, Y; G) \to H_q(A, Y; G) \oplus H_q(B, Y; G) \to$$
$$\to H_q(X, Y; G) \to H_{q-1}(A \cap B, Y; G) \to \cdots,$$

respectively; the cohomology sequences are constructed similarly. All these sequences are exact.

C. An obvious consequence of the exactness of the Mayer-Vietoris sequences is the existence of canonical isomorphisms

$$\tilde{H}_q(X \vee Y; G) = \tilde{H}_q(X; G) \oplus \tilde{H}_q(Y; G),$$
$$\tilde{H}^q(X \vee Y; G) = \tilde{H}^q(X; G) \oplus \tilde{H}^q(Y; G).$$

These generalize (in an obvious way) to bouquets of arbitrary families of spaces:

$$\tilde{H}_q(\vee_\alpha X_\alpha; G) = \bigoplus_\alpha \tilde{H}_q(X_\alpha; G), \quad \tilde{H}^q(\vee_\alpha X_\alpha; G) = \prod_\alpha \tilde{H}^q(X_\alpha; G).$$

2.6. Dependence on the coefficients.

In this section we shall state the results for absolute homology and cohomology, but they are also valid without any change for relative and reduced homology and cohomology.

A. A homomorphism ϕ of an Abelian group G_1 into an Abelian group G_2 induces homomorphisms

$$\phi_* : H_q(X; G_1) \to H_q(X; G_2), \quad \phi^* : H^q(X; G_1) \to H^q(X; G_2)$$

for any space X. For $q = 0$, ϕ_* can be represented as the sum or product of the underlying collection of copies of the homomorphism ϕ.

B. Since $S(X)$ is a free complex, the constructions in 1.5.A can be applied to it, producing from a short exact sequence

$$0 \to G' \to G \to G'' \to 0,$$

the exact *coefficient sequences*

$$\cdots \to H_q(X; G') \to H_q(X; G) \to H_q(X; G'') \to H_{q-1}(X; G') \to \cdots,$$
$$\cdots \to H^q(X; G') \to H^q(X; G) \to H^q(X; G'') \to H^{q-1}(X; G') \to \cdots.$$

The connecting homomorphisms

$$H_q(X; G'') \to H_{q-1}(X; G'), \quad H^q(X; G'') \to H^{q-1}(X; G')$$

in these sequences are called the *Bockstein homomorphisms*. However, this name more often refers to the case when the sequence (9) has the form $0 \to \mathbb{Z}_m \to \mathbb{Z}_{m^2} \to \mathbb{Z}_m \to 0$ or $0 \to \mathbb{Z} \to \mathbb{Z} \to \mathbb{Z}_m \to 0$. The corresponding homomorphisms

$$\beta = \beta_m : H_q(X; \mathbb{Z}_m) \to H_{q-1}(X; \mathbb{Z}_m),$$
$$\tilde{\beta} = \tilde{\beta}_m : H_q(X; \mathbb{Z}_m) \to H_{q-1}(X; \mathbb{Z}),$$
$$\beta = \beta_m : H^q(X; \mathbb{Z}_m) \to H^{q-1}(X; \mathbb{Z}_m),$$
$$\tilde{\beta} = \tilde{\beta}_m : H^q(X; \mathbb{Z}_m) \to H^{q-1}(X; \mathbb{Z}),$$

used to be denoted by the symbols $\frac{1}{m}\partial$ and $\frac{1}{m}\delta$. These form a convenient reminder of their construction (we confine ourselves to the case of homology, as the cohomology case is similar). Let $c = \sum a_i s_i \in C_q(X; \mathbb{Z}_m)$ be a cycle representing the class $\gamma \in$

$H_q(X; \mathbb{Z}_m)$. The coefficients $a_i \in \mathbb{Z}_m$ can be written as the numbers $0, 1, \ldots, m-1$, so that the cycle c has the same form as an integral chain $\tilde{c} \in C_q(X)$. The latter need not be a cycle, but all the coefficients of its boundary $\partial \tilde{c}$ will be divisible by m (since c is a cycle). Divide them by m to obtain a chain $\tilde{a} = \frac{1}{m} \partial \tilde{c} \in C_{q-1}(X)$ which will be a cycle (since $\partial \partial = 0$), whose homology class does not depend on the choices made in the construction; this is the class $\tilde{\beta}(\gamma)$. The class $\beta(\gamma)$ is represented by the cycle $a \in C_{q-1}(X; \mathbb{Z}_m)$ obtained from a by reducing $\bmod m$.

There is an important extension of this construction. Suppose that the cycle a obtained in the preceding construction is homologous to 0: $a = \partial b$, $b \in C_q(X; \mathbb{Z}_m)$. Then the chain $\partial(\tilde{c} - m\tilde{b})$, where $\tilde{b}$ is obtained from b in the same way as $\tilde{c}$ was obtained from c, is divisible by m^2. Dividing we get $\tilde{a}' = \frac{1}{m^2} \partial(\tilde{c} - m\tilde{b}) \in C_{q-1}(X)$. The homology class of the cycle $a' \in C_{q-1}(X; \mathbb{Z}_m)$, obtained from $\tilde{a}'$ by reducing $\bmod m$, is determined up to addition of an arbitrary element of the image $\operatorname{Im} \beta$ (verify!). Hence there is a homomorphism

$$\beta_m^{(2)} : \operatorname{Ker}[\beta : H_q(X; \mathbb{Z}_m) \to H_{q-1}(X; \mathbb{Z}_m)] \to$$
$$\to \frac{H_{q-1}(X; \mathbb{Z}_m)}{\operatorname{Im}[\beta : H_q(X; \mathbb{Z}_m) \to H_{q-1}(X; \mathbb{Z}_m)]},$$

called the *secondary* (or *second*) *Bockstein homomorphism*. Similarly homomorphisms $\beta_m^{(3)}, \beta_m^{(4)}, \ldots$ can be defined, each of which acts from the kernel of the preceding homomorphism to its cokernel.

C. The results of 1.5.C applied to $S(X)$ give split exact sequences

$$0 \to H_q(X) \otimes G \to H_q(X; G) \to \operatorname{Tor}(H_{q-1}(X), G) \to 0,$$
$$0 \to H^q(X; \mathbb{Z}) \otimes G \to H^q(X; G) \to \operatorname{Tor}(H^{q-1}(X; \mathbb{Z}), G) \to 0,$$
$$0 \to \operatorname{Ext}(H_{q-1}(X), G) \to H^q(X; G) \to \operatorname{Hom}(H_{q-1}(X), G) \to 0.$$

In particular,

$$H_q(X; G) \cong [H_q(X) \otimes G] \oplus \operatorname{Tor}(H_{q-1}(X), G), \tag{10}$$
$$H^q(X; G) \cong [H^q(X; \mathbb{Z}) \otimes G] \oplus \operatorname{Tor}(H^{q-1}(X; \mathbb{Z}), G), \tag{11}$$
$$H^q(X; G) \cong \operatorname{Hom}(H_{q-1}(X), G) \oplus \operatorname{Ext}(H_{q-1}(X), G) \tag{12}$$

(non-canonical isomorphisms); formulae (10)–(12) are called *universal coefficient formulae*. In addition, for any field k,

$$H^q(X; k) \cong \operatorname{Hom}_k(H_q(X; k), k).$$

Note also that if the groups $H_q(X)$ are finitely generated, then it follows from (12) that $H_q(X)$ and $H^q(X; \mathbb{Z})$ have the same rank, and $H_q(X)$ and and $H^{q+1}(X; \mathbb{Z})$ have the same torsion. In particular, $H^1(X; \mathbb{Z})$ is torsion free.

D. In conclusion we give a method of restoring the integer homology (and cohomology) groups from the homology groups with coefficients in the cyclic groups of prime order, on which the action of the Bockstein homomorphisms is known.

Let X be a space with finitely generated homology, and let p be prime. Suppose that (where dim denotes the dimension over the field of p elements)

$$\dim H(X; \mathbb{Z}_p) = m,$$
$$\dim \mathrm{Ker}[\beta_p : H_q(X; \mathbb{Z}_p) \to H_{q-1}(X; \mathbb{Z}_p)] = m_1,$$
$$\dim \mathrm{Ker}[\beta_p^{(2)} : \mathrm{Ker}\,\beta_p \to H_{q-1}(X; \mathbb{Z}_p)/\mathrm{Im}\,\beta_p] = m_2,$$
$$\cdots$$
$$\dim \mathrm{Im}[\beta_p : H^{q+1}(X; \mathbb{Z}_p) \to H_q(X; \mathbb{Z}_p)] = n_1,$$
$$\dim \mathrm{Im}[\beta_p^{(2)} : \mathrm{Ker}\,\beta_p \to H_q(X; \mathbb{Z}_p)/\mathrm{Im}\,\beta_p] = n_2,$$
$$\cdots$$

The sequence $m_1, m_2, \ldots$ is non-increasing, and stabilizes: $m_k = m^0$ for sufficiently large k; the series $n_1 + n_2 + \ldots$ converges to some $n^0 \le m^0$. It turns out that

$$H_q(X) = \mathbb{Z}^{m^0 - n^0} \oplus \mathbb{Z}_p^{m - m_1} \oplus \mathbb{Z}_{p^2}^{m_1 - m_2} \oplus \mathbb{Z}_{p^3}^{m_2 - m_3} \oplus \cdots \oplus T,$$

where T is a finite group whose order is relatively prime to p.

§3. Homology of cellular spaces

3.1. The cellular complex.
A. Let X be a cellular space with fixed characteristic maps given for its cells. (It is sufficient to fix not the characteristic maps but the *orientations* of the cells that is, classes of characteristic maps under the following equivalence relation: two characteristic maps $f : D^q \to X$, $g : D^q \to X$ of a cell e are equivalent if the composite homeomorphism $\mathrm{Int}\,D^q \xrightarrow{f} e \xrightarrow{g^{-1}} \mathrm{Int}\,D^q$ preserves orientation.) An (integral) *cellular q-chain* of X is a finite linear combination $\sum a_i e_i$, where the a_i are integers, and the e_i are q-cells. The group of cellular q-chains of X is denoted by $\mathscr{C}_q(X)$. The boundary operator $\partial : \mathscr{C}_q(X) \to \mathscr{C}_{q-1}(X)$ is defined by

$$\partial \sum a_i e_i = \sum a_i \partial e_i, \qquad \partial e = \sum_\varepsilon [e : \varepsilon]\varepsilon,$$

where in the last sum, ε runs through the $(q - 1)$-cells of X, and $[e : \varepsilon]$ is the *incidence number* defined as follows. Let $f : D^q \to X$, $g : D^{q-1} \to X$ be the characteristic maps of e and ε; then the number $[e : \varepsilon]$ is the degree of the composite map

$$S^{q-1} \xrightarrow{f|S^{q-1}} \mathrm{sk}_{q-1} X \xrightarrow{\text{projection}} \mathrm{sk}_{q-1} X/(\mathrm{sk}_{q-1} X \setminus \varepsilon)$$
$$= \mathrm{Cl}\,\varepsilon/\mathrm{Fr}\,\varepsilon \xrightarrow{g^{-1}} D^{q-1}/\partial D^{q-1} = S^{q-1}.$$

Lemma. *The composition* $\partial\partial : C_q(X) \to C_{q-1}(X)$ *is trivial for any* q. $\qquad\square$

This lemma is not hard to prove directly, but it is much easier to deduce it from the homological interpretation of the groups $\mathscr{C}_q(X)$ and homomorphisms ∂ obtained below (in **B**).

Thus the groups $\mathscr{C}_q(X)$ and homomorphisms $\partial : \mathscr{C}_q(X) \to \mathscr{C}_{q-1}(X)$ form a complex, which is denoted by $\mathscr{C}(X)$ and called the *cellular complex of X*. The main result of this section is the construction of a homotopy equivalence between $\mathscr{C}(X)$ and the singular complex $S(X)$ of X (see 3.2). In particular, we have:

Theorem. *The homology and cohomology of $\mathscr{C}(X)$ with coefficients in an arbitrary Abelian group G are the same as the homology and cohomology of the space X with coefficients in G.* $\Box$

Corollary. *If cellular spaces are homotopy equivalent, then the homology and cohomology of their cellular complexes are the same.* $\Box$

In contrast to the singular complex, where the groups are as a rule, enormous, the cellular complex is entirely suitable for practical computation. Examples of such computations will be given below in 3.4. More complicated examples can be found in Part III, Chapters 3 and 4.

B. Proposition 1. *For any X and q,*

$$H_r(\mathrm{sk}_q X, \mathrm{sk}_{q-1} X) = \begin{cases} 0 & \text{for } r \neq q, \\ \mathscr{C}_q(X) & \text{for } r = q. \end{cases}$$

Since the quotient space $\mathrm{sk}_q X / \mathrm{sk}_{q-1} X$ is canonically homeomorphic to a bouquet of q-spheres indexed by the q-cells of X, Proposition 1 follows from the results of 2.4.**B**, 2.4.**D**, and 2.5.**C**.

Proposition 1 allows us to identify the group $\mathscr{C}_q(X)$ with $H_q(\mathrm{sk}_q X, \mathrm{sk}_{q-1} X)$.

Proposition 2. *Under this identification, the homomorphism $\partial : \mathscr{C}_q(X) \to \mathscr{C}_{q-1}(X)$ is identified with the homomorphism*

$$H_q(\mathrm{sk}_q X, \mathrm{sk}_{q-1} X) \to H_{q-1}(\mathrm{sk}_{q-1} X, \mathrm{sk}_{q-2} X)$$

from the homology sequence of the triple $(\mathrm{sk}_q X, \mathrm{sk}_{q-1} X, \mathrm{sk}_{q-2} X)$. $\Box$

The proof consists of a direct comparison of the definitions.

C. A cellular map f of a cellular space X to a cellular space Y induces a chain map $f_\#$ of $\mathscr{C}(X)$ to $\mathscr{C}(Y)$ in a natural way. A cellular homotopy between cellular maps $f, g : X \to Y$ induces a chain homotopy between $f_\#$ and $g_\#$; since homotopic cellular maps are cellular homotopic, the maps $f_\#$, $g_\#$ induced by homotopic cellular maps are homotopic.

3.2. Interrelations with the singular complex.

A. The theorem in 3.1.**A** is usually proved by means of Propositions 1 and 2 in 3.1.**B** and the sequences of a pair and a triple. We shall confine ourselves to the case of homology; the cohomology case is similar. Since $H_r(\mathrm{sk}_s X, \mathrm{sk}_{s-1} X; G) = 0$, it follows from the exactness of the homology sequences of the triples $(\mathrm{sk}_{n+1} X, \mathrm{sk}_n X, \mathrm{sk}_m X)$ and $(\mathrm{sk}_n X, \mathrm{sk}_m X, \mathrm{sk}_{m-1} X)$ with $n > q$, $q - 1 > m$ that

$$H_q(\mathrm{sk}_n X, \mathrm{sk}_m X; G) = H_q(X, \mathrm{sk}_m X; G) = H_q(X; G).$$

These equalities show that

$$H_q(X; G) = H_q(\mathrm{sk}_{q+1}X, \mathrm{sk}_{q-2}X; G).$$

The equality

$$H_q(X; G) = \frac{\mathrm{Ker}[\partial : \mathscr{C}_q(X) \otimes G \to \mathscr{C}_{q-1}(X) \otimes G]}{\mathrm{Im}[\partial : \mathscr{C}_{q+1}(X) \otimes G \to \mathscr{C}_q(X) \otimes G]}$$

is proved by means of the commutative diagram

$$H_q(\mathrm{sk}_{q-1}X, \mathrm{sk}_{q-1}X; G) = 0$$
$$\downarrow$$

$$H_{q+1}(\mathrm{sk}_{q+1}X, \mathrm{sk}_q X; G) \to H_q(\mathrm{sk}_q X, \mathrm{sk}_{q-2}X; G) \to H_q(\mathrm{sk}_{q+1}X, \mathrm{sk}_{q-2}X; G) \to H_q(\mathrm{sk}_{q+1}X, \mathrm{sk}_q X; G)$$

$$\searrow \partial \qquad \downarrow \qquad \qquad \| $$

$$H_q(\mathrm{sk}_q X, \mathrm{sk}_{q-1}X; G) \qquad\qquad 0$$

$$\downarrow \partial$$

$$H_{q-1}(\mathrm{sk}_{q-1}X, \mathrm{sk}_{q-1}X; G),$$

in which the rows and columns are exact.

B. This argument can be refined by proving the homotopy equivalence of the complexes $S(X)$ and $\mathscr{C}(X)$ directly. To do this it is convenient to introduce an intermediate complex $\hat{S}(X)$ consisting of the groups

$$\hat{C}_q(X) = \{c \in C_q(X) \mid c \in C_q(\mathrm{sk}_q X), \partial c \in C_{q-1}(\mathrm{sk}_{q-1}X)\}.$$

On one hand, $\hat{S}(X)$ is a subcomplex of $S(X)$ so that the inclusion $i : \hat{S}(X) \to S(X)$ is defined. On the other hand, a chain $c \in C_q(X)$ is a relative cycle of $\mathrm{sk}_q X$ modulo $\mathrm{sk}_{q-1}X$, so that it determines an element of $H_q(\mathrm{sk}_q X, \mathrm{sk}_{q-1}X) = \mathscr{C}_q(X)$; the resulting maps $\hat{C}_q(X) \to \mathscr{C}_q(X)$ form a chain map $p : \hat{S}(X) \to \mathscr{C}(X)$.

Proposition. *The maps i and p are homotopy equivalences.* $\qquad\qquad\square$

The proof of this proposition involves very similar ideas to those in **A**.

Corollary. *The complexes $S(X)$ and $C(X)$ are homotopy equivalent.*

C. The results of **A** and **B** have relative versions. If A is a cellular subspace of a cellular space X, then the complex $\mathscr{C}(X, A) = \mathscr{C}(X)/\mathscr{C}(A)$ is homotopy equivalent to $S(X, A)$; in particular, relative homology and cohomology can be computed by means of cellular chains. Moreover, the homology and cohomology sequences of cellular pairs and triples can be constructed by means of cellular chains and cochains, thus bypassing singular chains and cochains. (The same is true for the coefficient sequences.) Finally, we may define the reduced cellular complex $\tilde{\mathscr{C}}(X)$ (in exactly the same way as the reduced singular complex), and the complexes $\tilde{S}(X)$ and $\tilde{\mathscr{C}}(X)$ will be homotopy equivalent.

D. A generalization of a cellular complex is a complex with an arbitrary *homologically proper filtration*. Let X be a topological space. A sequence of subspaces X_q $(q \in \mathbb{Z})$ is called a *filtration* if (i) $X_q \subset X_{q+1}$ for each q; (ii) $X_0 = \emptyset$ for $q < 0$;

(iii) $\bigcup_{q=-\infty}^{\infty} X_q = X$; (iv) a set $F \supset X$ is closed if and only if $F \cap X_q$ is closed in X_q for each q. A filtration is called *homologically proper* if $H_r(X_q, X_{q-1}) = 0$ for any $r \neq q$ and $H_q(X_q, X_{q-1})$ is a free Abelian group for each q. The groups $H_q(X_q, X_{q-1})$ and the homomorphisms $\partial_* : H_q(X_q, X_{q-1}) \to H_{q-1}(X_{q-1}, X_{q-2})$ form a complex called the *complex of the homologically proper filtration* $\{X_q\}$. (A cellular complex is the complex of the homologically proper filtration $\{sk_q X\}$.)

Theorem. *The complex of any homologically proper filtration of a space X is homotopy equivalent to $S(X)$.*

The proof of this imitates the previous proof.

In this part, we shall only need to encounter non-cellular homologically proper filtrations once – in the construction of the Poincaré isomorphism in §8.

3.3. The simplicial case.

Let X be an ordered simplicial space. Each q-simplex σ of X is associated with the singular q-simplex of X given by the canonical homeomorphism f_σ of the standard simplex T^q onto σ. Clearly the correspondence $\sigma \mapsto f_\sigma$ defines an embedding of $\mathscr{C}(X)$ in $S(X)$.

Theorem. *This inclusion is a homotopy equivalence.* $\qquad\qquad\qquad\square$

In fact it is the composition of the map $\mathscr{C}(X) \to \hat{S}(X)$, which is a homotopy inverse of the map p in 3.2.**B**, and the inclusion $i : \hat{S}(X) \to S(X)$.

The complex $\mathscr{C}(X)$ of an ordered simplicial space X was invented two decades earlier than the singular complex and is often called the *classical complex*. Its q-chains are finite integral linear combinations $\sum a_i \sigma_i$ of q-simplexes of X, and the boundary operator is defined by the formula (which we already know in another context) $\partial \sum a_i \sigma_i = \sum a_i \partial \sigma_i$, $\partial\sigma = \sum_i (-1)^i \Gamma_i \sigma$, where $\Gamma_i \sigma$ is the ith face of the simplex σ (cf. 2.3.**C**, example 3°)

3.4. Examples of calculations.

(The cell decompositions used in this section are described in Part I, 7.3.)

A. The homology of spheres was computed in 2.4.**D**, but it can just as easily be found by means of cell decompositions. The result of course is the same.

B. The standard surface P_g can be decomposed into one 0-cell, $2g$ 1-cells and one 2-cell. All the incidence numbers are zero; for the 1-cells and 0-cell this is obvious (if the closure of a 1-cell is homeomorphic to a circle, then its boundary is 0), and for the 2-cell and the 1-cells it is clear from the representation of the surface as a folded $4g$-sided polygon: the map $S^1 \to S^1$, whose degree is by definition the incidence number that interests us, is constant outside two segments (sides of the $4g$-gon), but these segments pile up on S^1 in different directions. Thus the differential in the complex $\mathscr{C}(P_g)$ is identically zero, and we have:

$$H_q(P_g; G) \cong H^q(P_g; G) \cong \begin{cases} G & \text{if } q = 0, 2, \\ \underbrace{G \oplus \cdots \oplus G}_{2g} & \text{if } q = 1, \\ 0 & \text{otherwise.} \end{cases}$$

C. The standard surface Q_h can be decomposed into one 0-cell, h 1-cells, and one 2-cell. The incidence numbers between the 1-cells and the 0-cell are zero, and those between the (naturally oriented) 2-cell and the 1-cells are equal to 2. Hence it follows in particular that

$$H_q(Q_h) \cong \begin{cases} \mathbb{Z} & \text{if } q = 0, \\ \underbrace{\mathbb{Z} \oplus \cdots \oplus \mathbb{Z}}_{h-1} \oplus \mathbb{Z}_2 & \text{if } q = 1, \\ 0 & \text{otherwise}; \end{cases}$$

$$H_q(Q_h; \mathbb{Z}_2) \cong H^q(Q_h; \mathbb{Z}_2) \cong \begin{cases} \mathbb{Z}_2 & \text{if } q = 0, \\ \underbrace{\mathbb{Z}_2 \oplus \cdots \oplus \mathbb{Z}_2}_{n} & \text{if } q = 1, \\ 0 & \text{otherwise}; \end{cases}$$

D. The standard cell decomposition of real projective space $\mathbb{R}P^n$ $(1 \leq n \leq \infty)$ consists of cells e^q $(0 \leq q < n+1)$, $\dim e^q = q$, with

$$[e^q : e^{q-1}] = \begin{cases} 2 & \text{for even positive } q \leq n, \\ 0 & \text{otherwise}. \end{cases}$$

In particular, it follows from this that

$$H_q(\mathbb{R}P^n) \cong \begin{cases} \mathbb{Z} & \text{if } q = 0 \text{ or if } q = n \text{ and } n \text{ is odd}, \\ \mathbb{Z}_2 & \text{if } 0 < q < n \text{ and } q \text{ is odd}, \\ 0 & \text{otherwise}. \end{cases}$$

$$H^q(\mathbb{R}P^n; \mathbb{Z}) \cong \begin{cases} \mathbb{Z} & \text{if } q = 0 \text{ or if } q = n \text{ and } n \text{ is odd}, \\ \mathbb{Z}_2 & \text{if } 0 < q \leq n \text{ and } q \text{ is even}, \\ 0 & \text{otherwise}. \end{cases}$$

$$H_q(\mathbb{R}P^n; \mathbb{Z}_2) \cong H^q(\mathbb{R}P^n; \mathbb{Z}_2) \cong \begin{cases} \mathbb{Z}_2 & \text{if } 0 \leq q \leq n, \\ 0 & \text{otherwise}. \end{cases}$$

E. The standard cell decomposition of complex projective space $\mathbb{C}P^n$ $(1 \leq n \leq \infty)$ consists of cells e^q $(0 \leq q < n+1)$, $\dim e^q = 2q$. The incidence numbers are all zero, since there are no cells of adjacent dimensions. Hence

$$H_q(\mathbb{C}P^n; G) \cong H^q(\mathbb{C}P^n; G) = \begin{cases} G & \text{if } 0 \leq q \leq 2n \text{ and } q \text{ is even}, \\ 0 & \text{otherwise}. \end{cases}$$

3.5. Other applications.

A. If there are no q-cells in a cellular space X, then $H_q(X; G) = H^q(X; G) = 0$ for any Abelian group G. In particular, if X is n-dimensional, then $H_q(X; G) = H^q(X; G)$ for $q > n$ and any G.

B. If X is an open subset of $\mathbb{R}^n$ with $n > 0$, then $H_q(X; G) = H^q(X; G) = 0$ for $q \geq n$ and any G.

(It is sufficient to prove that these groups are trivial for any polyhedron Y contained in X. This is obvious for $q > n$; for $q = n$ it follows because the groups $H_{n+1}(\mathbb{R}^n, Y; G)$, $H^{n+1}(\mathbb{R}^n, Y; G)$ are trivial, in view of the exactness of the sequence of the pair $(\mathbb{R}^n, Y)$.)

C. If X is a finite cellular space, then the Euler characteristic of X may be calculated knowing only the numbers of cells of each dimension:

$$\mathrm{Eu}(X) = \sum (-1)^q c_q(X),$$

where $c_q(X)$ is the number of q-cells of X (see the theorem in 1.4.A).

§4. Homology and homotopy

4.1. Weak homotopy equivalence and homology.

A. Theorem. *If a continuous map $f : X \to Y$ is a weak homotopy equivalence (see Part I, 9.4), then for any q and G, the maps*

$$f_* : H_q(X; G) \to H_q(Y; G), \quad f^* : H^q(Y; G) \to H^q(X; G)$$

are isomorphisms. Moreover, the maps $f_\# : C_q(X) \to C_q(Y)$ form a homotopy equivalence of the singular complexes of X and Y. □

Corollary. *Let X and Y be connected topological spaces with base points x_0, y_0. If a map $f : (X, x_0) \to (Y, y_0)$ induces an isomorphism of all the homotopy groups, then it induces an isomorphism of all homology and cohomology groups with arbitrary coefficients.* □

The proof (of the homology version) depends on the following lemma.

Lemma. *If $\alpha \in H_q(X; G)$ is any homology class, then there exists a q-dimensional cellular space K, a homology class $\beta \in H_q(K; G)$ and a continuous map $f : K \to X$ such that $f_*(\beta) = \alpha$.* □

The theorem can be deduced in an obvious way from the lemma.

Proof of the lemma: choose a cycle $\sum_{i=1}^{N} g_i f_i$ ($f_i : T^q \to X$) representing α, consider the union $\coprod T_i^q$ of N copies of the standard simplex T^q and identify faces $S_1 \subset T_i^q$, $S_2 \subset T_j^q$ of the same dimension with each other if $f_i \mid S_1 = f_j \mid S_2$ (the identification is performed by the simplicial homeomorphism preserving the order of the vertices); the resulting cellular space is K, and β and f are defined in an obvious way. □

B. There is a relative version of the theorem in which the spaces X and Y are replaced by topological pairs. A similar assertion is valid for k-equivalence (see Part I, 9.4.**H**); k-equivalence between topological spaces or topological pairs induces isomorphisms of homology and cohomology with arbitrary coefficients in dimensions $< k$, and in dimension k it induces an epimorphism of homology and a monomorphism of cohomology.

4.2. The Hurewicz theorems.

A. The *Hurewicz homomorphism*

$$h : \pi_n(X, x_0) \to H_n(X)$$

for any space X with base point x_0 and any n is defined by $h(\alpha) = f_*([S^n])$, where f is any spheroid $S^n \to X$ representing α and $[S^n]$ is the canonical generator of the group $H_n(S^n) \cong \mathbb{Z}$ – see 2.4.**D**; h is well-defined because of the homotopy invariance of homology (see 2.1.**C**). The Hurewicz homomorphism is natural: the diagram

$$\pi_n(X, x_0) \xrightarrow{h} H_n(X)$$

$$f_* \downarrow \qquad\qquad \downarrow f_*$$

$$\pi_n(Y, y_0) \xrightarrow{h} H_n(Y)$$

is commutative for any continuous map $f : (X, x_0) \to (Y, y_0)$.

Theorem (Hurewicz). *Let X be a connected space with base point x_0.*

(i) *If $\pi_1(X, x_0) = \ldots = \pi_{n-1}(X, x_0) = 0$, where $n \geq 2$, then $H_1(X) = \ldots = H_{n-1}(X) = 0$, and*

$$h : \pi_n(X, x_0) \to H_n(X)$$

is an isomorphism.

(ii) *If X is simply connected and $H_2(X) = \ldots = H_{n-1}(X) = 0$, where $n \geq 2$, then $\pi_2(X, x_0) = \ldots = \pi_{n-1}(X, x_0) = 0$, and*

$$h : \pi_n(X, x_0) \to H_n(X)$$

is an isomorphism. □

In other words, for a simply connected space the first nontrivial homotopy and homology groups appear in the same dimension, and they are isomorphic.

Part (ii) of the Hurewicz theorem is derived in an obvious way from part (i). The most efficient proof of part (i) is the following. We first consider the case when X is a cellular space with a single vertex and with no cells of dimensions $1, \ldots, n-1$. For such a space the groups $H_1(X), \ldots, H_{n-1}(X)$ are trivial by the results of §2, and h is an isomorphism because the procedures for computing the nth homology group and the nth homotopy group, described in §2 and in Part I, §11, are absolutely identical. The case of a general cellular space can be reduced to the case already considered in view of the homotopy invariance of homology and homotopy groups and the results of Part I, 9.2.**B**; finally the general case reduces to the cellular case in view of the weak homotopy invariance of homology and homotopy groups (see 4.1 and Part I, 9.4) and the existence of a cellular approximation to an arbitrary topological space (see Part I, 9.5).

B. The *relative Hurewicz homomorphism*

$$h : \pi_n(X, A, x_0) \to H_n(X, A)$$

is defined in a similar way to the absolute case: a class $\alpha \in \pi_n(X, A, x_0)$ is mapped to $f_*([D^n])$, where $f : (D^n, S_{n-1}) \to (X, A)$ is a spheroid representing α, and $[D^n]$ is the canonical generator of the group $H_n(D^n, S^{n-1}) = \mathbb{Z}$ (this group is related to $H_{n-1}(S^{n-1}) = \mathbb{Z}$ by the isomorphism $\partial_* : H_n(D^n, S^{n-1}) \to H_{n-1}(S^{n-1})$). The relative Hurewicz homomorphism is natural in the same sense as the absolute one.

Theorem (Hurewicz). *Let (X, A) be a topological pair with X and A connected and simply connected, and with base point $x_0 \in A$.*

(i) *If $\pi_2(X, A, x_0) = \ldots = \pi_{n-1}(X, A, x_0) = 0$, where $n \geq 3$, then $H_1(X, A) = \ldots = H_{n-1}(X, A) = 0$ and*

$$h : \pi_n(X, A, x_0) \to H_n(X, A)$$

is an isomorphism.

(ii) *If* $\pi_2(X, A, x_0) = 0$ *and* $H_3(X, A) = \ldots = H_{n-1}(X, A) = 0$, *where* $n \geq 3$, *then* $\pi_3(X, A) = \ldots = \pi_{n-1}(X, A) = 0$ *and* $h : \pi_n(X, A, x_0) \to H_n(X, A)$ *is an isomorphism.*   □

The proof is similar to that of the absolute Hurewicz theorem.

4.3. The theorems of Poincaré and Hopf.

A. Theorem (Poincaré). *For any connected space* X *with base point* x_0, *the Hurewicz homomorphism*

$$h : \pi_1(X, x_0) \to H_1(X)$$

is an epimorphism, whose kernel is the commutator subgroup $[\pi_1(X, x_0), \pi_1(X, x_0)]$ *of* $\pi_1(X, x_0)$. *Thus*

$$H_1(X) \cong \pi_1(X, x_0)/[\pi_1(X, x_0), \pi_1(X, x_0)],$$

that is, the first homology group is the fundamental group made Abelian.   □

The proof is similar to that of the Hurewicz theorem.

B. Theorem (Hopf). *Let* X *be a connected topological space with base point* x_0 *and fundamental group* $\pi_1(X, x_0) = \pi$, *and let* K *be a* $K(\pi, 1)$ *space. Then there is an exact sequence*

$$\pi_2(X, x) \xrightarrow{h} H_2(X) \xrightarrow{\psi^*} H_2(K) \longrightarrow 0,$$

where ψ *is the (unique up to homotopy) continuous map* $X \to K$ *that induces the identity map of the fundamental group.*   □

Remarks. 1°. The group $H_2(K)$ is determined by π in a purely algebraic way. Namely, if $\pi = F_1/F_2$ where F_1 is a free group, then

$$H_2(K) \cong (F_2 \cap [F_1, F_1])/[F_1, F_2].$$

2°. Hopf's theorem has the following generalization. Suppose that $\pi_1(X, x_0) = \pi$, and $\pi_2(X, x_0) = \ldots = \pi_{n-1}(X, x_0) = 0$. Then $\psi_* : H_q(X) \to H_q(K)$ is an isomorphism for $q < n$, and the sequence

$$\pi_n(X, x_0) \longrightarrow H_n(X) \xrightarrow{\psi_*} H_n(K) \longrightarrow 0$$

is exact.

3°. The following exact sequence is a refinement of the preceding one:

$$H_{n+1}(X) \xrightarrow{\psi_*} H_{n+1}(K) \longrightarrow \pi_n(X, x_0)_\pi \xrightarrow{h} H_n(X) \xrightarrow{\psi_*} H_n(K) \to 0,$$

where $\pi_n(X, x_0)_\pi$ denotes the factor group of $\pi_n(X, x_0)$ by the normal subgroup generated by elements of the form

$$\sigma - \sigma\alpha, \quad \alpha \in \pi_n(X, x_0), \quad \sigma \in \pi = \pi_1(X, x_0).$$

4.4. Whitehead's theorem.

The following theorem is deduced in a standard way from Hurewicz's theorem.

Theorem (Whitehead). *Let X and Y be simply connected topological spaces with base points x_0, y_0, and $f : (X, x_0) \to (Y, y_0)$ be a continuous map that induces an epimorphism $\pi_2(X, x_0) \to \pi_2(Y, y_0)$. Then the following statements are equivalent.*

(i) *The homomorphism $f_* : \pi_q(X, x_0) \to \pi_q(Y, y_0)$ is an isomorphism for $q < n$ and an epimorphism for $q = n$.*

(ii) *The homomorphism $f_* : H_q(X) \to H_q(Y)$ is an isomorphism for $q < n$ and an epimorphism for $q = n$.* □

Corollary. *If a map $f : X \to Y$ of a simply connected space into another simply connected space induces an epimorphism $\pi_2(X) \to \pi_2(Y)$ and induces an isomorphism $H_q(X) \cong H_q(Y)$ for all q, then f is a weak homotopy equivalence (even a homotopy equivalence if X and Y are cellular).* □

4.5. Some instructive examples.

A. The spaces S^2 and $\mathbb{C}P^\infty \times S^3$ have the same homotopy groups, but different homology groups; the same is true for $S^m \times \mathbb{R}P^n$ and $S^n \times \mathbb{R}P^m$ with $m \neq 1, n \neq 1$, $m \neq n$. On the other hand, the spaces $S^1 \vee S^1 \vee S^2$ and $S^1 \times S^1$ have the same homology groups but different homotopy groups. Thus in Whitehead's theorem it is essential that the isomorphism between the homotopy or homology groups of X and Y does not merely exist, but is actually established by some continuous map.

B. The Hopf map $S^3 \to S^2$ induces the trivial homomorphism on the (reduced) homology groups, but induces a nontrivial map of the homotopy groups. On the other hand, the projection $S^1 \times S^1 \to (S^1 \times S^1)/(S^1 \vee S^1) = S^2$ induces the trivial homomorphism in all homotopy groups, but a nontrivial map on the homology groups (including the reduced groups).

C. The compositions

$$S^1 \times S^1 \times S^1 \xrightarrow{\text{projection}} (S^1 \times S^1 \times S^1)/\text{sk}_2(S^1 \times S^1 \times S^1) = S^3 \xrightarrow{\text{Hopf}} S^2,$$

$$S^{2n-2} \times S^3 \xrightarrow{\text{projection}} (S^{2n-2} \times S^3)/(S^{2n-2} \vee S^3) = S^{2n+1} \xrightarrow{\text{Hopf}} \mathbb{C}P^n$$

induce trivial maps both in the (reduced) homology groups and in the homotopy groups, but neither of these maps is homotopic to a constant map.

§5. Homology and fixed points

5.1. Lefschetz's theorem.

A. Let X be a topological space with the group $H_*(X)$ finitely generated, and $f : X \to X$ a continuous map. Let t_q denote the trace of the operator

$$f_* : H_q(X)/\text{tors} \to H_q(X)/\text{tors}$$

(or, which comes to the same thing, the trace of the operator $f_* : H_q(X; k) \to H_q(X; k)$, where k is any field of characteristic 0, say $\mathbb{Q}$, $\mathbb{R}$, or $\mathbb{C}$). The alternating sum

$$\text{Le}(f) = \sum (-1)^q t_q$$

(which in accordance with the general theory of §1 should be called the Lefschetz number of the map $f_\# : S(X) \to S(X)$) will be called the *Lefschetz number of the map* f. From the theorem of 1.4.**B** it follows that if X is a finite cellular space, the Lefschetz number can be found from the induced cellular chain map:

Proposition.

$$\text{Le}(f) = \sum (-1)^q \text{Tr}[f_\# : \mathscr{C}_q(X) \to \mathscr{C}_q(X)]. \quad \square$$

Note. Clearly Le(id) is just the Euler characteristic Eu(X) of X. Since Le(f) is homotopy invariant, the same is true for Le(f) if $f \sim$ id.

Theorem. *Let X be a compact triangulable space. If the continuous map f : $X \to X$ has no fixed points, then* Le(f) = 0. $\quad \square$

To prove this it is sufficient to triangulate X so finely that the diameters of the simplexes are considerably smaller than the lower bound of the distance between x and $f(x)$ (in some arbitrary metric on X). Then we find a simplicial approximation g to f; g also has no fixed points, and moreover, no simplex σ intersects its image $g(\sigma)$. Hence the matrix of the map $g_\# \mathscr{C}_q(X) \to \mathscr{C}_q(X)$ has zero diagonal, and hence zero trace. Consequently Le(g) = 0, so that Le(f) = 0. $\quad \square$

(The requirement that X is triangulable can be relaxed: the theorem is true for compact absolute neighbourhood retracts.)

Example. The following necessarily have fixed points: every continuous map of a ball of any dimension into itself (*Brouwer's theorem*); every continuous map of four-dimensional real projective space into itself; every map of an odd-dimensional sphere into itself that is not homotopic to the identity; every continuous map of an even-dimensional sphere into itself with degree not equal to -1.

B. The preceding theorem shows that a continuous map between sufficiently "good" topological spaces with nonzero Lefschetz number must have a fixed point. We shall now refine this formulation and show that if certain natural conditions are satisfied the Lefschetz number can be interpreted as the number of fixed points counted with appropriate multiplicities.

A fixed point of a map $f : X \to X$ is called *regular* if it has a neighbourhood homeomorphic to $\mathbb{R}^n$ for some n that contains no other fixed points. If X is compact then the number of fixed points is finite, provided they are all regular. The class of maps all of whose fixed points are regular may seem special, but in many important cases such maps turn out to be typical.

Let x be a regular fixed point of the continuous map $f : X \to X$ and ϕ : $(U, x) \to (\mathbb{R}^n, 0)$ be a homeomorphism, where U is a neighbourhood of x. Let $S \subset \mathbb{R}^n$ be a sphere with centre 0 small enough for $f(\phi^{-1}(S)) \subset U$. Let ε be the radius of S, and for $s \in S$ put $g(s) = s - \phi \circ f \circ \phi^{-1}(s) \in \mathbb{R}^n \setminus 0$, and consider the map $h : S \to S$, $s \to \varepsilon g(s)/\|g(s)\|$. The degree of this map is called the *index* of the fixed point x, denoted by $\text{ind}_f(x)$. It is easy to see that this is well-defined, that is, $\text{ind}_f(x)$ does not depend on the choice of U, ϕ and S.

Let us point out a special case in which it is particularly easy to calculate the index. Suppose that X is a smooth manifold and f a smooth map (or, more generally, that there is a pair of open sets $V \subset U$ in X, where U has the structure of a smooth manifold, V contains all the fixed points of f, $f(V) \subset U$ and f is smooth on V). Then at each fixed point x of f the differential $d_x f$ is defined as a linear transformation of the tangent space. The point x is called *non-degenerate* if 1 is not an eigenvalue of the differential of f. It is easy to see that a non-degenerate fixed point must be regular and have index ± 1; more precisely: the index is 1 if there is an even number of real eigenvalues of $d_x f$ greater than 1, and the index is -1 if there is an odd number of such eigenvalues. An important example: f is close to the identity diffeomorphism of the manifold X, defined by a vector field v; then the fixed points of f correspond to the singular points of v, the non-degenerate fixed points correspond to the non-degenerate singular points and the corresponding indices are the same.

The fundamental result of this section is the following

Theorem (Lefschetz). *Let X be a compact triangulable space (or compact absolute neighbourhood retract), and let $f : X \to X$ be a continuous map all of whose fixed points are regular. Then*

$$\mathrm{Le}(f) = \sum_{\mathrm{Fix}\, f} \mathrm{ind}_f(x)$$

(Fix f denotes the set of all fixed points of f). $\Box$

There are several ways of proving this theorem; we outline one of the most efficient proofs. We begin by transforming X and f in such a way that (i) neither side of the equality to be proved is altered; (ii) all the fixed points become regular; (iii) the inverse image of each fixed point consists of precisely that point; (iv) the index of a fixed point having a neighbourhood homeomorphic to $\mathbb{R}^n$ becomes equal to $(-1)^n$. This may be done in various ways, for example as follows. First we simply perturb the map near each fixed point, so that the fixed points become regular, but the sum of their indices is unchanged (this is easy to prove). Then we form the product of X and the interval $I = [0, 1]$ and replace f by the map $(x, t) \to (f(x), \phi_x(t))$, where ϕ_x is the linear map of I onto an interval $I_x \subset I$ of half the length, such that $I_x = [1/8, 5/8]$ if x is a fixed point, and $I_x = [3/8, 7/8]$ if x lies outside the union of the small neighbourhoods of the fixed points. The fixed points of this latter map are the points $(x, 1/4)$, where x is a fixed point of f, and it is clear that this map satisfies condition (iii). Finally, if a map satisfies conditions (i)–(iii), it is easy to arrange that condition (iv) is also satisfied: it is sufficient to blow up the neighbourhoods of those fixed points where (iv) is not satisfied by taking their product with an interval and defining the map as the composition of the projection of the new X onto the old X with the old map. The next step is to deform our map in the neighbourhood of each fixed point so that all the fixed points become "repelling"; for example, are given by the formula $u \to 2u$ in a neighbourhood $U \approx \mathbb{R}^n$ (a deformation of a map near a fixed point is the same thing as a deformation of a vector field, and vector fields given near a singular point can be deformed into each other without the appearance of new singular points if and only if they have the same index). Finally we remove neighbourhoods of all the fixed points from X, and obtain a new map g which has no fixed points at all.

From the previous theorem, $\mathrm{Le}(g) = 0$, and at the same time it is clear that $\mathrm{Le}(g)$ and $\mathrm{Le}(f)$ differ by precisely the right-hand side of the equality to be proved.

Corollary (Hopf's theorem). *The sum of the indices of the singular points of a vector field on a (not necessarily orientable) closed manifold is equal to the Euler characteristic of the manifold.*   □

C. We shall give a useful generalization of Lefschetz's theorem. Suppose we are concerned with the following situation. We are given a triangulated space X and a continuous map $f : X \to X$; in addition we are given a pair of open sets $V \subset U \subset X$ and the structure of a smooth manifold on U, with $f(V) \subset U$, f smooth on V, and all the fixed points of f contained in V. Further, the set of fixed points is the union of closed non-intersecting submanifolds X_i (in general of different dimensions) of V, where the differential of f at any point of X_i has 1 as an eigenvalue of multiplicity precisely dim X_i We define the index $\mathrm{ind}_f(X_i)$ of the fixed manifold X_i to be equal to 1, if there is an even number of real eigenvalues of this differential that are greater than 1 (this is obviously independent of the choice of point in X_i), and equal to -1 if there is an odd number of such eigenvalues.

Theorem.

$$\mathrm{Le}(f) = \sum_i \mathrm{ind}_f(X_i)\mathrm{Eu}(X_i). \ \square$$

To prove this the map f is subjected to deformations that change it only in the neighbourhood of the manifolds X_i, so that the new map maps X_i into X_i as before, but the resulting map $X_i \to X_i$ has only non-degenerate fixed points (the algebraic number of which is exactly equal to $\mathrm{Eu}(X_i)$).

To conclude we give a very important application of Lefschetz's theorem in **B**.

Theorem. *Let X be a compact triangulated space whose homology with coefficients in $\mathbb{Q}$ is the same as that of S^n with n even (for example, $X = S^n$ with n even). If G is a group acting freely and continuously on X, then G is 0 or $\mathbb{Z}_2$.*   □

Proof. Let $1 \neq g \in G$. Since the transformation produced on X by g (which we shall denote also by g) has no fixed points, $\mathrm{Le}(g) = 0$ and $g_* : H_n(X; \mathbb{Q}) \to H_n(X; \mathbb{Q})$ is multiplication by -1. If $1 \neq g' \in G$, then $g'_* : H_n(X; \mathbb{Q}) \to H_n(X; \mathbb{Q})$ is also multiplication by -1, so that $(g^{-1}g')_* : H_n(X; \mathbb{Q}) \to H_n(X; \mathbb{Q})$ is the identity map. Hence $g^{-1}g' = 1$ and $g' = g$.   □

This argument does not work for odd-dimensional spheres, but the following result can be proved.

Theorem. *If a finite Abelian group G acts freely and continuously on an odd-dimensional sphere S^n and the quotient space S^n/G is triangulable, then the group G is cyclic.*   □

A stronger result can be obtained for Euclidean space.

Theorem. *A nontrivial finite group cannot act freely on Euclidean space.*   □

Corollary. *If p is prime, then any continuous map $f : \mathbb{R}^n \to \mathbb{R}^n$ with $f^p = \mathrm{id}$ has a fixed point.*   □

It turns out that a map $f : \mathbb{R}^n \to \mathbb{R}^n$ with $f^m = \mathrm{id}$ must have a fixed point also when m is a prime power, but need not have fixed points if m is the product of two distinct primes (see Borel (1960)).

5.2. Smith theory.

A. Let X be a compact topological space and $t : X \to X$ an involution (that is, a map whose square is equal to the identity). We shall assume that X has a triangulation in which t is simplicial (for example, a suitable case is when X is a smooth manifold and t is a diffeomorphism). Let F denote the set of fixed points of t, and X' be the quotient space $X/(x = tx)$. The projection $X \to X'$ maps F homeomorphically onto a subset of X', which we again denote by F.

Theorem. *There are exact sequences*

$$\cdots \longrightarrow H_r(X', F; \mathbb{Z}_2) \oplus H_r(F; \mathbb{Z}_2) \xrightarrow{s_r} H_r(X; \mathbb{Z}_2) \xrightarrow{m_r}$$
$$\xrightarrow{m_r} H_r(X', F; \mathbb{Z}_2) \xrightarrow{i_r} H_{r-1}(X', F; \mathbb{Z}_2) \oplus H_{r-1}(F; \mathbb{Z}_2) \longrightarrow \cdots,$$

$$\cdots \longrightarrow H^{r-1}(X', F; \mathbb{Z}_2) \oplus H^{r-1}(F; \mathbb{Z}_2) \xrightarrow{i^r} H^r(X', F; \mathbb{Z}_2) \xrightarrow{m^r}$$
$$\xrightarrow{m^r} H^r(X; \mathbb{Z}_2) \xrightarrow{s^r} H^r(X', F; \mathbb{Z}_2) \oplus H^r(F; \mathbb{Z}_2) \longrightarrow \cdots. \ \square$$

These sequences are called the *Smith sequences*. We shall describe the construction of the homomorphisms s, m, and i in the Smith homology sequence (the cohomology Smith homomorphisms are constructed similarly).

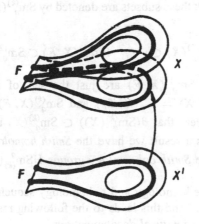

Fig. 2

The homomorphism s_r is composed of the homomorphism $H_r(F; \mathbb{Z}_2) \to H_r(X; \mathbb{Z}_2)$ induced by the inclusion $F \to X$, and the homomorphism of "taking the inverse image" $H_r(X', F; \mathbb{Z}_2) \to H_r(X; \mathbb{Z}_2)$ (Fig. 2): given a $\mathbb{Z}_2$-homology class of the space X' mod F, we choose a representative cycle in such a way that each of its singular simplexes intersects F in a face; then in X we take the sum of all the inverse images of the simplexes of this cycle (we need not worry about orientation since the coefficients are in $\mathbb{Z}_2$). The homomorphism m is the composition of the projection and reduction mod F. Finally, the homomorphism i_r is composed of the homomorphism $\partial_* : H_r(X', F; \mathbb{Z}_2) \to H_{r-1}(F; \mathbb{Z}_2)$ and the map $H_r(X', F; \mathbb{Z}_2) \to H_{r-1}(X', F; \mathbb{Z}_2)$, which is described at the chain level as

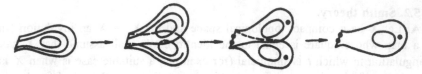

Fig. 3

the result of four successive operations: lifting, taking the boundary, lowering and reduction mod F (Fig. 3).

B. To explain the construction of the Smith sequences it will be convenient for us to consider a generalization of them, which also often turns out to be useful. Let p be a prime, X a compact topological space, $t : X \to X$ a transformation of order p ($t^p = $ id), F the fixed-point set, $X' = X/(x = tx)$. As before, we assume that X has a triangulation in which t is simplicial.

The transformation t induces a transformation of singular chains:

$$t_\# : C_q(X; \mathbb{Z}_p) \to C_q(X; \mathbb{Z}_p), \quad t_\# : C_q(X, F; \mathbb{Z}_p) \to C_q(X, F; \mathbb{Z}_p).$$

In $C_q(X; \mathbb{Z}_p)$ and $C_q(X, F; \mathbb{Z}_p)$ we consider the subsets consisting of the chains c such that $(\mathrm{id} - t_\#)^k c = 0$; these subsets are denoted by $\mathrm{Sm}_q^{(k)}(X)$ and $\mathrm{Sm}_q^{(k)}(k)(X, F)$ respectively. Clearly

$$\mathrm{Sm}_q^{(1)}(X) \subset \mathrm{Sm}_q^{(2)}(X) \subset \cdots, \quad \mathrm{Sm}_q^{(1)}(X, F) \subset \mathrm{Sm}_q^{(2)}(X, F) \subset \cdots,$$

so that $\mathrm{Sm}_q^{(1)}(X)$ and $\mathrm{Sm}_q^{(1)}(X, F)$ are just the sets of invariant chains. It is easy to see that $\mathrm{Sm}_q^{(k)}(X) = C_q(X; \mathbb{Z}_p)$ and $\mathrm{Sm}_q^{(k)}(X, F) = C_q(X, F; \mathbb{Z}_p)$ for $k \geq p$. Finally, it is clear that $\partial(\mathrm{Sm}_q^{(k)}(X)) \subset \mathrm{Sm}_q^{(k)}(X)$ and $\partial(\mathrm{Sm}_q^{(k)}(X, F)) \subset \mathrm{Sm}_q^{(k)}(X, F)$, so that as a result we have the *Smith homology groups* $\mathrm{HSm}_q^{(k)}(X)$ and $\mathrm{HSm}_q^{(k)}(X, F)$. The *Smith cohomology groups* $\mathrm{HSm}_{(k)}^q(X)$ and $\mathrm{HSm}_{(k)}^q(X, F)$ are defined similarly.

The analogues of the homology and cohomology sequences of the pair (X, F) in the Smith theory split, and this leads to the following result.

Proposition. *There are natural decompositions*

$$\mathrm{HSm}_q^{(k)}(X) = \mathrm{HSm}_q^{(k)}(X, F) \oplus H_q(F; \mathbb{Z}_p),$$
$$\mathrm{HSm}_{(k)}^q(X) = \mathrm{HSm}_{(k)}^q(X, F) \oplus H^q(F; \mathbb{Z}_p). \quad \square$$

We introduce obvious isomorphisms which will be useful later:

$$\mathrm{HSm}_q^{(1)}(X, F) = H_q(X', F; \mathbb{Z}_p), \quad \mathrm{HSm}_{(1)}^q(X, F) = H^q(X', F; \mathbb{Z}_p).$$

We now turn to the construction of the generalized Smith sequences. We fix numbers k and l with $1 \leq k < l \leq p$. Clearly

$$\mathrm{Sm}_q^{(l)}(X)/\mathrm{Sm}_q^{(k)}(X) = \mathrm{Sm}_q^{(l-k)}(X, F)$$

(the identification is established by the map $c \mapsto (\mathrm{id} - t_\#)^k c$). This equality leads (by the scheme in 1.3) to exact sequences

$$\cdots \to \mathrm{HSm}_q^{(k)}(X) \to \mathrm{HSm}_q^{(l)}(X) \to$$
$$\to \mathrm{HSm}_q^{(l-k)}(X, F) \to \mathrm{HSm}_{q-1}^{(k)}(X) \to \cdots,$$
$$\cdots \to \mathrm{HSm}_{(k)}^{q-1}(X) \to \mathrm{HSm}_{(l-k)}^q(X, F) \to$$
$$\to \mathrm{HSm}_{(l)}^q(X) \to \mathrm{HSm}_{(k)}^q(X) \to \cdots.$$

These are called the *Smith (k, l)-sequences* mod p.

Of course the most interesting cases are those in which the Smith homology and cohomology groups appearing in the Smith sequence can be expressed in terms of the ordinary homology and cohomology groups. This has only been done completely for $p = 2$. In this case we must have $k = 1, l = 2$, and our sequences reduce to those in **A**. For $p > 2$ the most interesting sequences are those with $(k, l) = (p - 1, p)$ and $(k, l) = (1, p)$; we write these out, replacing the Smith homology and cohomology groups by the ordinary ones where possible:

$$\cdots \to \mathrm{HSm}_q^{(p-1)}(X) \to H_q(X; \mathbb{Z}_p) \to$$
$$\to H_q(X', F; \mathbb{Z}_p) \to \mathrm{HSm}_{q-1}^{(p-1)}(X) \to \cdots,$$

$$\cdots \to \mathrm{HSm}_{(p-1)}^{q-1}(X) \to H^q(X', F; \mathbb{Z}_p) \to$$
$$\to H^q(X; \mathbb{Z}_p) \to \mathrm{HSm}_{(p-1)}^q(X) \to \cdots;$$

$$\cdots \to H_q(X', F; \mathbb{Z}_p) \oplus H_q(F; \mathbb{Z}_p) \to H_q(X; \mathbb{Z}_p) \to$$
$$\to \mathrm{HSm}_q^{(p-1)}(X, F) \to H_{q-1}(X', F; \mathbb{Z}_p) \oplus H_{q_1}(F; \mathbb{Z}_p) \to \cdots,$$

$$\cdots \to H^{q-1}(X', F; \mathbb{Z}_p) \oplus H^{q-1}(F; \mathbb{Z}_p) \to \mathrm{HSm}_{(p-1)}^q(X, F) \to$$
$$\to H^q(X; \mathbb{Z}_p) \to H^q(X', F; \mathbb{Z}_p) \oplus H^q(F; \mathbb{Z}_p) \to \cdots$$

C. The following statements can be deduced from the exactness of the Smith sequences in the standard way.

1°. $\mathrm{Eu}(X) - \mathrm{Eu}(F) = p(\mathrm{Eu}(X') - \mathrm{Eu}(F))$; in particular, $\mathrm{Eu}(X) \equiv \mathrm{Eu}(F) \bmod p$.

2°. For any q

$$\dim H_q(X', F; \mathbb{Z}_p) + \sum_{r=q}^{\infty} \dim H_r(F; \mathbb{Z}_p) \le \sum_{r=q}^{\infty} \dim H_r(X; \mathbb{Z}_p);$$

in particular,

$$\dim H_*(F; \mathbb{Z}_p) \le \dim H_*(X; \mathbb{Z}_p).$$

3°. If $H_r(X; \mathbb{Z}_p) = 0$ for $r \ge q$, then $H_r(F; \mathbb{Z}_p) = 0$ for $r \ge 0$. If $\tilde{H}_*(X; \mathbb{Z}_p) = 0$, then $\tilde{H}_*(F; \mathbb{Z}_p) = 0$.

$4°$. If $H_q(X; \mathbb{Z}_p) \cong H_q(S^n; \mathbb{Z}_p)$ for all q, then there exists k such that $H_q(F; \mathbb{Z}_p) \cong H_q(S^k, \mathbb{Z}_p)$ for all q; if p is odd, then $k \equiv n \bmod 2$.

$5°$. If $p = 2$ and $\dim H_*(F; \mathbb{Z}_2) = \dim H_*(X; \mathbb{Z}_2)$ then the endomorphism t_* : $H_*(X; \mathbb{Z}_2) \to H_*(X; \mathbb{Z}_2)$ is the identity and in the Smith $\mathbb{Z}_2$-homology sequence the homomorphisms s_r are epimorphisms, the homomorphisms m_r are trivial, and the homomorphisms i_r are monomorphisms.

$6°$. For any q and n

$$\sum_{k=0}^{2n}(-1)^k H_{q+k}(X; \mathbb{Z}_p)$$

$$\leq \sum_{k=0}^{2n}(-1)^k[\dim H_{q+k}(X'; \mathbb{Z}_p) + (p-1)\dim H_{q+k}(X', F; \mathbb{Z}_p)];$$

in particular

$$\dim H_q(X; \mathbb{Z}_p) \leq \dim H_q(X'; \mathbb{Z}_p) + (p-1)\dim H_q(X', F; \mathbb{Z}_p).$$

§6. Other homology and cohomology theories

6.1. The Eilenberg-Steenrod axioms.

If the calculations of the homology and cohomology of cellular spaces in §3 are analysed, it may be noticed that they only use a restricted range of properties of singular homology and cohomology and in fact do not appeal to the details of the singular construction. This suggests the idea that the homology and cohomology of cellular spaces could be defined axiomatically and that the construction with singular simplexes and chains is only one of the possible ways of realizing the system of axioms; the calculations of §3 will then provide a uniqueness theorem for these axiomatic homology and cohomology theories. Moreover, the uniqueness theorem will only include the cellular case; beyond the bounds of cellular spaces other methods of constructing homology theories may lead to different results, which could be vital for geometric applications.

In this section we shall describe the axiomatic approach to homology and cohomology theory, and also mention some other versions of this theory.

A. We say that we are given a *homology theory* if for each finite cellular pair (X, A) there is given a sequence of groups $h_q(X, A)$, $q = 0, 1, 2, \ldots$, and homomorphisms $\partial_* : h_q(X, A) \to h_{q-1}(A)$ (for brevity the pair $(B, \emptyset)$ is denoted just by B), $q = 1, 2, \ldots$, and for each continuous map $f : (X, A) \to (Y, B)$ of one cellular pair into another there is given a sequence of homomorphisms $f_* : h_q(X, A) \to h_q(Y, B)$, so that these groups and homomorphisms satisfy the following system of axioms.

Axiom 1. $(\mathrm{id})_* = \mathrm{id}$.

Axiom 2. $(fg)_* = f_* g_*$.

Axiom 3. For any continuous map $f : (X, A) \to (Y, B)$ and any $q \geq 1$, the diagram

$$h_q(X, A) \xrightarrow{\partial_*} h_{q-1}(A)$$

$$f_* \downarrow \qquad\qquad \downarrow f_*$$

$$h_q(X, B) \xrightarrow{\partial_*} h_{q-1}(B)$$

is commutative.

Axiom 4 (*axiom of exactness*). For any cellular pair (X, A) the sequence

$$\cdots \longrightarrow h_q(A) \xrightarrow{i_*} h_q(X) \xrightarrow{j_*} h_q(X, A) \xrightarrow{\partial_*} h_{q-1}(A) \longrightarrow \cdots$$

is exact, where $i : A \to X$, $j : (X, \emptyset) \to (X, A)$ are the inclusions.

Axiom 5 (*homotopy axiom*). If the maps $f, g : (X, A) \to (Y, B)$ are homotopic then the maps $f_*, g_* : h_q(X, A) \to h_q(Y, B)$ are the same for each q.

Axiom 6 (*excision axiom*). Let (X, A) be a finite cellular pair, and let U be an open subset of X consisting of whole cells and contained in A. Then the homomorphism $h_q(X \setminus U, A \setminus U) \to h_q(X, A)$ induced by the inclusion is an isomorphism for all q.

Axiom 7 (*dimension axiom*). $h_q(\mathrm{pt}) = 0$ for $q \neq 0$.

Axiom 6 is equivalent to the following axiom.

Axiom 6' (*factorization axiom*). If (X, A) is a finite cellular pair, then the homomorphism $h_q(X, A) \to h_q(X/A, \mathrm{pt})$ induced by the projection is an isomorphism for all q.

As we have already said the following theorem was essentially proved in §3.

Theorem. *If h is a homology theory, then for any finite cellular pair (X, A) there is an isomorphism $h_q(X, A) = H_q(X, A; G)$, where $G = h_0(\mathrm{pt})$. This isomorphism is compatible with the maps f_* and ∂_*.* $\qquad\square$

B. An obvious modification of what has been said in **A** gives the cohomology version of the Eilenberg-Steenrod axioms 1–7 and 6'. The uniqueness theorem for a cohomology theory satisfying these axioms is stated in precisely the same way as for homology.

C. The class of pairs (X, A) considered may be extended without detriment to **A**, by admitting pairs of CNRS spaces, as well as finite cellular pairs. We may also axiomatize homology theory for arbitrary cellular pairs that are not necessarily finite, but then we must add the following axiom to axioms 1–7 (as before axiom 6 will be equivalent to axiom 6').

Axiom 8 (*axiom of continuity*). Let (X, A) be a cellular pair, and $\{X_\alpha\}$ the directed family of all finite subspaces of X. Then

$$h_q(X, A) = \varinjlim_\alpha h_q(X_\alpha, X_\alpha \cap A).$$

If a homology theory satisfies this additional axiom, then the uniqueness theorem holds for arbitrary cellular spaces. A similar assertion is true for cohomology.

It will be convenient for us not to include axiom 8 among the Eilenberg-Steenrod axioms, so that when we speak of an Eilenberg-Steenrod theory, we shall always mean a homology or cohomology theory defined on finite cellular spaces and satisfying axioms 1–7.

6.2. An alternative construction of the Eilenberg-Steenrod homology and cohomology theory: the Aleksandrov-Čech theory.

A. An open covering of the pair (X, A) is an open covering $\{U_i \mid i \in I\}$ of X together with a subset J of I such that $\bigcup_{j \in J} U_j \supset A$. We fix such a covering and denote it by U. For $q = 0, 1, \ldots$ let $I^{(q)}$ denote the collection of ordered subsets $(i_0, i_1, \ldots, i_q)$ of $q + 1$ elements of I such that the intersection $\bigcap_{s=0}^{q} U_{i_s}$ is nonempty, and we consider the maps $\gamma_{q,s} : I^{(q)} \to I^{(q-1)}$ $(q = 0, 1, \ldots; s = 0, 1, \ldots, q)$ defined by

$$\gamma_{q,s}(i_0, \ldots, i_q) = (i_0, \ldots, \hat{i}_s, \ldots, i_q)$$

where the notation $\hat{i}_s$ means that this element is omitted. Also let $J^{(q)}$ denote the subset of $I^{(q)}$ consisting of those $(i_0, \ldots, i_q)$ for which $i_0 \in J, \ldots, i_q \in J$. Finally we fix a coefficient group G and consider the following groups:

$C_q(\mathcal{U}; G)$ is the group of formal finite linear combinations $\sum g_i S_i$, where $g_i \in G$, $S_i \in I^{(q)}$, factored by the subgroup generated by (i) linear combinations of the form $\sum g_i S_i$ with $S_i \in J^{(q)}$; (ii) expressions of the form $g(i_0, \ldots, i_q) - \operatorname{sgn} \sigma g(i_{\sigma(0)}, \ldots, i_{\sigma(q)})$, where σ is a permutation and $\operatorname{sgn} \sigma = \pm 1$ is its sign. $C^q(\mathcal{U}; G)$ is the collection of functions $\phi : I^{(q)} \to G$ such that

$$\phi(i_0, \ldots, i_q) = 0 \text{ for } (i_0, \ldots, i_q) \in J^{(q)},$$

$$\phi(i_0, \ldots, i_q) = \operatorname{sgn} \sigma \phi(i_{\sigma(0)}, \ldots, i_{\sigma(q)}).$$

These groups are called the *group of q-chains* and the *group of q-cochains* respectively of the covering $\mathcal{U}$ with coefficients in g. The boundary and coboundary operators

$$\partial : C_q(\mathcal{U}; G) \to C_{q-1}(\mathcal{U}; G) \text{ and } \delta : C^q(\mathcal{U}; G) \to C^{q+1}(\mathcal{U}; G)$$

are defined by

$$\partial[g(i_0, \ldots, i_q)] = \sum_s (-1)^s g \gamma_{q,s}(i_0, \ldots, i_q),$$

$$\delta\phi(i_0, \ldots, i_{q+1}) = \sum_s (-1)^s \phi(\gamma_{q+1,s}(i_0, \ldots, i_{q+1})).$$

The homology and cohomology groups of the resulting complexes are denoted by $H_q(\mathcal{U}; G)$ and $H^q(\mathcal{U}; G)$ respectively.

B. The following step consists of passage to the limit in the set of all coverings. Let

$$\mathcal{U} = \{\{U_i \mid i \in I\}, \ J \subset I\}, \quad \mathcal{U}' = \{\{U_{i'} \mid i' \in I'\}, \ J' \subset I'\}$$

be two coverings of the pair (X, A). The covering $\mathcal{U}'$ is said to *refine* the covering $\mathcal{U}$ (written $U' \prec U$) if there is a map $\alpha : I' \to I$ such that $\alpha(J') \subset J$ and $U_{i'}' \subset U_{\alpha(i')}$ for any $i' \in I$. The map α induces homomorphisms that commute with ∂ and δ:

$$\alpha_* : C_q(\mathcal{U}'; G) \to C_q(\mathcal{U}; G) \text{ and } \alpha^* : C^q(\mathcal{U}; G) \to C^q(\mathcal{U}'; G):$$

$$a_*[g(i'_0, \ldots, i'_q)] = \begin{cases} g(\alpha(i'_0), \ldots, \alpha(i'_q)) & \text{if } \alpha(i'_0), \ldots, \alpha(i'_q) \\ & \text{are all different,} \\ 0 & \text{otherwise.} \end{cases}$$

$$[\alpha^*\phi](i'_0, \ldots, i'_q) = \begin{cases} \phi(\alpha(i'_0), \ldots, \alpha(i'_q)) & \text{if } \alpha(i'_0), \ldots, \alpha(i'_q) \\ & \text{are all different,} \\ 0 & \text{otherwise.} \end{cases}$$

(here it is important that $U_{i'_0} \cap \ldots \cap U_{i'_q}$ nonempty implies that $U_{\alpha(i'_0)} \cap \ldots \cap U_{\alpha(i'_q)}$ is nonempty). Consequently there are homomorphisms

$$H_q(\mathcal{U}'; G) \to H_q(\mathcal{U}; C) \text{ and } H^q(\mathcal{U}; G) \to H^q(\mathcal{U}'; C),$$

and it is easy to prove that they depend only on $\mathcal{U}$ and $\mathcal{U}'$ (and not on the choice of α). Moreover it is very easy to see that the partial order of the set of coverings is directed, that is, for any coverings $\mathcal{U}_1$ and $\mathcal{U}_2$ there is a covering $\mathcal{V}$ such that $\mathcal{V} \prec \mathcal{U}_1$ and $\mathcal{V} \prec \mathcal{U}_2$. This allows us to take limits and we put

$$\check{H}_q(X, A; G) = \lim_{\leftarrow \mathcal{U}} H_q(U; G),$$

$$\check{H}^q(X, A; G) = \lim_{\leftarrow \mathcal{U}} H^q(U; G).$$

The groups $\check{H}_q(X, A; G)$ and $\check{H}^q(X, A; G)$ are called the *Aleksandrov-Čech homology and cohomology groups of the pair* (X, A).

(Hence an element of the group $\check{H}(X, A : G)$ is a collection of elements $\alpha_{\mathcal{U}}$ of the groups $H_q(\mathcal{U}; G)$ given for all $\mathcal{U}$ in such a way that if $\mathcal{U}' \prec \mathcal{U}$, then $\alpha_{\mathcal{U}'} \to \alpha_{\mathcal{U}}$ under the resulting homomorphism $H_q(\mathcal{U}'; G) \to H_q(\mathcal{U}; G)$. An element of $H^q(X, A; G)$ is defined by an element γ of a single group $H^q(\mathcal{U}; G)$, but here $\gamma_1 \in H^q(\mathcal{U}_1; G)$ and $\gamma_2 \in H^q(\mathcal{U}_2; G)$ define the same element in $H^q(X, A; G)$ if there exists $\mathcal{V}$ such that $\mathcal{V} \prec \mathcal{U}_1$, $\mathcal{V} \prec \mathcal{U}_2$ and γ_1 and γ_2 have the same image in $H^q(\mathcal{V}; G)$. Incidentally the notation $\check{H}$ for Aleksandrov-Čech homology and cohomology is in general use, and derives from the initial letter of the name Čech.)

C. The definition of Aleksandrov-Čech homology and cohomology may seem very cumbersome, but in fact each of the two stages (**A** and **B**) can be clarified considerably. We begin with the fact that a covering $\mathcal{U}$ of the pair (X, A) can be naturally associated with a pair consisting of a triangulated space and a subspace. Namely, the set I will be the set of vertices (0-simplexes) of our space, and a finite subset $(i_0, \ldots, i_q) \subset I$ is the vertex set of a q-simplex if and only if the intersection $U_{i_0} \cap \ldots \cap U_{i_q}$ is nonempty. The resulting space is called the *nerve* of the covering $\{U_i\}$. The union of the simplexes all of whose vertices belong to J forms a subspace B of Y. It is easy to see that the homology and cohomology of $\mathcal{U}$ is the same as that of the pair (Y, B) calculated by means of a cell complex. On the other hand, if the covering $\{U_i\}$ is sufficiently fine, and the set J is chosen in a sufficiently economical way (this means that the union $\bigcup_{j \in J} U_j$ is sufficiently close to A), then the pair (Y, B) is a good approximation to (X, A). In addition, if all the nonempty finite intersections $U_{i_0} \cap \ldots \cap U_{i_q}$ are contractible, then it can be shown that (Y, B) is homotopy equivalent to (X, A) (convince yourself of this in the example when $(X, A) = (S^1, \text{pt})$, the covering $\mathscr{U}$ consists of three intervals, and the set J consists of a single element – see Fig. 4). Thus with an appropriate choice of covering the passage to the limit is not necessary: the Aleksandrov-Čech homology and cohomology is the same as the homology and cohomology of this covering.

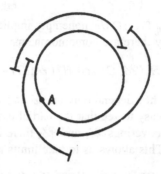

Fig. 4

D. Theorem. *The Aleksandrov-Čech theory satisfies the Eilenberg-Steenrod axioms 1–7, as well as axiom 8. Hence for any (finite or infinite) cellular pair (X, A) and any Abelian group G,*

$$\check{H}_q(X, A; G) = H_q(X, A; G) \text{ and } \check{H}^q(X, A; G) = H^q(X, A; G). \ \square$$

The proof of this theorem can be found in Eilenberg and Steenrod (1952), Chapter IX.

Remarks. 1°. For certain pairs and even certain spaces the singular homology and cohomology may be different from the Aleksandrov-Čech homology and cohomology. The simplest example can be constructed from the graph of the function $\sin 1/x$ in the plane (see Fig. 13 in Part I, 9.4.F; for the space X shown there the one-dimensional singular homology and cohomology is trivial, while the corresponding Aleksandrov-Čech groups are nontrivial). In some cases this difference is to the advantage of the Aleksandrov-Čech theory; for example, the Smith theory (see 5.2) turns out to be applicable to a wider class of $\mathbb{Z}_p$-spaces if the homology and cohomology groups in the Smith sequence are defined in the Aleksandrov-Čech sense.

2°. A considerably stronger continuity property than that of Axiom 8 in 6.1 holds for Aleksandrov-Čech homology and cohomology: the relations $\check{H}_q(\varinjlim(X_\alpha, A_\alpha);$ $G) = \varinjlim\limits_{\alpha} \check{H}^q(X_\alpha, A_\alpha; G)$ and $\check{H}^q(\varinjlim(X_\alpha, A_\alpha); G) = \varinjlim\limits_{\alpha} \check{H}^q(X_\alpha, A_\alpha; G)$ are true for arbitrary direct spectra of topological pairs; simple examples show that this is not true for the singular theory. On the other hand, for non-cellular pairs the Aleksandrov-Čech homology and cohomology sequences turn out to be in general only half-exact (that is, the composition of successive homomorphisms is trivial), but not exact; this is a deficiency of the Aleksandrov-Čech theory compared with the singular theory. In Chapter X of Eilenberg and Steenrod (1952) it is explained that the properties of exactness and continuity contradict each other if they are understood too widely, so that the singular theory and the Aleksandrov-Čech theory choose, each in their own way, the lesser of two evils (lack of exactness or lack of continuity).

6.3. Extraordinary theories.

A. Since the end of the 1950s "extraordinary" homology and cohomology theories have played an important role in topology; these differ from the usual theories in that they satisfy an axiom system omitting the dimension axiom (recall that from the outset we have not included the continuity axiom in our system). In addition when speaking of extraordinary homology and cohomology theories it is advisable to assume that the homology and cohomology groups are indexed by all the integers, not just the non-negative ones; this changes nothing in the Eilenberg-Steenrod theory since it follows from the axioms that the groups indexed by the negative integers are all trivial.

Thus an extraordinary homology theory assigns to each finite cellular space X a sequence of groups $h_n(X)$, $n \in \mathbb{Z}$, and to each finite cellular pair (X, A) a sequence of groups $h_n(X, A)$, $n \in \mathbb{Z}$, such that $h_n(X, \emptyset) = h_n(X)$; the induced homomorphisms and connecting homomorphisms are defined, the sequences of a pair are exact, and the homotopy and excision axioms are satisfied. The groups $h_n(\mathrm{pt})$, where pt is a one-point space, can in principle be arbitrary. Extraordinary cohomology theories (which in fact are much more important than homology theories) are defined similarly.

The following theorem replaces the uniqueness theorem for extraordinary homology and cohomology theories.

Theorem. *Let h and k be two extraordinary homology or cohomology theories and let $\phi : h \to k$ be a morphism of the first to the second (to give a morphism from one theory to another means to give for each cellular space and each cellular pair a map of the groups of one theory to the corresponding groups of the other theory in such a way that these maps commute with the induced and connecting homomorphisms). If this morphism is an isomorphism for a one-point space, then it is an isomorphism for any cellular space and any cellular pair.* □

The proof consists of some obvious steps. First of all, because the maps that make up the morphism ϕ commute with the suspension isomorphisms (which exist in any theory), it follows that ϕ defines isomorphisms for spheres. Then in an obvious way we can pass to (finite) bouquets of spheres. Then by considering the h- and k-sequences of a pair of the form $(\mathrm{sk}_n X, \mathrm{sk}_{n-1} X)$ and using the fact that $\mathrm{sk}_n X / \mathrm{sk}_{n-1} X$ is a bouquet of spheres, we prove that ϕ defines an isomorphism between the h- and k-groups for any finite cellular space X. The extension to cellular pairs follows from the factorization axiom or by means of the 5-lemma.

Remark. For the theories h and k to be isomorphic it is not sufficient that there exist abstract isomorphisms $h_n(pt) \cong k_n(\mathrm{pt})$ (not induced by a morphism $\phi : h \to k$); examples will appear below.

B. An instructive example of an extraordinary homology theory is provided by the cubical singular theory. A *singular q-dimensional cube* of a topological space

X is a continuous map $i^q \to X$, where $I^q = \{(x_1, \ldots, x_q) \in \mathbb{R}^q \mid 0 \le x_i \le 1\}$ is the standard q-cube. A formal linear combination $\sum k_i f_i$ of singular q-cubes with integer coefficients is called a *cubical singular q-chain* of X. The cube I^q has $2q(q-1)$-dimensional faces:

$$I_{i,\varepsilon}^{q-1}\{(x_1, \ldots, x_q) \in I^q \mid x_i = \varepsilon\} \quad (i = 1, \ldots, q; \ \varepsilon = 0, 1),$$

which can be naturally identified with the cube I^{q-1}. For a singular cube f we put

$$\Gamma_{i,\varepsilon} f = f\big|_{I_{i,\varepsilon}^{q-1}}, \quad \partial f = \sum_{i,\varepsilon} (-1)^i (-1)^\varepsilon \Gamma_{i,\varepsilon} f,$$

and for a chain $c = \sum k_i f_i$,

$$\partial c = \sum k_i \partial f_i.$$

It is easy to verify that $\partial\partial = 0$, so that we obtain the *cubical singular homology group* $H_q^\square(X)$ of X. The whole theory of singular homology set out in §2 can be carried over word for word to the cubical case. In particular, relative cubical homology is defined, the exactness of the sequence of a pair can be proved, as well as homotopy invariance and the excision theorem (in precisely the same formulations as in §2). However, computation of the homology of a point gives

$$H_q^\square(\text{pt})) = \begin{cases} \mathbb{Z} & \text{for } q \ge 0, \\ 0 & \text{for } q < 0 \end{cases}$$

(for each $q \ge 0$ there is exactly one cubical singular simplex and its boundary is equal to 0). Hence the cubical singular theory is an extraordinary homology theory. (Cubical singular homology and cohomology theories with arbitrary coefficients can be defined similarly, and they too are extraordinary.)

The cubical singular theory appeared long before the beginning of the boom in extraordinary theories, and its extraordinariness was regarded as an annoying defect rather than an interesting property. A method of avoiding this defect was devised: for this it was necessary to factor the group of singular cubical chains by the subgroup generated by "degenerate" singular cubes; a singular cube $f : I^q \to x$ with $q > 0$ is called degenerate if $f(x_1, \ldots, x_q)$ does not depend on the coordinate x_q. After factoring out this subgroup the theory becomes equivalent to the usual singular theory. However, singular cubical homology theory is of no interest as an extraordinary theory; more interesting examples of extraordinary theories will appear below in **C** and **D**.

C. Of the extraordinary theories, K-theory has the most striking applications (both inside and outside topology), and indeed its study stimulated the consideration of general extraordinary homology and cohomology theories. There are several versions of K-theory; we shall consider the *complex* or *unitary K-functor*.

Let X be a topological space. Let $F_n(X)$, $n \ge 0$, denote the set of classes of equivalent n-dimensional complex vector bundles over X, and put $F(X) = \sqcup_n F_n(X)$. The set $F(X)$ forms an Abelian semigroup with zero relative to the addition in the bundles, but of course is not a group: we cannot for example subtract a bundle from a bundle of higher dimension. We form the corresponding *Grothendieck group* $K(X)$. Thus, $K(X)$ is the set of classes of formal differences $a - b$ where $a, b \in F(X)$ and $a - b \sim c - d$ if there exists $e \in F(X)$ such that $a \oplus d \oplus e = b \oplus c \oplus e$. The operation in $K(X)$ is denoted by $+$. The elements of the group $K(X)$ are called *virtual bundles*. Note that an ordinary bundle defines a virtual bundle ($\xi \mapsto \xi - 0$), but nonequivalent bundles of the same dimension may define equivalent virtual bundles (for example, if $\xi_1 \not\sim \xi_2$, but $\xi_1 \oplus \eta \sim \xi_2 \oplus \eta$ for some η, then $\xi_1 - 0 \sim \xi_2 - 0$). The map dim : $F(X) \to \mathbb{Z}_+$ extends to a homomorphism $K(X) \to \mathbb{Z}$, also denoted by dim. (Note that the dimension of a virtual bundle may be negative.) The kernel of this homomorphism is denoted by $\tilde{K}(X)$, and it is clear that $K(X) = \mathbb{Z} \oplus \tilde{K}(X)$. The group $\tilde{K}(X)$ has a particularly simple meaning when X is a finite cellular space; in this case $\tilde{K}(X)$ is the group of classes of stably equivalent complex vector bundles over X.

Given a continuous map $f : X \to Y$, the induced bundle operation enables us to construct a map $f^* : F(Y) \to F(X)$ and a homomorphism $f^* : K(Y) \to K(X)$.

For a topological pair (X, A) and an integer $n \leq 0$, we put

$$K^n(X, A) = \tilde{K}(\Sigma^{-n}(X/A)),$$

where Σ is the suspension. We emphasize that $X/\emptyset$ is pt $\sqcup X$, so that

$$K^n(X) = \tilde{K}(\Sigma^{-n}(\text{pt} \sqcup X)) = \tilde{K}(S^{-n} \vee \Sigma^{-n}X),$$

and in particular

$$K^0(X) = \tilde{K}(\text{pt} \sqcup X) = K(\text{pt}) \oplus \tilde{K}(X) = \mathbb{Z} \oplus \tilde{K}(X) = K(X)$$

(a natural isomorphism). A key feature in the construction of K-theory is the definition of $K^n(X, A)$ for $n > 0$, which is done by means of *Bott periodicity*. This is in fact an assertion about the homotopy groups of classical Lie groups, but it can also be expressed in our present language.

Theorem (Bott). *If X is a cellular space, then*

$$\tilde{K}(X) \cong \tilde{K}(\Sigma^2 X);$$

this isomorphism can be made natural with respect to X. $\square$

It follows from this theorem that for any cellular pair (X, A)

$$K^n(X, A) = K^{n-2}(X, A),$$

and this allows us to extend the definition of the functor K^n to the case $n > 0$:

$$K^n(X, A) = K^{n-2m}(X, A), \text{ where } m > n/2.$$

If we restrict ourselves to cellular pairs it is easy to show that the collection $\{K^n(X, A)\}$ satisfies the Eilenberg-Steenrod (cohomology) axioms, except, of course, the dimension axiom:

$$K^n(\text{pt}) = \begin{cases} \mathbb{Z} & \text{for even } n, \\ 0 & \text{for odd } n. \end{cases}$$

(We note that attempts to extend K-theory to arbitrary cellular pairs that are not necessarily finite encounter difficulties: there are several competing methods of defining the K-functor on infinite cellular pairs, but none of them gives a theory satisfying the Eilenberg-Steenrod axioms 1–6 and 8; see Atiyah (1965).)

Thus K-theory is an extraordinary cohomology theory. We shall not go more deeply into its structure, which the reader can learn from the books Karoubi (1978), Mishchenko (1984), Fomenko and Fuchs (1989), Atiyah (1965). We only remark that K-theory is different from the theory defined by

$$h^n(X, A) = \bigoplus_{k \in \mathbb{Z}} H^{n+2k}(X, A; \mathbb{Z}),$$

although $h^n(\text{pt}) = K^n(\text{pt})$ (for example, $h^0(\mathbb{R}P^4) = \mathbb{Z} \oplus \mathbb{Z}_2$, but $k^0(\mathbb{R}P^4) = \mathbb{Z} \oplus \mathbb{Z}_4$. An example of this kind was promised in **B**.

We note also that real or orthogonal K-theory can be defined in exactly the same way but with real instead of complex vector bundles, and has a similar structure but is rather more complicated.

D. The so-called *bordisms* provide an example of an extraordinary homology theory. Its definition is very natural so it is not surprising that Poincaré invented it before homology. However, Poincaré's invention was before its time, and the definition was discarded as clearly unsuccessful. Topologists returned to it half a century later when they were at last in a position to understand its true significance.

Let X be a topological space. An *n-dimensional singular (oriented) manifold* in X is a pair consisting of an n-dimensional closed oriented smooth manifold M and a continuous map $f : M \to X$. Two singular n-manifolds (M_1, f_1) and (M_2, f_2) are said to be *bordant* if there exists an $(n+1)$-dimensional compact oriented smooth manifold N, a map $F : N \to X$ and a diffeomorphism $\phi : M_1 \sqcup M_2 \to \partial N$ (where ∂N is the boundary of N), preserving orientation on M_2 and reversing it on M_1, such that $(F \circ \phi)|_{M_1} = f_1$ and $(F \circ \phi)|_{M_2} = f_2$. The set of classes of bordant singular n-manifolds is denoted by $\Omega_n(X)$. There is a natural group structure on $\Omega_n(X)$: the sum of the singular manifolds (M_1, f_1), (M_2, f_2) is by definition $(M_1 \sqcup M_2, f)$, where $f|_{M_i} = f_i$, the negative corresponds to change of orientation, and the zero element is the empty singular manifold. It is easy to show that the groups $\Omega_0(X)$ and $\Omega_1(X)$ are the same as $H_0(X)$ and $H_1(X)$; in fact, $\Omega_n(X) = H_n(X)$ for $n > 4$. But for $n = 4$ a difference appears: suppose, for example, that X is a point; take $M = \mathbb{C}P^2$ and $f : M \to X$ the unique map. Then the singular manifold (M, f) is not bordant to 0 simply because $\mathbb{C}P^2$ is not the boundary of any manifold (see §8 below); hence $\Omega_4(X) \neq 0$ (in fact $\Omega_4(X) = \mathbb{Z}$), while $H_4(X) = 0$. Thus the bordism groups differ from the homology groups, but an extraordinary homology theory can be obtained from them: it is sufficient to put (for cellular X, A) $\Omega_n(X, A) = 0$ for $n < 0$, $\Omega_n(X, A) = \tilde{\Omega}_n(X/A) = \text{Ker}(\Omega_n(X/A) \to \Omega_n(\text{pt}))$ for $n \geq 0$.

Other versions of bordism theory may be obtained by taking non-oriented manifolds or manifolds with additional structures, for example, almost complex manifolds. There are also numerous theories of bordisms with singularities. There are many books on the subject, for example Snaith (1979) and Stong (1968).

E. There is a general method of constructing extraordinary homology and co-homology theories, which in fact enable any such theory to be constructed. The basis of this method is the concept of a *spectrum* (or Ω-spectrum).

By a spectrum we mean a sequence of cellular spaces W_n ($n = 1, 2, \ldots$) with base points and continuous maps $h_n : W_n \to \Omega W_{n+1}$ ($n = 1, 2, \ldots$) that map base points to base points (Ω is the operator that forms the loop space). Let Y be a space with base point. For $k \in \mathbb{Z}$ consider the sequence

$$\cdots \longrightarrow (Y, \Omega^{n-k} W_n) \xrightarrow{\phi_n} \pi(Y, \Omega^{n+1-k} W_n) \longrightarrow \cdots,$$

where π denotes the set of pointed homotopy classes of maps (the sequence begins with $n = \max(1, k)$), and ϕ_n takes the class of a map $g : Y \to \Omega^{n-k} W_n$ to the class of the composition

$$Y \xrightarrow{g} \Omega^{n-k} W_n \xrightarrow{\Omega^{n-k} h_n} \Omega^{n+1-k} W_n.$$

For $n \geq k + 1$ the set $\pi(Y, \Omega^{n-k} W_n)$ is a group, and for $n \geq k + 1$ an Abelian group (see Part I, 2.12). Hence the limit

$$\tilde{H}^k(Y; W) = \varinjlim_n [\pi(Y, \Omega^{n-k} W_n), \phi_n]$$

is an Abelian group for any k. For a finite cellular pair (X, A) we put

$$H^k(X, A; W) = \tilde{H}^k(X/A; W).$$

A map $f : Y_1 \to Y_2$ obviously induces a map (homomorphism)

$$f^* : \pi(Y_2, \Omega^{n-k} W_n) \longrightarrow \pi(Y_1, \Omega^{n-k} W_n),$$

which allows us to construct the induced maps for the groups $H^k(X, A; W)$. The connecting homomorphism

$$H^k(A; W) \longrightarrow H^{k+1}(X, A; W)$$

is induced by the composite map

$$X/A \sim X \cup CA \longrightarrow (X \cup CA)/X =$$
$$= \Sigma A \longrightarrow S^1 \vee \Sigma A = \Sigma(\mathrm{pt} \sqcup A) : \pi(\mathrm{pt} \sqcup A, \Omega^{n-k} W_n) =$$
$$= \pi(\Sigma(\mathrm{pt} \sqcup A), \Omega^{n-1-k} W_n) \longrightarrow \pi(X/A, \Omega^{n-(k+1)} W_n)$$

[here we use the universal equality $\pi(A, \Omega B) = \pi(\Sigma A, B)$]. The groups $H^k(X, A; W)$ are easily seen to constitute an extraordinary cohomology theory; in particular

$$H^k(\mathrm{pt}; W) = \varinjlim_n \pi(\mathrm{pt} \sqcup \mathrm{pt}, \Omega^{n-k} W_n) = \varinjlim_n (S^0, \Omega^{n-k} W_n) =$$

$$= \varinjlim_n \pi(\Sigma^{n-k} S^0, W_n) = \varinjlim_n \pi(S^{n-k}, W_n) = \varinjlim_n \pi_{n-k}(W_n).$$

These groups are *called cohomology groups with coefficients in the spectrum* $W = \{W_n, h_n\}$. Homology with coefficients in a spectrum is also defined:

$$H_k(X, A; W) = \varinjlim_n \pi_{n+k}((X/A)\#W_n).$$

This is an extraordinary homology theory with

$$H_k(\text{pt}; W) = \varinjlim_n \pi_{n+k}(W_n).$$

It can be shown that any kind of homology and cohomology can be represented as homology and cohomology with coefficients in an appropriate spectrum. Thus ordinary homology and cohomology with coefficients in G is homology and cohomology with coefficients in the Eilenberg-MacLane spectrum $\{K(G, n), h_n\}$, where $h_n : K(G, n) \to \Omega K(G, n + 1)$ is the standard homotopy equivalence. The spectrum corresponding to K-theory has the form

$$U, BU, U, BU, U, BU, \ldots$$

where $U = U(\infty) = \bigcup_n U(n)$ is the infinite unitary group, BU its classifying space; the map $U \to \Omega BU$ in the spectrum is the standard homotopy equivalence, but the map $BU \to \Omega U \to \Omega^2 BU$ is defined in a more delicate way using Bott periodicity. Bordism theory corresponds to the Thom spectrum

$$MSO(1), MSO(2), MSO(3), \ldots,$$

where $MSO(n)$ is the Thom space of the universal vector $SO(n)$-bundle. The map

$$MSO(n) \longrightarrow \Omega MSO(n + 1),$$

that is,

$$MSO(n) \longrightarrow MSO(n + 1)$$

is defined as follows: $MSO(n)$ is the Thom space of the sum of the universal vector $SO(n)$-bundle and a 1-dimensional trivial bundle, and this sum is canonically mapped to the universal vector $SO(n + 1)$-bundle. We emphasize that bordism is the extraordinary homology theory associated with the Thom spectrum; the corresponding cohomology theory, called *cobordism*, plays an extremely important role in topology (see Stong (1968)). Among other spectra we mention the sphere spectrum

$$S^1, S^2, S^3, S^4, S^5, \ldots$$

(with the standard map $S^n \to \Omega S^{n+1}$). The corresponding extraordinary homology groups are the *stable homotopy groups* well known to topologists; the corresponding extraordinary cohomology groups are called the *stable cohomotopy groups*.

Further information about extraordinary homology and cohomology theories can be found in the literature cited here, to which we add the books Dold (1972) and Switzer (1975).

6.4. Homology and cohomology with local coefficients.

These homology and cohomology theories are neither Eilenberg-Steenrod theories nor extraordinary theories, although they have properties close to the axioms in 6.1. The difference is that they are related not just to a topological space, but to a topological space with an additional structure in the form of a "local system of Abelian groups".

A. A *local system* of Abelian groups on a topological space X is defined as follows. With each point $x \in X$ there is associated an Abelian group G_x, and for each path $s : I \to X$ joining $x \in X$ and $y \in Y$ there is an isomorphism $\phi_s : G_x \to G_y$, such that homotopic paths give rise to the same isomorphism, the product of paths corresponds to the product of isomorphisms (and hence the constant path corresponds to the identity isomorphism, and mutually inverse paths to mutually inverse isomorphisms). In particular, the automorphism $\phi_s : G_x \to G_x$ corresponding to a loop with origin at x defines an action of the group $\pi_1(X, x)$ on the group G_x. It turns out that for a (path-) connected space X all local systems are determined (up to equivalence) by this group and its action:

Proposition. *For a connected space X the preceding construction establishes a one-one correspondence*

$$\begin{array}{ccc} \text{equivalence classes of local} & & \text{Abelian groups endowed with} \\ \text{systems of Abelian groups on } X & \leftrightarrow & \text{an action of } \pi_1(X, x). \end{array} \quad \Box$$

Proof. $1°$. Let G be an Abelian group with an action of $\pi_1(X, x)$. For each point $y \in X$ fix (without regard to continuity) a path σ_y joining it to x. Then put $G_y = G$ for all $y \in X$ and define the isomorphism $\phi_s : G_{s(0)} \to G_{s(1)}$ as the transformation $G \to G$ produced by the class of the loop $\sigma_{s(0)} s \sigma_{s(1)}^{-1}$.

$2°$. Let $\{G_y, \phi_s\}$ and $\{H_y, \phi_s\}$ be two local systems such that for some fixed point $x \in X$ the groups G_x and H_x are related by an isomorphism $\alpha_x = \alpha$, compatible with the action of $\pi_1(X, x)$. For any point $y \in X$, let α_y denote the composition

$$G_y \xrightarrow{\phi_s^{-1}} G_x \xrightarrow{\alpha_x} H_x \xrightarrow{\psi_s} H_y,$$

where s is an arbitrary path from x to y. The independence of α_y from the path s follows from the compatibility of α_x with the action of $\pi_1(X, x)$. It is easy to verify that the α_y are isomorphisms, which establishes the equivalence of the local systems.

Examples. $0°$. The "trivial local system": $G_x \equiv G$, $\phi_s \equiv$ id.

$1°$. A large number of examples of local systems can be derived from the preceding proposition. For example, if $\pi_1(X) = \mathbb{Z}_2$ ($X = \mathbb{R}P^n$ with $n \geq 2$, say, or $SO(n)$ with $n \geq 3$), then we may take G to be $\mathbb{Z}$ and multiplication by -1 to be the automorphism $\mathbb{Z} \to \mathbb{Z}$ produced by the nonzero element of $\pi_1(X)$. The resulting local system is denoted by $\mathbb{Z}_T$ (T for "twisted", and the system is called the *twisted integers*).

$2°$. Let X be a connected smooth manifold. For $x \in X$ let $\mathbb{Z}_x$ denote the group (isomorphic to $\mathbb{Z}$) generated by the (two-element) set of orientations at x with the relation: the sum of the generators is equal to 0. A path joining points x and y establishes a correspondence between the orientations at these points, and hence defines an isomorphism $\mathbb{Z}_x \to \mathbb{Z}_y$. The resulting local system is called the *local system of orientations of the manifold X*, and is denoted by OGX. (To describe it in terms of the fundamental group, it consists of the group $\mathbb{Z}$ on which an element of the fundamental group determines the identity transformation if it preserves orientation, and multiplication by -1 if it reverses orientation.)

$3°$. Let X be the base of a Serre bundle with a connected and simply connected (or homotopically simple) fibre. Then the nth homotopy groups of the fibres for any $n \geq 2$ form a local system of Abelian groups on X. The same holds for any homology groups of the fibres, but in this case we do not require the fibres to be simply connected or even homotopically simple.

B. Suppose we are given a local system $\mathfrak{G} = \{G_x, \phi_s\}$ of Abelian groups on a space X. A singular q-chain with coefficients in the local system is defined to be a finite linear combination $\sum_i g_i f_i$, where $f_i : T^q \to X$ is a singular simplex, $g_i \in G_{f_i(C_q)}$, $c_q = (1/(q+1), \ldots, 1/(q+1))$, and the resulting group is denoted by $C_q(X; \mathfrak{G})$. The boundary operator $\partial : C_q(X; \mathfrak{G}) \to C_{q-1}(X; \mathfrak{G})$ is defined by

$$\partial\left(\sum_i g_i f_i\right) = \sum_i \sum_{r=0}^{q} \phi_{f_i \circ l_{q,r}}(g_i) \Gamma_r f_i,$$

where $l_{q,r}$ denotes the rectilinear path joining the centre c_q of the standard simplex T^q to the centre $c_{q,r} = (1/q, \ldots, 1/q, 0, 1/q, \ldots, 1/q)$ of its rth face. The chains with this boundary operator form a complex whose homology is called the *homology with coefficients in the local system* $\mathfrak{G}$, denoted by $H_q(X; \mathfrak{G})$. The cochain group $C^q(X; \mathfrak{G})$ and cohomology groups $H^q(X; \mathfrak{G})$ are defined similarly: a q-chain with coefficients in $\mathfrak{G}$ is a function c taking a singular simplex $f : T^q \to X$ to an element of the group $G_{f_i(C_q)}$, and the coboundary operator δ is defined by

$$\delta c(f) = \sum_{r=0}^{q+1} \phi_{f \circ l_{q+1,r}}^{-1}(c(\Gamma_r f)).$$

Homology and cohomology with coefficients in the trivial system $\{G, \text{id}\}$ is just homology and cohomology with coefficients in G, but with a nontrivial local system we obtain new homology and cohomology in general; for example:

Proposition. $H_0(X; \mathfrak{G})$ *is the factor group of* $G = G_x$ *by the subgroup generated by elements of the form* $g - \alpha g$ $(g \in G, \alpha \in \pi_1(X, x))$; $H^0(X; \mathfrak{G})$ *is the subgroup of* $G = G_x$, *consisting of those g for which* $\alpha g = g$ *for any* $\alpha \in (X, x)$. $\square$

The proof is obvious.

C. We shall indicate a method of describing homology and cohomology with local coefficients in an algebraically more convenient way. Recall that a local system (on a connected space X) can be specified by giving an Abelian group G and an action of the fundamental group $\pi_1(X, x)$ on it. The fundamental group also acts on the universal covering $\hat{X}$ of X, and hence on the groups of singular chains of $\hat{X}$. We put

$$C_q(X; \mathfrak{G}) = C_q(\hat{X}) \underset{\pi_1(X,x)}{\otimes} G,$$

$$C^q(X; \mathfrak{G}) = \text{Hom}_{\pi_1(X,x)}(C_q(\hat{X}), G)).$$

In other words, $C_q(X; \mathfrak{G})$ is the factor group of $C_q(\hat{X}; G) = C_q(\hat{X}) \otimes G$ by the subgroup generated by elements of the form $c \otimes \alpha g - \alpha^{-1} c \otimes g$ ($c \in C_q(\hat{X})$, $g \in G$, $\alpha \in \pi_1(X, x)$), and $C^q(X; \mathfrak{G})$ is the subgroup of $C^q(\hat{X}; G) = \mathrm{Hom}(C^q(\hat{X}), G)$ consisting of those homomorphisms $f : C_q(\hat{X}) \to G$, such that $f(\alpha c) = \alpha f(c)$ ($c \in C_q(\hat{X})$, $\alpha \in \pi_1(X, x)$). Moreover, the complexes $\{C_q(X; \mathfrak{G})\}$ and $\{C^q(X; \mathfrak{G})\}$ are respectively a factor complex and a subcomplex of the complexes $\{C(\hat{X}; G)\}$ and $\{C^q(\hat{X}; G)\}$ of singular chains and cochains of the space $\hat{X}$.

D. Homology and cohomology with local coefficients can be calculated, like ordinary homology and cohomology, by means of a cell decomposition. In fact, a cellular q-chain of a cellular space X with coefficients in a local system $\mathfrak{G} = \{G_x, \phi_s\}$ is a formal finite linear combination $\sum g_i \sigma_i$ of q-cells of X, where $g_i \in G_x$, $x \in \sigma_i$ (the groups associated with different points of the same cell can be regarded as identical, since these points can be joined inside the cell by a loop uniquely defined up to homotopy). Cellular q-cochains with coefficients in G are defined as functions on the set of q-cells whose value on a cell σ lies in the group G_x, $x \in \sigma$. The groups of cellular q-chains and q-cochains are denoted by $\mathscr{C}(X; \mathfrak{G})$ and $\mathscr{C}^q(X; \mathfrak{G})$. The boundary and coboundary operators

$$\partial : \mathscr{C}_q(X; \mathfrak{G}) \to \mathscr{C}_{q-1}(X; \mathfrak{G}),$$
$$\delta : \mathscr{C}^q(X; \mathfrak{G}) \to \mathscr{C}^{q1}(X; \mathfrak{G})$$

can be most simply described by representing $\mathscr{C}_q(X; \mathfrak{G})$ and $\mathscr{C}^q(X; \mathfrak{G})$ in the spirit of **C** as a factor group and a subgroup of the groups of cellular chains and cochains of the universal covering of X. There is also a direct description. Namely, for a q-cell $\sigma \subset X$ we fix a characteristic map $f : D^q \to X$ and assume that a point y in a $(q - 1)$-cell τ is a regular value for f. This means that y has a neighbourhood U in τ, whose inverse image in $S^{q-1} \subset D^q$ is the union of N balls homeomorphically mapped onto U (with orientation preserved or reversed). We join the point $f^{-1}(x) \in D^q$, where x is an arbitrarily chosen point of σ, by rectilinear paths to all the points of the set $f^{-1}(y)$; the images of these paths will be paths $s_1, \ldots, s_N$, joining x to y. Then for $g \in G_x$ we put

$$\partial(g\sigma) = \sum_{i=1}^{N} \varepsilon_i \phi_{s_i}(g)\tau,$$

where $\varepsilon_i = +1$ or -1 depending on whether f preserves or reverses orientation on the ith of the N balls that make up U. (For $q = 0, 1$, this definition requires a small modification which we leave to the reader.) The coboundary operator is defined similarly.

E. *Example.* Let $\mathfrak{G}$ be a local system of Abelian groups on projective space $\mathbb{R}P^n$, $n > 1$, defined by a group G on which the generator of the fundamental group acts by an automorphism t (if $n \geq 2$, then $t^2 = \mathrm{id}$. Then

$$\mathscr{C}_q(\mathbb{R}P^n; \mathfrak{G}) = \mathscr{C}^q(\mathbb{R}P^n; \mathfrak{G}) = G \text{ for } q = 0, 1, \ldots, n$$

(we use the standard cell decomposition of $\mathbb{R}P^n$), and the preceding description of the boundary and coboundary operators shows that the maps

$\partial : \mathscr{C}_q(\mathbb{R}P^n; \mho) \to \mathscr{C}_{q-1}(\mathbb{R}P^n; \mho)$ and $\delta : \mathscr{C}^{q-1}(\mathbb{R}P^n; \mho) \to \mathscr{C}^q(\mathbb{R}P^n; \mho)$

$(1 \le q \le n)$ are $G \xrightarrow{1-t} G$ for even q, and $G \xrightarrow{1+t} G$ for odd q. Hence

$$H_q(\mathbb{R}P^n; \mho) = \begin{cases} G/\mathrm{Im}(1-t) & \text{if } q = 0, \\ \mathrm{Ker}(1+(-1)^q t)/\mathrm{Im}(1-(-1)^q t) & \text{if } 0 < q < n, \\ \mathrm{Ker}(1+(-1)^n t) & \text{if } q = n; \end{cases}$$

$$H^q(\mathbb{R}P^n; \mho) = \begin{cases} \mathrm{Ker}(1-t) & \text{if } q = 0, \\ \mathrm{Ker}(1+(-1)^q t)/\mathrm{Im}(1+(-1)^q t) & \text{if } 0 < q < n, \\ G/\mathrm{Im}(1+(-1)^n t) & \text{if } q = n \end{cases}$$

(compare these results with those of **B** and 3.4.**D**). For example,

$$H_q(\mathbb{R}P^n; \mathbb{Z}_T) = \begin{cases} \mathbb{Z} & \text{if } q = n \text{ and } q \text{ is even}, \\ \mathbb{Z}_2 & \text{if } q < n \text{ and } q \text{ is even}, \\ 0 & \text{otherwise}; \end{cases}$$

$$H^q(\mathbb{R}P^n; \mathbb{Z}_T) = \begin{cases} \mathbb{Z} & \text{if } q = n \text{ and } q \text{ is even}, \\ \mathbb{Z}_2 & \text{if } q \le n \text{ and } q \text{ is odd}, \\ 0 & \text{otherwise}. \end{cases}$$

6.5. Cohomology with coefficients in a sheaf.

A. A *sheaf* of Abelian groups over a topological space X is a generalization of a local system on X. A sheaf is a topological space S, endowed with a projection $\pi : S \to X$ and a fibrewise group structure; the latter consists of a continuous fibrewise multiplication

$$\mu : \{(s_1, s_2) \in S \times S \mid \pi(s_1) = \pi(s_2)\} \to S, \quad \pi(\mu(s_1, s_2)) = \pi(s_1) = \pi(s_2),$$

and a fibrewise inverse map

$$\nu : S \to S, \quad \pi(\nu(s)) = \pi(s),$$

which make the *stalk* $\pi^{-1}(x) \in S$ into an Abelian group. (Sheaves of rings, algebras, modules, etc. can also be considered.) The following condition is imposed on the topological space X: the projection π is a local homeomorphism, that is, each point $s \in S$ has a neighbourhood T that is mapped homeomorphically by π onto its image $\pi(T)$. (We emphasize that S is not assumed to be Hausdorff, and indeed is not in the majority of significant examples.) If $U \subset X$ is open, then $\Gamma_U(S)$ denotes the set of *sections* of S over U, that is, continuous maps $\phi : U \to S$ such that $\pi(\phi(u)) = u$ for all $u \in U$. Clearly $\Gamma_U(S)$ is an Abelian group, and for an inclusion $V \subset U$ there is a "restriction homomorphism"

$$\phi_V^U : \Gamma_U(S) \to \Gamma_V(S),$$

satisfying the condition

$$(*) \quad \phi_W^V \phi_V^U = \phi_W^U.$$

It turns out that the groups $\Gamma_U(S)$ and homomorphisms ϕ_V^U completely determine the sheaf S. Moreover, a system of groups $\Gamma_U(S)$ and homomorphisms ϕ_V^U

satsifying the condition (∗) (which itself is called a *presheaf*) arises from some sheaf if and only if it satisfies the condition

(∗∗)   if $\alpha_1 \in \Gamma_{U_1}(S)$, $\alpha_2 \in \Gamma_{U_2}(S)$ and $\phi_{U_1 \cap U_2}^{U_1}(\alpha_1) = \phi_{U_1 \cap U_2}^{U_2}(\alpha_2)$,

then there is a unique element $\alpha \in \Gamma_{U_1 \cup U_2}(S)$ such that
$$\phi_{U_1}^{U_1 \cup U_2} = \alpha_1, \quad \phi_{U_2}^{U_1 \cup U_2} = \alpha_2.$$

If this condition is satisfied then the presheaf uniquely determines the sheaf. Because of this, a sheaf is often defined as a presheaf satisfying condition (∗∗).

(The stalk over a point $x \in X$ of the sheaf corresponding to the presheaf $\{P(U)\}$ is $\varinjlim_{U \ni x} P(U)$.)

Examples 0°. (The constant sheaf.) $\Gamma_U(S) = G$ for any connected U, $\phi_V^U = \mathrm{id}$ for any connected U, V. In other words, $S = X \times G$ (the group G is assumed to be discrete), $\pi : S \to X$ is the projection onto the first factor.

1°. (Locally constant sheaf.) There exists a covering $\mathscr{U}$ of X such that $\Gamma_U(S) \cong G$ for any connected U contained in an element of the covering; for two such sets $U, V \subset U$ the map ϕ_V^U is an isomorphism. In other words: S is a fibre bundle over X with (discrete) fibre G and structure group $\mathrm{Aut}\, G$, the automorphism group of G. It is easy to see that locally constant sheaves over a connected space X correspond precisely to local systems of Abelian groups on X.

2°. (Sheaves of germs.) X is a (topological, smooth, analytic) manifold and $\Gamma_U(S)$ is the group of (continuous, smooth, analytic) functions on U, or of sections of some vector bundle over X. The space of this sheaf is the space of germs of functions or sections of the type considered. If these are, say, C^∞ functions or sections, and the manifold X has positive dimension, then this space is automatically not Hausdorff: germs of functions or sections f, g at a point $x \in X$ cannot be separated in the sheaf space if $f = g$ on some open set whose closure contains x (for example, if $X = \mathbb{R}$, $f(t) = g(t)$ for $t \le x$ and $f(t) \ne g(t)$ for $t > x$).

3°. If Y is a subspace of X and S is a sheaf over Y, then a sheaf $\bar{S}$ over X can be defined by putting $\Gamma_U(\bar{S}) = \Gamma_{U \cap Y}(S)$. Note that the topological space $\bar{S}$ as a set is $S \cup (X - Y)$ with the identity projection over $X - Y$; the topology on $\bar{S}$ has a rather complicated structure: if $y \in Y$ lies in the closure of some open set in X that is disjoint from Y, then any neighbourhoods of the points of $\bar{S}$ that project to y will intersect.

In fact there is a great variety of sheaves and the examples mentioned are only an insignificant part of this diversity. There is an extensive literature on sheaves (and cohomology with coefficients in sheaves); see Godement (1958) and Hirzebruch (1966).

B. Only cohomology with coefficients in a sheaf occurs (homology can be defined with coefficients in "cosheaves" but this concept is not particularly important). The simplest way to define cohomology with coefficients in a sheaf is to apply a construction in the spirit of Aleksandrov-Čech (see 6.2). We fix a covering $\mathscr{U} = \{U_i \mid i \in I\}$ of the space X on which we are given a sheaf S of Abelian groups. A q-cochain of this covering with coefficients in S is a function taking a set $U_{i_1}, \ldots, U_{i_q}$ of pairwise distinct elements of $\mathscr{U}$ to an element of the group

$\Gamma_{U_{i_1} \cap \cdots \cap U_{i_q}}(S)$ that depends skew-symmetrically on $i_1, \ldots, i_q \in I$. The group of these cochains is denoted by $C^q(\mathscr{U}; S)$. The coboundary operator

$$\delta : C^q(\mathscr{U}; S) \to C^{q+1}(\mathscr{U}; S)$$

is defined by

$$\delta c(U_{i_1}, \ldots, U_{i_q}) = \sum_{s=0}^{q} (-1)^s \phi_{U_{i_0} \cap \cdots \cap U_{i_q}}^{U_{i_0} \cap \cdots \cap \hat{U}_{i_s} \cap \cdots \cap U_{i_q}} (c(U_{i_0}, \ldots, \hat{U}_{i_s}, \ldots, U_{i_q}).$$

Hence we obtain a complex $\{C^q(\mathscr{U}; S), \delta\}$ and cohomology groups $H^q(\mathscr{U}; S)$. The cohomology groups $H^q(X; S)$ are defined by the usual Aleksandrov-Čech passage to the limit with respect to the directed set of coverings:

$$H^q(X; S) = \varinjlim_{\mathscr{U}} H^q(\mathscr{U}; S).$$

If S is the constant sheaf with stalk G, then cohomology with coefficients in S is the same as the usual (Aleksandrov-Čech) cohomology with coefficients in G. If S is a locally constant sheaf then cohomology with coefficients in S is the same as cohomology with local coefficients.

As we have already said, thick books have been dedicated to the properties of sheaf cohomology, and we cannot go into this subject in any detail; we mention two of the most useful properties.

1°. (0-dimensional cohomology.) $H^0(X; S) = \Gamma_X(S)$, the group of global sections of the sheaf.

2°. (The coefficient sequence.) If S' is a subsheaf of S and S'' is the factor sheaf S/S' (these words have an obvious meaning), that is, if we are given a so-called short exact sequence

$$0 \to S' \to S \to S'' \to 0$$

of sheaves, then there is a long exact cohomology sequence

$$\cdots \to H^q(X; S') \to H^q(X; S) \to H^q(X; S'') \to H^{q+1}(X; S') \to \cdots$$

C. Several large classes of sheaves are known with the property that cohomology groups of positive dimension with coefficients in these sheaves are trivial. One such class is that of *fine sheaves*. A sheaf S over a space X is called *fine* if for any open $U \subset X$, section $\alpha \in \Gamma_U(S)$ and open set $V \subset X$ with cl $V \subset U$, there is a global section $\tilde{\alpha} \in \Gamma_X(S)$ coinciding with α on $V : \phi_V^X(\tilde{\alpha}) = \phi_V^U(\alpha)$.

Proposition. *If X is paracompact and S is a fine sheaf, then $H^q(X; S) = 0$ for $q > 0$.*   □

This is easily deduced from the definitions.

D. In conclusion we shall give a (rather banal) application of sheaf cohomology and the properties we have listed.

Let X be a n-dimensional smooth manifold, $\mathscr{A}^q(X)$ the sheaf of germs of exterior differential forms of degree q on X (that is, $\Gamma_U(\mathscr{A}(X))$ is the space of exterior q-forms on U), and $d : \mathscr{A}^q \to \mathscr{A}^{q+1}(X)$ the exterior differential. Let $A^q(X)$ be the space of exterior q-forms on the whole of X,

$$A^q(X) = \Gamma_X(\mathscr{A}^q(X));$$

the exterior differential $d : A^q(X) \to A^{q+1}(X)$ is also defined. Our aim is to prove the following theorem.

Theorem (de Rham).

$$H^q(X; \mathbb{R}) = \frac{\operatorname{Ker}[d : A^q(X) \to A^{q+1}(X)]}{\operatorname{Im}[d : A^{q-1}(X) \to A^1(X)]}. \quad \square$$

We shall prove this by combining the above-mentioned properties of sheaf cohomology with the following two facts:

1°. The sheaf $\mathscr{A}^q(X)$ is fine.

2°. The sequence

$$(*) \quad \mathscr{A}^0(X) \to \mathscr{A}^1(X) \to \mathscr{A}^2(X) \to \cdots$$

is exact.

(The first assertion is obvious, the second forms the content of the so-called Poincaré lemma, and is also proved quite easily.)

Let $\mathscr{F}^q(X)$ denote the kernel of $d : \mathscr{A}^q(X) \to \mathscr{A}^{q+1}(X)$; it is the sheaf of germs of closed q-forms. Note that $\mathscr{F}^0(X) = \mathbb{C}$ (the sheaf of germs of constants) is the constant sheaf with stalk $\mathbb{C}$, and that for $q \geq 1$, the sheaf $\mathscr{F}^q(X)$ is also the image of $d : \mathscr{A}^{q-1}(X) \to \mathscr{A}^q(X)$. The exactness of the sequence $(*)$ provides a chain of short exact sequences

$$0 \to \mathbb{C} \to \mathscr{A}^0(X) \to \mathscr{F}^1(X) \to 0,$$
$$0 \to \mathscr{F}^1(X) \to \mathscr{A}^1(X) \to \mathscr{F}^2(X) \to 0,$$
$$0 \to \mathscr{F}^2(X) \to \mathscr{A}^2(X) \to \mathscr{F}^3(X) \to 0.$$

Consider the fragments

$$H^{q-1}(X; \mathscr{A}^0(X)) \to H^{q-1}(X; \mathscr{F}^1(X)) \to H^q(X; \mathbb{C}) \to H^q(X; \mathscr{A}^0(X)),$$

$$\cdots$$

$$H^1(X; \mathscr{A}^{q-2}(X)) \to H^1(X; \mathscr{F}^{q-1}(X)) \to H^2(X; \mathscr{F}^{q-2}(X)) \to$$
$$\to H^2(X; \mathscr{A}^{q-2}(X)),$$

$$H^0(X; \mathscr{A}^{q-1}(X)) \to H^0(X; \mathscr{F}^q(X)) \to H^1(X; \mathscr{F}^{q-1}(X)) \to$$
$$\to H^1(X; \mathscr{A}^{q-1}(X))$$

of the corresponding cohomology sequences. Since $H^1(X; \mathscr{A}^r(X)) = \ldots = H^q(X; \mathscr{A}^r(X)) = 0$ and $H^0(X; \mathscr{A}^{q-1}(X)) = A^{q-1}(X)$, $H^0(X; \mathscr{F}^q(X)) = \mathscr{F}^q(X) = \operatorname{Ker}[d : A^q(X) \to A^{q+1}(X)]$, then it follows from the exactness of the above fragments that there are isomorphisms

$$H^q(X; \mathbb{C}) \cong H^{q-1}(X; \mathscr{F}^1(X));$$
$$H^{q-1}(X; \mathscr{F}^1(X)) \cong H^{q-2}(X; \mathscr{F}^2(X));$$

$$\cdots$$

$$H^2(X; \mathscr{F}^{q-2}(X)) \cong H^1(X; \mathscr{F}^{q-1}(X));$$

$$H^1(X; \mathscr{F}^{q-1}(X)) \cong \frac{H^0(X; \mathscr{F}^q(X))}{\mathrm{Im}[d_* : H^0(X; \mathscr{A}^{q-1}(X)) \to H^0(X; \mathscr{F}^q(X))]}$$
$$\cong \frac{\mathrm{Ker}(d : A_q(X) \to A_{q-1}(X))}{\mathrm{Im}(d : A_{q-1}(X) \to A_q(X))}.$$

6.6. Conclusion.

It should not be thought that all possible versions of homology and cohomology that topologists have associated with topological spaces have been presented in this section. It is enough to recall the Smith homology theory considered in the previous section, which incidentally has far-reaching generalizations (see Borel (1960)). The so-called *Deligne-Goresky-MacPherson homology* (intersection homology) is widely applied in topology, complex analysis, and representation theory (see Goresky and MacPherson (1980)). Another example of non-standard homology and cohomology theory is the "local theory" relating a topological space X and a point $x \in X$ to the limits

$$\lim_{\leftarrow U} H_q(U, U - x; G), \quad \lim_{\rightarrow U} H^q(U, U - x; G),$$

in which U runs through the set of neighbourhoods of x; these limits are denoted by $H_q^{\mathrm{loc}}(X, x; G)$ and $H_{\mathrm{loc}}^q(X, x; G)$, and called the local homology and cohomology groups of X at x with coefficients in G. We may also consider the homology and cohomology of various algebraic objects associated with a space X: the group of homeomorphisms of X, or, if X is a smooth manifold, the group of diffeomorphisms of X and the Lie algebras of vector fields on X. All these various homology theories remain outside the scope of our survey.

Chapter 2
Multiplicative theory

§7. Products

7.1. Introduction.

Although homology is more geometric than cohomology it plays a considerably more modest role in topology. There are many reasons for this, but the main one is that cohomology classes can be multiplied together, so that for any commutative ring G the sum $\oplus_q h^q(X; G) = H^*(X; G)$ becomes an associative skew-symmetric graded ring; there is nothing like this for homology (the reader may independently attach a precise meaning to this remark and prove it). We shall begin with the most transparent construction of the cohomology product.

A. Let G be a commutative ring, and X_1, X_2 be cellular spaces. Given cellular cochains $c_1 \in \mathscr{C}^{q_1}(X_1; G)$, $c_2 \in \mathscr{C}^{q_2}(X_2; G)$, we shall construct a cochain $c_1 \times c_2 \in \mathscr{C}^{q_1+q_2}(X_1 \times X_2; G)$: on the cell $\sigma \times \tau \in X_1 \times X_2$ it takes the value $(-1)^{q_1 q_2} c_1(\sigma) c_2(\tau)$ (multiplication in G). Verification shows that $\delta(c_1 \times c_2) = (\delta c_1) \times c_2 + (-1)^{q_1} c_1 \times \delta c_2$, so that if c_1 and c_2 are cocycles, then so is $c_1 \times c_2$. The cohomology class of the cocycle $c_1 \times c_2$ depends only on the cohomology classes of c_1 and c_2 so that we obtain a well-defined multiplication

$$[\gamma_1 \in H^{q_1}(X_1; G), \ \gamma_2 \in H^{q_2}(X_2; G)] \mapsto \gamma_1 \times \gamma_2 \in H^{q_1+q_2}(X_1 \times X_2; G).$$

There is also a similar multiplication for homology: if $a_1 = \sum_i g_{1i}\sigma_i \in \mathscr{C}_{q_1}(X_1; G)$, $a_2 = \sum_j g_{2j}\tau_j \in \mathscr{C}_{q_2}(X_2; G)$, then $a_1 \times a_2 = \sum_{i,j} g_{1i} g_{2j}(\sigma_i \times \tau_j) \in \mathscr{C}_{q_1+q_2}(X_1 \times X_2; G)$; here $\partial(a_1 \times a_2) = (\partial a_1) \times a_2 + (-1)^{q_1} a_1 \times \partial a_2$, so that we obtain a multiplication $[\alpha_1 \in H_{q_1}(X_1; G), \ \alpha_2 \in H_{q_2}(X_2; G)] \mapsto \alpha_1 \times \alpha_2 \in H_{q_1+q_2}(X_1 \times X_2; G)$.

The two products $\times$ are related by the formula $\langle \gamma_1 \times \gamma_2, \ \alpha_1 \times \alpha_2 \rangle = (-1)^{q_1 q_2} \langle \gamma_1, \alpha_1 \rangle \langle \gamma_2, \alpha_2 \rangle$.

B. The formula $(\sum_i k_i \sigma_i) \times (\sum_j l_j \tau_j) \mapsto \sum_{i,j} k_i l_j (\sigma_i \times \tau_j)$ defines an isomorphism between the complexes $\mathscr{C}(X_1) \otimes \mathscr{C}(X_2)$ (see 1.6) and $\mathscr{C}(X_1 \times X_2)$. The results of 1.6 therefore entail an exact sequence

$$0 \longrightarrow \bigoplus_{r+s=q} [H_r(X_1) \otimes H_s(X_2)] \longrightarrow$$
$$\longrightarrow H_q(X_1 \times X_2) \longrightarrow \bigoplus_{r+s=q-1} \mathrm{Tor}(H_r(X_1), H_s(X_2)) \longrightarrow 0 \tag{13}$$

and an isomorphism

$$H_q(X_1 \times X_2) = \bigoplus_{r+s=q} [(H_r(X_1) \otimes H_s(X_2))] \oplus_{r+s=q-1} \mathrm{Tor}(H_r(X_1), H_s(X_2)).$$

The last formula is called the *Künneth formula*, a name we have already met in 1.6. We have proved it for cellular spaces, but it is also valid for any spaces: this is deduced in the standard way from the existence of cellular approximations to topological spaces (see Part 1, 9.5).

To this we must add that the second arrow in the sequence (13) is just the homology $\times$-product.

C. Later we shall speak of an important distinction between homology and cohomology, namely, that the first is covariant and the second contravariant. For any space X we can define the *diagonal map* $\Delta : X \to X \times X$, $\Delta(x) = (x, x)$. This induces homomorphisms

$$\Delta_* : H_q(X; G) \to H_q(X \times X; G),$$
$$\Delta^* : H^q(X \times X; G) \to H^q(X; G).$$

The first is of no use to us in this survey, but the second defines a cohomology multiplication for X: for $\gamma_1 \in H^{q_1}(X; G)$, $\gamma_2 \in H^{q_2}(X; G)$ we put

$$\gamma_1 \cup \gamma_2 = \Delta^*(\gamma_{\times}\gamma_2) \in H^{q_1+q_2}(X; G)$$

(the classical notation $\cup$ ("cup product") is not particularly convenient, and we shall sometimes write simply $\gamma_1\gamma_2$ instead of $\gamma_1 \cup \gamma_2$).

7.2. Direct construction of the $\cup$-product.

We shall not prove that the product is independent of the cell structure. Instead, we introduce another method of constructing the $\cup$-product which is in some respects the opposite of the previous one: we begin with an independent definition of the $\cup$-product, by means of which we define the $\times$-product, and then deduce our original description of the $\cup$-product.

A. Let X be any topological space and G a commutative ring, and let $c_1 \in C^{q_1}(X; G)$, $c_2 \in C^{q_2}(X; G)$. Define the cochain $c_1 \cup c_2 \in C^{q_1+q_2}(X; G)$ by

$$[c_1 \cup c_2](f) = c_1(f_{0...q_1})c_2(f_{q_1...q_1+q_2}),$$

where $f : T^{q_1+q_2} \to X$, and $f_{i_0...i_r}$ $(0 \le i_0 < i_1 < \cdots < i_r \le q_1 + q_2)$ denotes the singular r-simplex obtained by restriction of the singular simplex f to the r-dimensional face of $T^{q_1+q_2}$ spanned by the vertices numbered $i_0, \ldots, i_r$.

Proposition (properties of the cochain $\cup$-product). *Let* $c_1 \in C^{q_1}(X; G)$, $c_2 \in C^{q_2}(X; G)$. *Then:*

(i) $\delta(c_1 \cup c_2) = (\delta c_1) \cup c_2 + (-1)^{q_1}c_1 \cup \delta c_2$;

(ii) $c_1 \cup (c_2 \cup c_3) = (c_1 \cup c_2) \cup c_3$ $(c_3 \in C^{q_3}(X; G))$;

(iii) *let* $\omega = \{\omega_X\}$ *be the neutral transformation introduced in 2.3.C; for any singular $(q_1 + q_2)$-chain a we have the equation*

$$[c_1 \cup c_2](a) = (-1)^{q_1 q_2}[c_2 \cup c_1](\omega_X a);$$

(iv) *if* $g : Y \to X$ *is a continuous map, then* $g^{\#}(c_1 \cup c_2) = (g^{\#}c_1) \cup (g^{\#}c_2)$; *if* $h : G \to H$ *is a ring homomorphism, then* $h \circ (c_1 \cup c_2) = (h \circ c_1) \cup (h \circ c_2)$. □

The proof is obvious.

Note. The non-commutativity of the cochain $\cup$-product is not an accidental defect of the definition, but the appearance of a deep general regularity which has important consequences (see "Homotopy theory" (1981)).

It follows from (i) that the $\cup$-product of cocycles is a cocycle, whose cohomology class depends only on the cohomology classes of the factors. Hence the cochain $\cup$-product induces a cohomology $\cup$-product

$$[\gamma_1 \in H^{q_1}(X; G), \ \gamma_2 \in H^{q_2}(X; G)] \mapsto \gamma_1 \cup \gamma_2 \in H^{q_1+q_2}(X; G).$$

Theorem (properties of the cohomology $\cup$-product). *Let* $\gamma_1 \in H^{q_1}(X; G)$, $\gamma_2 \in H^{q_2}(X; G)$. *Then*

(i) $\gamma_1 \cup (\gamma_2 \cup \gamma_3) = (\gamma_1 \cup \gamma_2) \cup \gamma_3$ $(\gamma_3 \in H^{q_3}(X; G))$;

(ii) $\gamma_1 \cup \gamma_2 = (-1)^{q_1 q_2}\gamma_2 \cup \gamma_1$;

(iii) *for a continuous map* $g : Y \to X$ *and a ring homomorphism* $h : G \to H$

$$g^*(\gamma_1 \cup \gamma_2) = (g^*\gamma_1) \cup (g^*\gamma_2), \quad h_*(\gamma_1 \cup \gamma_2) = (h_*\gamma_1) \cup (h_*\gamma_2);$$

(iv) *if X is connected and $\gamma \in H^0(X; G) = G$, then $\gamma \cup \gamma_1 = \gamma\gamma_1$.* $\qquad\square$
This is a consequence of the previous proposition.

The construction described can be generalized in various ways. There is for example a relative version of the product

$$H^{q_1}(X, A; G) \times H^{q_2}(X, B; G) \to H^{q_1+q_2}(X, A \cup B; G).$$

Also, if instead of a ring structure we are given a pairing $G_1 \times G_2 \to G$, then a $\cup$-product is defined

$$H^{q_1}(X; G_1) \times H^{q_2}(X; G_2) \to H^{q_1+q_2}(X; G).$$

The latter can be generalized to the case of local coefficients. Without going into details (which the reader can supply if required) we mention the products

$$H^{q_1}(X; \mathbb{Z}) \times H^{q_2}(X; \mathfrak{G}) \to H^{q_1+q_2}(X; \mathfrak{G}),$$

where $\mathfrak{G}$ is any local system, and $H^{q_1}(X; \text{or} X) \times H^{q_2}(X; \text{or} X) \to H^{q_1+q_2}(X; \mathbb{Z})$, where X is a manifold and or X is an oriented local system.

B. As before, suppose that G is a ring, and X_1 and X_2 are topological spaces. Let $\gamma_1 \in H^{q_1}(X_1; G)$, $\gamma_2 \in H^{q_2}(X_2; G)$. Put

$$\gamma_1 \times \gamma_2 = (p^*\gamma_1) \cup (p^*\gamma_2) \in H^{q_1+q_2}(X_1 \times X_2; G)$$

(p_1 and p_2 are the projections of the product onto its factors). We leave the reader to construct the relative version of the $\times$-product and to prove that it is natural in all possible ways.

C. Theorem. *This definition of the $\times$-product is equivalent to that given at the beginning of this section.* $\qquad\square$

To prove this it turns out to be sufficient to calculate the $\times$-product explicitly in a simple special case. Namely, the $\times$-product of the standard generators of the groups

$$H^{q_1}(T^{q_1}, \partial T^{q_1}; \mathbb{Z}) = \mathbb{Z}, \quad H^{q_2}(T^{q_2}, \partial T^{q_2}; \mathbb{Z}) = \mathbb{Z}$$

is the standard generator of the group

$$H^{q_1+q_2}(T^{q_1} \times T^{q_2}, (\partial T^{q_1} \times T^{q_2}) \cup (T^{q_1} \times \partial T^{q_2}); \mathbb{Z}) =$$
$$= H^{q_1+q_2}(T^{q_1} \times T^{q_2}, \partial(T^{q_1} \times T^{q_2}); \mathbb{Z}) = \mathbb{Z}$$

(the standard simplexes and their products are homeomorphic to balls). The proof of this fact and the derivation of the theorem from it are standard (for details see Fomenko and Fuchs (1989)).

7.3. Application: the Hopf invariant.

To show at once how powerful a tool the multiplicative structure of cohomology is in algebraic topology, we shall now use it to prove a highly non-trivial result on the homotopy groups of spheres, even though in practice we are able to calculate the cohomology product of hardly any spaces.

Theorem. *The group $\pi_{4n-1}(S^{2n})$ is infinite for any $n \geq 1$. Furthermore, the Whitehead square $[\iota_{2n}, \iota_{2n}]$ of the generator ι_{2n} of $\pi_{2n}(S^{2n})$ has infinite order.*   □

This theorem is proved with the aid of the *Hopf invariant*, an integer associated with a class $\phi \in \pi_{4n-1}(S^{2n})$. It is defined as follows. Consider a spheroid $f : S^{4n-1} \to S^{2n}$ in the class ϕ and form the space $X_\phi = S^{2n} \cup_f D^{4n}$ (the mapping cone of f). The space X_ϕ up to homotopy equivalence depends only on ϕ (which justifies the notation). It has a natural cell structure with three cells of dimensions 0, $2n$, and $4n$. Hence

$$H^q(X_\phi; \mathbb{Z}) = \begin{cases} \mathbb{Z} & \text{for } q = 0, 2n, 4n, \\ 0 & \text{for } q \neq 0, 2n, 4n. \end{cases}$$

The groups $H^{2n}(X_\phi; \mathbb{Z})$ and $H^{4n}(X_\phi; \mathbb{Z})$ have natural generators (determined by the canonical orientations of S^{2n} and D^{4n}), which we denote by a and b. Since $\dim a^2 = 4n$ ($a^2 = a \cup a$), $a^2 = hb$ for some $h \in \mathbb{Z}$. The number $h = h(\phi)$ is called the *Hopf invariant* of the class ϕ. (The Hopf invariant has numerous generalizations, some of which are very useful; see for example Postnikov (1985).) The theorem to be proved is covered by two lemmas, which easily follow from the naturality of the $\cup$-product (for details see Fomenko and Fuchs (1989)):

Lemma 1. *The Hopf invariant is additive.*   □

Lemma 2. *The Hopf invariant is non-trivial; in particular, $h([\iota_{2n}, \iota_{2n}]) = 2$.*   □

7.4. Other products.

A. The $\cap$-product. This is a mixed operation involving homology and cohomology. Let $a = \sum g_i f_1 \in C_{q_1}(X; G)$, $c \in C^{q_2}(X; G)$, $q_1 \geq q_2$. Let f_i' denote the q_2-dimensional face of the singular simplex f_i spanned by the first $q_2 + 1$ vertices, and f_i'' its $(q_1 - q_2)$-dimensional face spanned by the last $q_1 - q_2 + 1$ vertices, and put

$$a \cap c = \sum g_i c(f_i') f_i'' \in C_{q_1-q_2}(X; G).$$

Lemma.
$$\partial(a \cap c) = \partial a \cap c + (-1)^{q_1} a \cap \partial c. \ \square$$

It follows from this lemma that the chain-cochain $\cap$-product induces a homology-cohomology $\cap$-product

$$[\alpha \in H_{q_1}(X; G), \ \gamma \in H^{q_2}(X; G)] \mapsto \alpha \cap \gamma \in H_{q_1-q_2}(X; G).$$

Among the properties of the $\cap$-product we mention its naturality and mixed associativity:

$$\alpha \cap (\gamma_1 \cup \gamma_2) = (\alpha \cap \gamma_1) \cap \gamma_2.$$

B. The Pontryagin-Samelson product. If there is a product on X making it into a topological group, or even an H-space, then a homology product can be defined for X. The definition is obvious: if $\mu : X \times X \to X$ is the product on X and $\alpha_1 \in H_{q_1}(X; G)$, $\alpha_2 \in H_{q_2}(X; G)$, where G is a ring, then $\alpha_1 \alpha_2 = \mu_*(\alpha_1 \times \alpha_2)$.

§8. Homology and manifolds

8.1. Introduction.

Among the elementary geometric tools which homology theory uses the most effective are those provided by the topology of manifolds and we shall touch on this subject here.

The fundamental object in this section will be a smooth manifold endowed with a smooth triangulation. Note that every smooth manifold has a smooth triangulation (a proof of this can be found in Milnor (1963) and Whitney (1957)). In fact many of the results of this section are valid for wider classes of spaces; we shall list the most important of these.

A *PL-manifold of dimension n* is an n-dimensional topological manifold with a triangulation in which the links of the q-simplexes ($q = 0, 1, \ldots, n - 1$) are combinatorially equivalent to $(n - q - 1)$-spheres. (Recall that the link of a simplex σ is defined to be $\bigcup_{\tau \supset \sigma} \Gamma_{\bar{\sigma}} \tau$, where the summation is over all simplexes of the triangulation that contain σ, and $\Gamma_{\bar{\sigma}} \tau$ denotes the face of τ spanned by the vertices not appearing in σ.) A *homology n-manifold* is an n-dimensional simplicial space in which the link of each q-simplex is a homology $(n - q - 1)$-sphere. (We also speak of homology manifolds with respect to one or another domain of coefficients.) A *pseudomanifold of dimension n* is an n-dimensional simplicial space satisfying the three conditions: every simplex is a face of some n-simplex (dimensional homogeneity); every $(n - 1)$-simplex is a face of exactly two n-simplexes (non-branching); any two n-simplexes are joined by a chain of n-simplexes in which adjacent simplexes have a common $(n - 1)$-dimensional face (strong connectedness). Each of these three definitions has a modification for manifolds with boundary (that is, we can define PL-manifolds, homology manifolds, and pseudomanifolds with boundary) and orientations can be defined naturally in all cases. The word "closed" applied to all these classes of spaces means "compact without boundary".

Smooth triangulated manifolds can be referred to all these classes, when the concepts of differential and simplicial orientation coincide.

For the rest of this section, X denotes a connected smooth triangulated manifold.

8.2. The fundamental class.

(For the rest of this subsection we assume that X is a pseudomanifold.)

A. Theorem.

$$H_n(X) = \begin{cases} \mathbb{Z} & \text{if } X \text{ is closed and oriented,} \\ 0 & \text{otherwise;} \end{cases}$$

$$H_n(X; \mathbb{Z}_2) = \begin{cases} \mathbb{Z}_2 & \text{if } X \text{ is closed,} \\ 0 & \text{otherwise.} \ \square \end{cases}$$

For consider an n-chain $c = \sum k_i \sigma_i$ of the classical complex of X with coefficients in $\mathbb{Z}$ or $\mathbb{Z}_2$. If the simplexes σ_i and σ_j have a common $(n - 1)$-dimensional face τ, then τ appears in c with coefficient $\pm(k_i - k_j)$ if the orientations of σ_i and σ_j are compatible, and with coefficient $\pm(k_i + k_j)$ if the orientations are not compatible. Suppose that $\partial c = 0$. Because X is non-branching and strongly connected (see 8.1) this is only possible if all the coefficients k_i are the same to within a change of sign. If the coefficients are taken in $\mathbb{Z}_2$ it follows that c is either 0 or $\sum \sigma_i$ where the sum is taken over all n-simplexes of the triangulation. The last expression will be a cycle if two conditions are satisfied: (i) the number of simplexes is finite, that is, the manifold is compact; (ii) X is without boundary. As there are no boundaries (there being no $(n+1)$-chains) this proves the theorem for coefficients in $\mathbb{Z}_2$. If the coefficients are in $\mathbb{Z}$, then a cycle has the form $\sum(\pm k)\sigma_i$; by changing the orientation of some of the n-simplexes, we can make the cycle take the form $k \sum \sigma_i$. Then the orientations of all the simplexes will be compatible, and in addition to the two previous conditions we obtain the condition: (iii) X is orientable. This completes the proof of the theorem.

B. This proof in fact contains an explicit construction for the generators of the groups $H_n(X)$ and $H_n(X; \mathbb{Z}_2)$ if they are nontrivial. Each generator is the homology class of the cycle $\sum \sigma_i$, where all the coefficients are equal to $1 \in \mathbb{Z}_2$ or $1 \in \mathbb{Z}$, and in the second case all the simplexes are oriented compatibly with a chosen orientation of the manifold. This homology class is called the *fundamental class of the manifold* (or pseudomanifold). We might say very expressively, if not very precisely, that the fundamental class is the homology class of the manifold itself, regarded as one of its own cycles. The fundamental class of a manifold X is usually denoted by $[X]$ (if we want to emphasize that we are concerned with the $\mathbb{Z}_2$-class, we write $[X]_2$).

C. If X and Y are manifolds or pseudomanifolds with the same dimension n having fundamental classes (integral or mod 2), and $f : X \to Y$ is a continuous map, then the integer (or residue mod 2) k, defined by $f_*[X] = k[Y]$, is called the *degree* of the map f. In the differentiable case the degree of f may be calculated as the number of inverse images of a regular value $y \in Y$ of f, where in the integral case the inverse images are counted with sign $+$ or $-$ depending on the sign of the Jacobian. If f is a simplicial map, then the degree may be found as the number of n-simplexes of X mapped onto a fixed n-simplex of Y. Finally, the degree of a map $S^n \to S^n$ can also be defined as the element of the group $\pi_n(S^n) = \mathbb{Z}$ determined by the map.

The degree of f is denoted by $\deg f$ (or $\deg_2 f$ in the $\mathbb{Z}_2$ case).

D. Everything that has been said in **A–C** can be carried over to the case of manifolds with boundary; the homology of X is replaced by the homology of the pair $(X, \partial X)$ and the degree is only computed for maps that take the boundary into the boundary.

E. We say that a closed smooth manifold Y lying in a topological space X *realizes* the class $\alpha \in H_k(X; \mathbb{Z}_2)$ or $H_k(X)$, if $i_*[Y] = \alpha$, where $i : Y \to X$ is the inclusion. We most often speak of such realizations when X is also a smooth manifold, and Y is a smooth submanifold. For example, the generators of the groups $H_q(\mathbb{R}P^n; \mathbb{Z}_2)$, $H_q(\mathbb{R}P^n)$, $H_q(\mathbb{C}P^n)$ can be realized by projective subspaces of the spaces $\mathbb{R}P^n$, $\mathbb{C}P^n$.

It turns out that not every homology class of a smooth manifold can be realized by a smooth submanifold. There is an extensive theory on this question with some brilliant results (see Thom (1954)); for example, for any homology class there is a multiple of this class that is realized by a smooth submanifold. However, there is a much simpler fact: every homology class can be realized by a smooth submanifold with singularities of codimension ≥ 2. (The precise definition: a closed submanifold Y of a smooth manifold X is called an (irreducible) smooth submanifold with singularities of codimension ≥ 2, if there are smooth submanifolds $Z_1, \ldots, Z_l$ in X such that $Y \cap [X - (Z_1 \cup \cdots \cup Z_l)]$ is a connected smooth submanifold of the manifold $X - (Z_1 \cap \cdots \cap Z_l)$ of dimension at least 2 greater than that of each of the manifolds $Z_1, \ldots, Z_l$.) Submanifolds with singularities of codimension ≥ 2 are pseudomanifolds, so that orientations are defined for them, and they have a fundamental class. As has already been said, every homology class of a smooth manifold can be realized by a smooth submanifold of codimension ≥ 2.

8.3. The Poincaré isomorphisms.

(The results of this subsection as well as those of 8.4 and 8.5 are valid when X is a homology manifold.)

A. The main result of the homology theory of manifolds is the following theorem.

Theorem. *If X is orientable, then*

$$H_q(X) \cong H^{n-q}(X; \mathbb{Z}) \text{ for any } q.$$

In the general case,

$$H_q(X; \mathbb{Z}_2) \cong H^{n-q}(X; \mathbb{Z}_2) \text{ for any } q. \ \square$$

We shall make an important refinement to this formulation, which consists of giving explicit formulae for the isomorphisms in the theorem. The resulting canonical isomorphisms are called the *Poincaré isomorphisms*. The formulae are very simple: the Poincaré isomorphisms

$$D : H^{n-q}(X; \mathbb{Z}) \to H_q(X),$$
$$D : H^{n-q}(X; \mathbb{Z}_2) \to H_q(X; \mathbb{Z}_2),$$

are given (in the connected case) by the formula

$$D(\alpha) = [X] \cap \alpha,$$

where $[X]$ is the fundamental class and $\cap$ denotes the $\cap$-product (see 7.4.A).

B. The proof of the theorem, however, does not include a map D defined in this way. It consists of an entirely different construction of the necessary isomorphisms. We consider the barycentric subdivision of our triangulation of the manifold X and for a simplex σ of the original triangulation, σ^* denotes the union of the simplexes of the subdivision that intersect σ precisely in its centre. The set σ^* is called the *barycentric star* of σ. σ^* may be more formally described as follows. The vertices of the barycentric subdivision coincide with the centres of the simplexes of the original triangulation, and hence are in one-one correspondence with these simplexes. The simplexes of the barycentric subdivision are chains $\sigma_r \supset \sigma_{r-1} \supset \cdots \supset \sigma_0$ of simplexes of the original triangulation (the simplex τ corresponds to the chain of simplexes whose centres are the vertices of τ). The barycentric star σ^* consists of the simplexes of the subdivision corresponding to chains $\cdots \supset \sigma$ ending in the simplex σ.

Example. Consider a triangulation of part of the plane (Fig. 5). The barycentric star of a 2-simplex is its centre. The barycentric star of an edge is a broken line with two segments leading from the centre of one of the triangles adjoining the given edge to the centre of the other triangle adjoining the edge. The barycentric star of a vertex is a polygonal region surrounding the vertex.

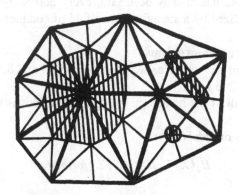

Fig. 5

The following properties of barycentric stars are obvious. (i) The barycentric star of a q-simplex is homeomorphic to an $(n-q)$-ball. (ii) The open barycentric stars form a cell decomposition of the manifold under consideration (an open barycentric star is a barycentric star from which the smaller barycentric stars contained in it have been removed). (In the case of a homology manifold this decomposition need not be cellular but it is still suitable for calculating homology in view of what was said in 3.2.C). (iii) If the manifold is orientable, then the orientation of a simplex naturally corresponds to that of its barycentric stars. (iv) The correspondence $\sigma \leftrightarrow \sigma^*$ is compatible with the boundary and coboundary operators in the following way. With a classical q-cochain $c = \{\sigma_i \mapsto k_i\}$ of the original triangulation we associate the cellular (or as we say, stellar) $(n-q)$-chain $Dc = \sum k_i \sigma_i^*$ of the star decomposition. (Here in the oriented case $k_i \in \mathbb{Z}$, and σ_i and σ_i^* denote oriented simplexes and stars; in the non-oriented case $k_i \in \mathbb{Z}_2$). Then $D\delta c = \pm \partial Dc$ (as usual, we do not specify the sign).

The last equation shows that the complexes from which the homology and cohomology of X are calculated are exactly the same:

$$
\begin{array}{ccccccc}
\mathscr{C}_n(X) & \xrightarrow{\partial} & \mathscr{C}_{n-1}(X) & \xrightarrow{\partial} \cdots \xrightarrow{\partial} & \mathscr{C}_1(X) & \xrightarrow{\partial} & \mathscr{C}_0(X) \\
\simeq \uparrow D & & \simeq \uparrow D & & \simeq \uparrow D & & \simeq \uparrow D \\
\mathscr{C}^0(X) & \xrightarrow{\delta} & \mathscr{C}^1(X) & \xrightarrow{\delta} \cdots \xrightarrow{\delta} & \mathscr{C}^{n-1}(X) & \xrightarrow{\delta} & \mathscr{C}^n(X)
\end{array}
$$

(the upper row is the star chain complex, and the lower row the classical cochain complex; in both cases the coefficients are in $\mathbb{Z}$ or $\mathbb{Z}_2$, but we omit them). This proves the theorem.

C. Corollary. *The Euler characteristic of a closed odd-dimensional manifold is equal to 0.* □

To prove this it is better to use the Poincaré isomorphism mod 2, since this also holds in the non-oriented case:

$$\text{Eu}(X) = \sum(-1)^q \dim H_q(X; \mathbb{Z}_2) =$$
$$= \sum(-1)^q \dim H_{n-q}(X; \mathbb{Z}_2) = -\text{Eu}(X).$$

D. The reader may convince himself that the isomorphism D constructed in **B** can in fact be expressed by means of the $\cap$-product, as was said in **A**. We also recommend the reader to consider all the information he has about the homology and cohomology of manifolds (including the theorem in 8.2) and convince himself in each case that it is consistent with the Poincaré isomorphism.

8.4. Intersection numbers and Poincaré duality.

A. From 2.6 we know the relations between the homology and cohomology of an arbitrary topological space. This allows us to transform the canonical isomorphisms in 8.2 into non-canonical isomorphisms

$$H_q(X; \mathbb{Z}_2) \cong H_{n-q}(X; \mathbb{Z}_2)$$

in all cases, and

the free part of $H_q(X) \cong$ the free part of $H_{n-q}(X)$,
the torsion of $H_q(X) \cong$ the torsion of $H_{n-q-1}(X)$

in the oriented case. It turns out that these *are* in fact natural isomorphisms, and are a manifestation of a completely natural duality, called *Poincaré duality*. We shall be concerned with the third isomorphism a little later, but now we say something about the first two.

B. Poincaré duality is based on the concept of the *intersection number*. Let $c_1 = \sum k_i \sigma_i$ be a q-chain of the classical complex of the manifold X, and $c_2 = \sum l_j \sigma_j^*$ an $(n-q)$-chain of the star complex. The number

$$\phi(c_1, c_2) = \sum \delta_{ij} k_i l_j = \pm \langle D^{-1} c_1, c_2 \rangle$$

is called the *intersection number* of the chains c_1 and c_2. Clearly $\langle \partial c', c'' \rangle = \pm \langle c', \partial c'' \rangle$ (it is sufficient to verify this when $c' = \sigma$ and $c'' = \tau^*$, and in this case both sides equal ± 1 if τ is a face of σ, and equal 0 otherwise). Hence the intersection number of two cycles depends only on the homology classes of these cycles. Hence we obtain a bilinear intersection number for homology classes

$$[\alpha_1 \in H_q(X), \ \alpha_2 \in H_{n-q}(X)] \mapsto \phi(\alpha_1, \alpha_2) \in \mathbb{Z},$$

and it is easy to show that $\phi(\alpha_1, \alpha_2) = (-1)^{q(n-q)}\phi(\alpha_2, \alpha_1)$.

In the non-oriented case an intersection number taking values in $\mathbb{Z}_2$ can be defined in an obvious way for $\mathbb{Z}_2$-chains and $\mathbb{Z}_2$-homology classes. There are also intersection numbers corresponding to an arbitrary pairing $G_1 \times G_2 \to G$.

C. A remarkable property of the intersection number is its easily visualized geometrical meaning. Clearly the barycentric star of a simplex intersects that simplex "transversally", and does not intersect any other simplexes of the same dimension. Hence $\phi(c_1, c_2)$ is a suitable expression for the "algebraic number of points of intersection" of the chains c_1 and c_2. Moreover it turns out that the intersection number can be defined in a purely "differential" way.

In fact, suppose that homology classes $\alpha_1 \in H_q(X)$ and $\alpha_2 \in H_{n-q}(X)$ are realized by oriented smooth submanifolds $Y_1 \subset X$, $Y_2 \in X$ in general position (that is, intersecting in a finite number of points and transversally at them). A sign can be assigned to the points of intersection of Y_1 and Y_2 in a natural way (the orientations of Y_1 and Y_2 at a point of intersection naturally determine an orientation of the surrounding manifold X, which may or may not coincide with the given orientation of X). The assertion states that $\phi(\alpha_1, \alpha_2)$ is the algebraic number of points of intersection. Similar assertions hold for homology classes mod 2 (no orientations are needed here) and for realizations of homology classes of manifolds with singularities.

Note that the requirement of general position is not a serious obstacle in using the above rule for computing intersection numbers: if the submanifolds Y_1 and Y_2 (even with singularities) realize the required homology classes, then they can be brought into general position by a small perturbation of either of them.

Example. The natural generators y_r, y_{n-r} of $H_{2r}(\mathbb{C}P^n)$, $H_{2n-2r}(\mathbb{C}P^n)$ have intersection number 1: they are realized by projective subspaces of $\mathbb{C}P^n$ of (complex) dimensions r and $n - r$, which (in general position) intersect in a single point. We make an important remark about the sign. If X is a complex manifold, then X has a natural "complex" orientation, and if Y_1 and Y_2 are complex submanifolds of X having complementary dimensions and also endowed with complex orientations, then the points of their transverse intersection always contribute $+1$ to the intersection number. Therefore $\phi(y_r, y_{n-r}) = 1$, not -1.

D. We now turn to our original definition of intersection number. The following fundamental theorem can be deduced from it in an obvious way.

Theorem. *Let X be a closed oriented smooth manifold. (i) For any homomorphism $\alpha : H_q(X) \to \mathbb{Z}$ there is a homology class $\alpha' \in H_{n-q}(X)$ such that $\phi(\alpha', \beta) = \alpha(\beta)$ for any $\beta \in H_q(X)$. (ii) The class α' is uniquely determined by the homomorphism α up to addition of an element of finite order. A similar assertion is valid in the non-oriented case for homology and intersection numbers mod 2, and in this situation α' is determined uniquely by α.* $\square$

Hence the intersection numbers determine a non-degenerate duality between the free parts of the groups $H_q(x)$ and $H_{n-q}(X)$ in the oriented case, and also between $H_q(X; \mathbb{Z}_2)$ and $H_{n-q}(X; \mathbb{Z}_2)$ in all cases. This duality too is called Poincaré duality. (It is astonishing that complete confusion reigns in the literature in the use of the terms "Poincaré isomorphism" and "Poincaré duality", although in other cases mathematicians have developed a sensitivity towards the distinction between a space and its adjoint, etc.).

E. As an application the reader may give a new proof of Lefschetz's theorem on the algebraic number of fixed points of a map $f : X \to X$ (see 5.1.**B**) in the case when X is a compact smooth manifold (perhaps with boundary) and f is a smooth map. In the closed case the proof consists of computing the intersection number of the graph of f (lying in $X \times X$) and the diagonal $X \subset X \times X$; the general case can be reduced to the closed case by means of the operation of doubling.

8.5. Linking coefficients.

We now turn to the duality between the torsion of the groups $H_q(X)$ and $H_{n-q-1}(X)$. It is constructed by means of the *linking coefficients*. This term is used in topology in several different senses among which there are two basic ones.

A. The first sense (the classical one) has nothing to do with Poincaré duality but is related to the so-called Alexander-Pontryagin duality which we shall meet at the end of this section. It consists of the following. Suppose that in an oriented n-manifold we are given two (integral) cycles c_1 and c_2 of dimensions q and $n-q-1$, homologous to 0 and having disjoint carriers (the *carrier* of a chain is the union of the images of its singular simplexes). We choose a chain d with boundary c_1 and calculate the intersection number of the chains d and c_2 (we may assume that all chains consist of smooth simplexes in general position; we may also assume that c_1 and d are classical chains of some triangulation, and c_2 is a chain of the corresponding star complex). This number is called the *linking coefficient* of c_1 and c_2, and is denoted by $\lambda(c_1, c_2)$. It is easy to verify that $\lambda(c_1, c_2)$ is well defined and does not depend on the choice of d. It is also easy to verify that

$$\lambda(c_1, c_2) = -(-1)^{\dim c_1 \, \dim c_2} \lambda(c_2, c_1).$$

The linking coefficient is a homology invariant in the sense that if the cycle c_2 is homologous to a cycle c_2' in the complement of the carrier of c_1, then $\lambda(c_1, c_2) = \lambda(c_1, c_2')$.

The linking coefficient has an easily visualized geometrical meaning. For example, it is defined for two oriented non-intersecting and non-self-intersecting closed curves in $\mathbb{R}^n$ (or S^n) and is equal to the algebraic number of points of intersection of one curve with a surface spanned by the other curve. The linking coefficient provides the first very crude obstruction to separating two non-intersecting curves in space without passing one through the other. (Of course curves with linking coefficient zero may still be linked, that is, cannot be separated; see Fig. 6.)

Since the linking coefficient is defined only for cycles homologous to 0, we must be careful in speaking of the induced homology linking coefficient. For example, if A and B are two non-intersecting subsets of an n-dimensional manifold X, then the linking coefficient defines a bilinear form

$$\mathrm{Ker}[H_p(A) \to H_p(X)] \times \mathrm{Ker}[H_q(B) \to H_q(X)] \to \mathbb{Z}, \quad p+q = n-1.$$

In the important case when X is a sphere we obtain the bilinear form

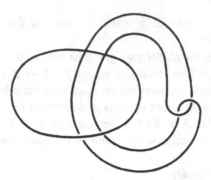

Fig. 6

$$H_p(A) \times H_q(B) \to \mathbb{Z}, \qquad p+q = n-1, \qquad p > 0, \qquad q > 0.$$

B. Another theory of linking coefficients having a direct relation to Poincaré duality can be constructed in the following situation. Let X be a closed oriented smooth n-manifold and let $\alpha \in H_q(X)$, $\beta \in H_{n-q-1}(X)$ be elements of finite order. Suppose also that a and b are cycles representing α and β. Assume that $Na = \partial c$. We define the *linking coefficient* $l(\alpha, \beta)$ as the rational number $\phi(c, b)/N$ reduced mod 1. Thus $l(\alpha\beta) \in \mathbb{Q}/\mathbb{Z}$.

The reader may verify that this is well defined (for this it is essential that β has finite order), and also prove the following assertions: (I) if $N\alpha = 0$, $M\beta = 0$ and $K = \mathrm{hcf}(M, N)$, then $Kl(\alpha, \beta) = 0$; (ii) $l(\alpha, \beta) = \pm l(\beta, \alpha)$ (make the signs precise).

Theorem. *With a class* $\alpha \in \mathrm{Tors}\, H_q(X)$ *let us associate the function* $\beta \mapsto l(\alpha, \beta)$. *Then we obtain an isomorphism*

$$\mathrm{Tors}\, H_q(X) \cong \mathrm{Hom}(\mathrm{Tors}\, H_{n-q-1}(X), \mathbb{Q}/\mathbb{Z}). \quad \square$$

This is the secondary Poincaré duality.

8.6. Inverse homomorphisms.

(The basic result of this section is valid for homology manifolds; the later ones can in general be extended to PL-manifolds.)

A. Let X and Y be connected closed oriented smooth manifolds, in general of different dimensions m and n, and let $f : X \to Y$ be a continuous map. The presence of the Poincaré isomorphism enables us to construct in terms of f maps of the homology and cohomology groups going in the "wrong direction":

$$f^! : H_q(Y) \xrightarrow{D^{-1}} H^{n-q}(Y; \mathbb{Z}) \xrightarrow{f^*} H^{n-q}(X; \mathbb{Z}) \xrightarrow{D} H_{m-n+q}(X),$$

$$f_! : H^q(X; \mathbb{Z}) \xrightarrow{D} H_{m-q}(X) \xrightarrow{f_*} H_{m-q}(Y) \xrightarrow{D^{-1}} H^{n-m+q}(Y).$$

Both homomorphisms change dimension by $m - n$: $f^!$ "increases" dimension by $m - n$ (we put the word "increases" in inverted commas since $m - n$ might be negative or zero), and $f_!$ "decreases" dimension by $m - n$.

B. Let us say a little about the cohomology homomorphism $f_!$. It is the simplest manifestation of an unusually general and important construction known as the "direct image". Its obvious analytic meaning (the intuitive sources of cohomology lie in analysis rather than in geometry) in any case when f is the projection of a smooth bundle, is expressed by the words "fibrewise integration", appealing to the de Rham theory (see 6.5.**D**).

C. The homomorphism $f^! : H_q(Y) \to H_{m-n+q}(X)$ is called the *Hopf inverse homomorphism*. Its geometric meaning is sufficiently clearly illustrated by the following theorem.

Theorem. *Suppose that the homology class $\alpha \in H_q(Y)$ is realized by an oriented submanifold $Z \subset Y$ and that f is transversally regular with respect to Z. Then $f^{-1}(Z) \subset X$ is an oriented submanifold realizing the class $f^!(\alpha)$.* $\square$

Let us clarify the meaning of the words in the theorem. A map $f : X \to Y$ is said to be *transversally regular* with respect to Z if for any point $x \in f^{-1}(Z)$ there is a small disc embedded in X with centre x and dimension $\dim Y - \dim Z$ that is mapped diffeomorphically by f onto a disk lying transversally to Z in Y. This is a condition involving general position; if it is not satisfied it may be achieved by a small perturbation of f (or of Z). It is easy to see that if f is transversally regular with respect to Z, then $f^{-1}(Z)$ is a submanifold having the same codimension in X as Z has in Y. If X, Y, and Z are oriented, then $f^{-1}(Z)$ also acquires a canonical orientation. (Note that if the codimension of Z in Y is greater than the dimension of X then being transversally regular reduces to saying that $f(X)$ and Z do not intersect.)

As regards the proof of the theorem we shall confine ourselves to the remark that if all the differential data are replaced by the corresponding conditions in terms of triangulations (X and Y are triangulated, f is simplicial, the class α is represented by a star cycle), then it becomes more or less obvious. Note too that the statement of the theorem does not say which of X and Y has the greater dimension: it is valid in both cases although they have different geometric content.

It is important that an analogous theorem is valid for manifolds with singularities of codimension ≥ 2.

D. Proposition. (i) $\langle \alpha, f^! \beta \rangle = \langle f^! \alpha, \beta \rangle$. (ii) *Let X and Y be closed smooth manifolds of the same dimension and let $f : X \to Y$ be a map of degree d. Then the compositions*

$$H_q(Y) \xrightarrow{f^!} H_q(X) \xrightarrow{f_*} H_q(Y),$$
$$H^q(Y; Z) \xrightarrow{f^*} H^q(X; Z) \xrightarrow{f_!} H^q(Y; Z)$$

are the same as multiplication by d. $\square$

Corollary. *If f is a map of degree ± 1, then f_* is an epimorphism and f^* a monomorphism. Generalization: if f is a map of degree $d \neq 0$, then each homology class of Y multiplied by d belongs to the image of f_* and each homology class of Y belonging to the kernel of f^* is annihilated by multiplication by d.* $\square$

For example, there is no map $S^2 \to S^1 \times S^1$ of nonzero degree (but there exists a map $S^1 \times S^1 \to S^2$ of degree 1: the quotient map of $S^1 \times S^1$ factored by $S^1 \vee S^1$).

Everything said here has an obvious $\mathbb{Z}_2$ analogue.

8.7. The relation with the ∪-product.
(The content of this section can in general be extended to PL-manifolds.)

A. Theorem. *Suppose that Y_1 and Y_2 are closed oriented smooth submanifolds in general position in a closed oriented smooth manifold X. This means that Y_1 and Y_2 intersect transversely in a smooth submanifold $Z \subset X$, whose dimension k is related to the dimensions m_1, m_2, and n of Y_1, Y_2, and X by $k = m_1 + m_2 - n$. Suppose also that $\alpha_1 \in H^{n-m_1}(X; \mathbb{Z})$, $\alpha_2 \in H^{n-m_2}(X; \mathbb{Z})$, $\beta \in H^{2n-m_1-m_2}(X; \mathbb{Z})$ are cohomology classes such that the homology classes $D\alpha_1$, $D\alpha_2$, $D\beta$ realize the manifolds Y_1, Y_2, Z. Then*

$$\alpha_1 \cup \alpha_2 = \beta. \quad \square$$

A similar result is true in the non-oriented case with coefficients in $\mathbb{Z}_2$.

(Note that the condition that Y_1 and Y_2 are in general position is equivalent to saying that the inclusion $i_1 : Y_1 \to X$ is transversally regular with respect to Y_2 or equivalently, that $i_2 : Y_2 \to X$ is transversally regular with respect to Y_1.)

The theorem is a direct corollary of the preceding theorem:

$$D(\alpha_1 \cup \alpha_2) = [X] \cap (\alpha_1 \cup \alpha_2) = ([X] \cap \alpha_1) \cap \alpha_2 = (i_{1*}[Y]) \cap \alpha_2 =$$
$$= i_{1*}([Y_1] \cap i_1^* \alpha_2) = i_{1*}(Di_1^* \alpha_2) = i_{1*}(Di_1^* D^{-1}(i_{2*}[Y_2])) =$$
$$= i_1^*(i_1^!(i_{2*}[Y_2])) = i_{1*}[i^{-1}(Y_2)] = i_*[Z] = D\beta$$

(where i is the inclusion of Z in X).

B. Theorem A provides a very powerful means of computing the multiplicative structure of the cohomology of various manifolds (and of other spaces, in view of the naturality of the product).

Example. If $q_1 + q_2 \leq n$, then the product of the canonical generators of the groups $H^{2q_1}(\mathbb{C}P^n; \mathbb{Z})$ and $H^{2q_2}(\mathbb{C}P^n; \mathbb{Z})$ is equal to the canonical generator of $H^{2(q_1+q_2)}(\mathbb{C}P^n; \mathbb{Z})$: in fact the first two generators are mapped by the Poincaré isomorphism to homology classes represented by projective subspaces of dimensions $(n - q_1)$ and $(n - q_2)$; their intersection in general position is an $(n - q_1 - q_2)$-dimensional projective space. Thus the ring $H^*(\mathbb{C}P^n; \mathbb{Z})$ has the following structure: it contains $1 \in H^0(\mathbb{C}P^n; \mathbb{Z})$ and a multiplicative generator $x \in H^2(\mathbb{C}P^n; \mathbb{Z})$. The remaining cohomology groups are generated by powers of this generator: $x^2 \in H^4(\mathbb{C}P^n; \mathbb{Z})$, $x^3 \in H^6(\mathbb{C}P^n; \mathbb{Z})$, and so on. If n is finite, then $x^{n+1} = 0$. In more algebraic language we may say that $H^*(\mathbb{C}P^n; \mathbb{Z})$ is a polynomial ring in one variable x, factored by the ideal generated by x^{n+1}:

$$H^*(\mathbb{C}P^n; \mathbb{Z}) = \mathbb{Z}[x]/(x^{n+1}), \quad \dim x = 2.$$

Similarly,

$$H^*(\mathbb{R}P^n; \mathbb{Z}_2) = \mathbb{Z}_2[x]/(x^{n+1}), \quad \dim x = 1,$$
$$H^*(\mathbb{H}P^n; \mathbb{Z}) = \mathbb{Z}[x]/(x^{n+1}), \quad \dim x = 4,$$
$$H^*(\mathbb{C}aP^2; \mathbb{Z}) = \mathbb{Z}[x]/(x^3), \quad \dim x = 8.$$

In all cases except $\mathbb{R}P^n$ we may replace $\mathbb{Z}$ (on both sides of the equality) by an arbitrary commutative ring with identity.

C. Theorem **A** shows that there is a rich multiplicative structure in the cohomology of a manifold (there are many nonzero products). This thesis is illustrated by the following proposition.

Corollary to Theorem A. *If α_1, α_2 are cohomology classes of complementary dimensions of a closed oriented manifold X, then*

$$\langle (\alpha_1 \cup \alpha_2, [X]) \rangle = \phi(D\alpha_1, D\alpha_2). \quad \square$$

(The non-oriented $\mathbb{Z}_2$ analogue is also valid.)

D. Since the intersection number is a non-degenerate bilinear form on the homology of a manifold (Poincaré duality), the value of the product on the fundamental class is a non-degenerate bilinear form on the cohomology (cohomological Poincaré duality). This duality in homology and cohomology is of particular interest in the middle dimension.

Theorem. *Let X be a closed oriented manifold of even dimension $2k$. Then the bilinear form $(\alpha_1, \alpha_2) \mapsto \phi(\alpha_1, \alpha_2)$ defined on the free part of $H^k(X; \mathbb{Z})$ is given by an (integer) matrix with determinant ± 1. This matrix is symmetric for even k and skew-symmetric for odd k. In particular, the kth Betti number is even if k is odd. The analogous fact for non-oriented manifolds is false (the Klein bottle).* $\square$

8.8. Generalizations of the Poincaré isomorphism and duality.

A. *The noncompact case.* If a (connected) manifold X is not compact (but as before has no boundary) then it has an infinite triangulation. There is still a correspondence between the simplexes and the barycentric stars, but there is no longer an isomorphism between the chains and cochains, since chains are finite linear combinations of simplexes while cochains are arbitrary functions on stars. Two ways out of this situation suggest themselves, and both are acceptable: to allow infinite chains or to restrict ourselve to finite cochains. Both modifications lead to specific homology and cohomology theories which can be validly defined on arbitrary topological spaces. The *open homology* $H_q^{\text{open}}(X; G)$ of a topological space X is defined as the homology of the complex $\{C_q^{\text{open}}(X; G), \partial\}$, where $C_q^{\text{open}}(X; G)$ is, in general, the group of infinite linear combinations $\sum g_i f_i$ of singular q-simplexes with the property that for any compact subset $K \subset X$ the intersection $f_i(T^q) \cap K$ is nonempty for only finitely many suffixes i; the boundary is defined by the usual formula. The *compact cohomology* $H_{\text{comp}}^q(X; G)$ of a space is defined by means of the complex $\{C_{\text{comp}}^q(X; G), \partial\}$, where $C_{\text{comp}}^q(X; G)$ is the group of functions on singular q-simplexes that have the property that there is a compact subset K (depending on the function) such that the value of the function is 0 on a singular simplex f for which $f(T^q) \cap K$ is empty. Defects of both theories are the absence of homotopy invariance and even of the induced homomorphisms f_*, f^* (these are only defined when the map f is proper, that is, the inverse image of a compact set is compact).

By applying the homology and cohomology theories described above we can immediately construct two different generalizations of the Poincaré isomorphism:

Theorem. *If X is a smooth oriented, in general noncompact, n-dimensional manifold without boundary, then*

$$H_q(X) \cong H^{n-q}_{comp}(X; \mathbb{Z}),$$
$$H^{open}_q(X) \cong H^{n-q}(X; \mathbb{Z}).$$

Similar isomorphisms hold for coefficients in $\mathbb{Z}_2$ without assuming orientability. □

The intersection numbers and linking coefficients between ordinary and open cycles provide a duality corresponding to these Poincaré isomorphisms. (Examine the case when X is the product of a closed manifold and a line.) In general more or less all of the preceding theory carries over to our new situation, but we shall not go into details. However, we shall single out the important case when open homology and compact cohomology reduce to the usual homology and cohomology.

We first note that if X is a compact space and A a closed subset, then there are natural (specify in what sense) isomorphisms

$$H^{open}_q(X \setminus A; G) \cong H_q(X, A; G),$$
$$H^q_{comp}(X \setminus A; G) \cong H^q(X, A; G).$$

In particular, if X is locally compact and $X^\bullet$ is its one-point compactification, then

$$H^{open}_q(X; G) \cong H_q(X^\bullet; G), \quad H^q_{comp}(X; G) \cong H^q(X^\bullet; G).$$

This assertion enables us to deduce an important corollary of the preceding theorem without bringing in exotic homology or cohomology.

Corollary. *Let X be a compact topological space, and A a closed subset such that $X - A$ is a smooth oriented n-manifold without boundary. Then*

$$H_q(X - A) \cong H^{n-q}(X, A; \mathbb{Z}),$$
$$H_q(X, A) \cong H^{n-q}(X - A; \mathbb{Z}).$$

A similar result is true with coefficients in $\mathbb{Z}_2$ without assuming orientability. □

As an exercise the reader should devise a construction for these isomorphisms with the aid of the relative $\cap$-product.

B. *The case of a manifold with boundary.* In the important special case when X is a connected compact oriented manifold with boundary and A is its boundary, the preceding isomorphisms take the form

$$H_q(X) \cong H^{n-q}(X, \partial X; \mathbb{Z}),$$
$$H_q(X, \partial X) \cong H^{n-q}(X; \mathbb{Z})$$

(the open set $X - \partial X$ does not differ homotopically from X).

It is easy to see that these isomorphisms together with the Poincaré isomorphisms for ∂X form an isomorphism between the homology and cohomology sequences of the pair $(X, \partial X)$, that is, there is a commutative ladder

$$\cdots \to \quad H_q(\partial X) \quad \to \quad H_q(X) \quad \to H_q(X, \partial X) \to \quad H_{q-1}(\partial X) \quad \to \cdots$$
$$\quad \quad \| \wr \quad \quad \quad \| \wr \quad \quad \quad \| \wr \quad \quad \quad \| \wr$$
$$\cdots \to H^{n-1-q}(\partial X; \mathbb{Z}) \to H^{n-q}(X, \partial X; \mathbb{Z}) \to H^{n-q}(X; \mathbb{Z}) \to H^{n-q}(\partial X; \mathbb{Z}) \to \cdots$$

We suppose that dim $X = 2k + 1$, and consider ⦂   is commutative
ladder

$$H_{k+1}(X, \partial X) \xrightarrow{\partial_*} H_k(\partial X) \xrightarrow{i_*} H_k(X) \longrightarrow \cdots$$

$$\| \wr \qquad\qquad \| \wr$$

$$H^k(\partial X; \mathbb{Z}) \xrightarrow{\delta^*} H^{k+1}(X, \partial X; \mathbb{Z}) \longrightarrow \cdots$$

(where i is the inclusion of ∂X in X). The homomorphisms δ^* and ∂_* are adjoint with respect to the pairing $\langle , \rangle$, and hence the ranks of the groups Coker $\partial_* = H_k(\partial X)/\text{Im } \partial_*$ and Ker δ^* are the same. Hence the ranks of the groups $H_k(\partial X)/\text{Im } \partial_*$ and Ker $i_* = \text{Im } \partial_*$ are the same. Hence we have the important corollary:

$$\text{rank Im } \partial_* = \frac{1}{2}\text{rank } H_k(\partial X).$$

Further corollaries. (i) *If an even-dimensional oriented manifold is the boundary of a compact manifold, then its middle Betti number is even.* (For non-oriented manifolds this is not true, of course, as is shown once again by the example of the Klein bottle; however, it is true that for any even-dimensional boundary of a compact manifold the dimension of the middle homology group with coefficients in $\mathbb{Z}_2$ is even. It follows from this for example that neither the real nor the complex projective plane can be the boundary of a compact manifold; many other similar examples can be found among the manifolds known to us.)

(ii) *The Euler characteristic of the boundary of a compact manifold is even.* (If the dimension of the boundary is even, this follows from the previous corollary, while if it is odd, the Euler characteristic of the boundary is equal to 0 from what was said in 8.3.C.)

(iii) *If $\alpha, \beta \in H_k(\partial X)$ lie in the image of ∂_*, then their intersection number is* 0. (This has a visual proof (see Fig. 7), but can easily be proved formally: $\phi(\partial_* \gamma, \partial_* \delta) = \phi(\gamma, i_* \partial_* \delta) = \phi(\gamma, 0) = 0$.)

The last fact is particularly important when k is even. We have already remarked above that if the middle homology group is even-dimensional, then there is a unimodular symmetric bilinear form defined on its free part. We now see that when the manifold is a boundary this form has a null space of half the dimension, which is only possible when the form has signature 0. The result is as follows.

Theorem. *If an orientable manifold with dimension divisible by 4 is the boundary of a compact orientable manifold, then the "intersection number form" on the free part of its middle homology group has signature* 0.

C. *The Alexander-Pontryagin isomorphism and duality.* Let A be a closed subset of the sphere S^n. Then from previous results there are isomorphisms

$$H_q(S^n - A) \cong H^{n-q}(S^n, A; \mathbb{Z}),$$
$$H_q(S^n, A) \cong H^{n-q}(S^n - A; \mathbb{Z}).$$

But, since the sphere has trivial homology and cohomology in almost all dimensions, the homology and cohomology sequences of the pair (S^n, A) give rise to isomorphisms

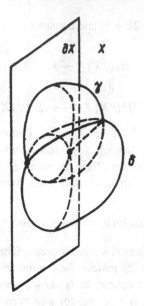

Fig. 7

$$\left.\begin{array}{c} H_q(S^n, A) \cong H_{q-1}(A), \\ H^q(S^n, A; \mathbb{Z}) \cong H^{q-1}(A; \mathbb{Z}) \end{array}\right\} \text{ for } 1 < q < n.$$

Thus

$$\left.\begin{array}{c} H_p(A) \cong H^r(S^n - A; \mathbb{Z}), \\ H^p(A; \mathbb{Z}) \cong H_r(S^n - A) \end{array}\right\} \text{ for } p + r = n - 1, \, p > 0, \, r > 0$$

(the conditions $p > 0$, $r > 0$ can be omitted if we use reduced homology and cohomology). These are called the *Alexander-Pontryagin isomorphisms*. Like the Poincaré isomorphisms they can be represented in terms of homology:

the free part of $H_p(A) \cong$ the free part of $H_{n-1-p}(S^n - A)$,

the torsion of $H_p(A) \cong$ the torsion of $H_{n-2-p}(S^n - A)$

and in this form they are instances of *Alexander-Pontryagin duality*, analogous to Poincaré duality. The duality between the free parts is realized by the classical linking coefficients (see above), and the duality between the torsions by the secondary linking coefficients which we have not considered, but whose theory the reader can construct for himself if desired.

D. *The integral non-orientable Poincaré isomorphisms*. These have the form

$$H_q(X; \mathbb{Z}) \cong H^{n-q}(X; \text{or } X),$$
$$H^q(X; \mathbb{Z}) \cong H_{n-q}(X; \text{or } X).$$

Here X is a closed, in general non-orientable smooth n-manifold, and or X is the local system of orientations (see example 2° in 6.4.A).

There are analogues of the intersection numbers and linking coefficients in the non-orientable situation. We leave details to the reader.

Finally we note that there is no reason why what is said here cannot be combined with that in **A** so that we obtain an integral Poincaré duality in the non-orientable non-compact case.

Chapter 3
Obstructions, characteristic classes and cohomology operations

§9. Obstructions

9.1. Obstructions to extending a continuous map.

We often meet the problem of extending a continuous map $A \to Y$ to a larger space X containing A. This is the problem (in many versions) which is considered in obstruction theory. We begin with a technically important special case.

A. Let X be a cellular space and Y a homotopically simple topological space. We shall use the abbreviated notation X^n for the n-skeleton of X.

We consider the problem of extending a given continuous map $f : X^n \to Y$ to a continuous map $f : X^{n+1} \to Y$.

Let $e \subset X$ be an $(n + 1)$-cell, $h : D^{n+1} \to X$ the corresponding characteristic map. Since $h(S^n) \subset X^n$, we obtain a map $f_e : S^n \to X$, $f_e(s) = f(h(s))$. It is completely obvious that f can be extended to $X^n \cup e$ if and only if f_e can be extended to a map $D^{n+1} \to X$, that is, when f_e is homotopic to 0, that is, when f_e represents $0 \in \pi_n(Y)$ (recall that Y is homotopically simple).

Clearly f can be extended to X^{n+1} if and only if it can be extended to $X^n \cup e$ for each $(n + 1)$-cell e. By constructing for each cell e a map $f_e : S^n \to X$ as above, and denoting the corresponding element of $\pi_n(Y)$ by ϕ_e, we can make the following, essentially tautological, assertion: a continuous map $f : X^n \to Y$ can be extended to a continuous map $X^{n+1} \to Y$ if and only if all the ϕ_e are zero.

The correspondence $e \to \phi_e$ can be regarded as a cellular cochain of X with coefficients in $\pi_n(Y)$. This cochain is denoted by c_f and is called the *obstruction to extending f to X^{n+1}*. Thus

$$c_f \in \mathscr{C}^{n+1}(X; \pi_n(Y)).$$

B. Everything said so far has been a triviality; here is the first significant assertion.

Theorem. *The obstruction is a cocycle:* $\delta c_f = 0$. $\qquad\square$

This is a variation on the theme $\partial\partial = 0$ (we need to show that $c_f(\partial a) = 0$ for any chain a, but the cochain c_f itself is defined by means of boundaries); for a precise proof see Fomenko and Fuchs (1989).

. The cohomology class of c_f will be denoted by C_f. In fact $C_f \in H^{n+1}(X; \pi_n(Y))$ is often called the obstruction (and c_f is then called the obstruction cocycle).

C. Theorem. $C_f = 0$ *if and only if* $f : X^n \to Y$ *can be extended to* X^{n+1} *by first altering* f *on* X^n, *but leaving it unchanged on* X^{n-1}.   □

(This theorem can be applied successively to extend a map from one skeleton to the next. In fact, if the map is given on X^n, then there is an obstruction in $(n + 1)$-dimensional cohomology to extension to X^{n+1} with a possible correction on X^n not affecting X^{n-1}. If such an extension exists then there is an obstruction in $(n+2)$-dimensional cohomology to extension to X^{n+2}, and so on. But it must be remembered that each successive obstruction depends on the choice of the extension at all the previous stages, and this lack of uniqueness accumulates from skeleton to skeleton.)

D. The proof of theorem **C** depends on the concept of difference cochain, which is also important in itself.

Suppose that $f, g : X^n \to Y$ are two maps that coincide on X^{n-1}. Consider an arbitrary n-cell $e \subset X$ with characteristic map $h : D^n \to X$. The maps $f \circ h$, $g \circ h : D^n \to Y$ agree on S^{n-1} (because $h(S^{n-1}) \subset X^{n-1}$ and f and g coincide on X^{n-1}), and together define a map $k_e : S^n \to Y$ (the domain of definition of f is identified with the upper hemisphere, and that of g with the lower hemisphere). We define the *difference cochain*

$$d_{f,g} \in \mathscr{C}^n(X; \pi_n(Y))$$

as the cochain whose value on e is the class of the spheroid k_e. Clearly $d_{f,g} = 0$ if and only if f and g are joined by a homotopy that is fixed on X^{n-1}. In the case that is important for us when both maps f, g are defined from the start on the whole of X (and coincide on X^{n-1}), the last assertion may be modified (in view of Borsuk's theorem) as follows: there is a homotopy of f, fixed on X^{n-1}, and joining f and g on X^n.

The following properties of the difference cochain are more or less obvious.

Proposition. (i) $d_{f,g} + d_{g,h} = d_{f,h}$ and $d_{f,g} = -d_{g,f}$.

(ii) $\delta d_{f,g} = c_g - c_f$.

(iii) *For any map* $f : X^n \to Y$ *and any cochain* $d \in C^n(X; \pi_n(Y))$ *there is a map* $g : X^n \to Y$ *coinciding with* f *on* X^{n-1} *and such that* $d_{f,g} = d$.   □

E. *Proof of theorem* **C**. If $C_f = 0$, then $c_f = \delta d$, so by (iii) there is a map $g : X^n \to Y$ coinciding with f on X^{n-1} and such that $d_{f,g} = -d$. But then $c_g = c_f + \delta d_{f,g} = 0$, that is, g can be extended to X^{n+1}. Conversely, if there is a map $g : X^n \to Y$ that can be extended to X^{n+1} and coinciding with f on X^{n-1}, then $c_g = 0$, $c_f = c_f - c_g = -\delta d_{f,g}$ and $C_f = 0$.   □

9.2. The relative case.

Let A be a cellular subspace of X and let a map f be defined on $A \cup X^n$. The obstruction cochain to extending such a map to $A \cup X^{n+1}$ lies in $\mathscr{C}^{n+1}(X, A; \pi_n(Y))$; it is a cocycle and the corresponding cohomology class $C_f \in H^{n+1}(X, A; \pi_n(Y))$ is called the *obstruction*. The theory of these relative obstructions runs completely parallel to the absolute prototype; in particular difference cochains can be defined and precise analogues (both in formulation and proof) can be proved of all the assertions in the previous subsection. We single out the following important absolute consequence of the relative theory.

Let $f, g : X \to Y$ be two continuous maps coinciding on X^n (the condition that they coincide can be replaced by the condition that they are homotopic, but with the proviso that a definite homotopy between $f|_{X^n}$ and $g|_{X^n}$ is given). We consider the problem of constructing a homotopy between f and g that is fixed (or coincides with the given homotopy) on X^n. This is equivalent to the problem of extending to $X \times I$ the map given on $(X \times 0) \cup (X \times 1) \cup (X^n \times I)$ by

$$(x, t) \mapsto \begin{cases} f(x) & \text{for } t = 0 \text{ or } x \in X^n, \\ g(x) & \text{for } t = 1 \text{ (or } x \in X^n). \end{cases}$$

(In the case when $f|_{X^n}$ and $g|_{X^n}$ are homotopic the formula is slightly different; we leave the details to the reader.) The obstruction to extending this map to $(X \times 0) \cup (X \times 1) \cup (X^{n+1} \times I)$ lies in $\mathscr{C}^{n+2}(X \times I, (X \times 0) \cup (X \times 1); \pi_{n+1}(Y)) = \mathscr{C}^{n+1}(X; \pi_{n+1}(Y))$; it is easy to see that it is equal to $d_{f,g}$. Now $\delta d_{f,g} = c_g - c_f = 0$, since f and g are defined on the whole of X. By applying the relative version of the theorem of 9.1.C to this situation, we obtain the following result.

Theorem. *If $f, g : X \to Y$ are maps coinciding on X^n, then $d_{f,g}$ is a cocycle, and its cohomology class $D_{f,g} \in H^{n+1}(X; \pi_{n+1}(Y))$ is zero if and only if $f \mid X^{n+1}$ and $g \mid X^{n+1}$ can be joined by a homotopy that is fixed on X^{n-1}.* $\square$

(We may make a remark similar to that after the statement of the theorem in 9.1.C.)

9.3. Application: cohomology and maps into $K(\pi, n)$ spaces.

Recall that the construction of the space $K(\pi, n)$ begins by taking a bouquet of n-spheres in one-one correspondence with the elements of some system of generators of the group π, and that this bouquet will be the n-skeleton of the space $K(\pi, n)$ under construction (see Part I, 11.7). If with each n-cell (a sphere of the bouquet) we associate the corresponding generator, we obtain a cochain $c \in \mathscr{C}^n(K(\pi, n); \pi)$.

Lemma. *The cochain c is a cocycle.*

Proof. The $(n+1)$-cells correspond to relations between the chosen generators of π; if a cell σ corresponds to the relation $\sum k_i g_i = 0$, where g_i are the generators, then the coefficient $[\sigma : e]$ is equal to the coefficient in this relation of the generator corresponding to e. If e_i denotes the cell corresponding to the generator g_i we obtain

$$\delta c(\sigma) = \sum [\sigma : e_i] c(e_i) = \sum k_i g_i = 0.$$

The cohomology class $F_\pi \in H^n(K(\pi, n); \pi)$ of the cocycle c is called the *fundamental cohomology class* of $K(\pi, n)$. Another construction of F_π can be derived by means of the universal coefficient formula: since $H_{n-1}(K(\pi, n)) = 0$,

$$H^n(K(\pi, n); \pi) = \text{Hom}(H_n(K(\pi, n)), \pi),$$

but by the Hurewicz theorem $H_n(K(\pi, n)) = \pi$. The fundamental class F_π corresponds to the identity homomorphism $\text{id}_\pi \in \text{Hom}(H_n(K(\pi, n)), \pi)$. Incidentally this construction can be applied to any $(n-1)$-connected space X, giving a fundamental class $F_X \in H^n(X; \pi_n(X))$. (We leave the reader to prove that the two definitions of the fundamental class are equivalent.)

We now turn to the main object of this subsection. Let X be any topological space and $f : X \to K(\pi, n)$ a continuous map. With the map f we associate the cohomology class $f^* F_\pi \in H^n(X; \pi)$. Since this class depends only on the homotopy class of f, we obtain a map

$$\pi(X, K(\pi, n)) \to H^n(X; \pi), \quad [f] \mapsto f^* F_\pi.$$

Theorem. *If X is cellular, then this map is a one-one correspondence.*   □

Proof. We first show that for any $\alpha \in H^n(X; \pi)$ there is a map $f : X \to K(\pi, n)$ with $f^* F_\pi = \alpha$. Choose a cocycle $\alpha \in \mathscr{C}^n(X; \pi)$ in the class α, and let f map the skeleton X^{n-1} to a point. Then f must be a spheroid on each oriented n-cell e, and we require this spheroid to belong to the class $a(e) \in \pi = \pi_n(K(\pi, n))$. Clearly the cochain a is the difference cochain between f and the constant map const that takes everything to a point. Hence by the results of 9.1

$$0 = \delta a = \delta d_{\text{const}, f} = c_f - c_{\text{const}} = c_f$$

($c_{\text{const}} = 0$, since const can be extended to the whole of X). Thus f can be extended to X^{n+1}. There is no obstruction to extending f to $X^{n+2}, X^{n+3}, \ldots$, since the corresponding obstructions lie in the groups $\mathscr{C}^{n+2}(X; \pi_{n+1}(K(\pi, n)) = 0)$, $\mathscr{C}^{n+3}(X; \pi_{n+2}(K(\pi, n)) = 0)$, etc. Hence we arrive at a map $f : X \to K(\pi, n)$, and by construction $f^* F_\pi = \alpha$.

We now show that if $f^* F_\pi = g^* F_\pi$ then $f \sim g$. We shall join f and g by a homotopy by successively computing difference cochains. These will lie in zero groups until we reach a difference cochain in $C^n(X; \pi_n(K(\pi, n)) = \pi)$. This difference cochain appears at the moment when f and g are already joined by a homotopy on X^{n-1}. In the first part of the proof we saw that $f^* F_\pi$ and $g^* F_\pi$ are the cohomology classes $D_{\text{const}, f}$ and $D_{\text{const}, g}$ of $d_{\text{const}, f}$ and $d_{\text{const}, g}$. Hence $D_{f, g} = D_{f, \text{const}} + D_{\text{const}, g} = D_{\text{const}, g} - D_{\text{const}, f} = g^* F_\pi - f^* F_\pi = 0$. Consequently f can be joined to g on X^n by a homotopy that is fixed on X^{n-2}. In the succeeding steps the difference cochains are again trivial (they lie in zero groups), and we successively join f to g by homotopies on $X^{n+1}, X^{n+2}, \ldots$. Hence $f \sim g$.   □

Note. The one-one correspondence $H^n(X; \pi) \approx \pi(X, K(\pi, n))$ is a one-one correspondence between groups: $\pi(X, K(\pi, n))$ is a group, since $K(\pi, n) \sim \Omega(K(\pi, n + 1))$ is an H-space. In fact this correspondence is a group isomorphism.

Corollary to the theorem. *If X is cellular there is an isomorphism $H^1(X; \mathbb{Z}) \cong \pi(X, S^1)$ (the Abelian group structure on the right-hand side arises because S^1 has a natural topological Abelian group s structure).*

9.4. Another application: Hopf's theorems.

A. Theorem (Hopf). *For any n-dimensional cellular space X there is a one-one correspondence*

$$H^n(X; \mathbb{Z}) \approx \pi(X, S^n), \quad [f] \leftrightarrow f^*(s),$$

where s is the canonical generator of the group $H^n(S^n; \mathbb{Z}) = \mathbb{Z}$. □

This classical theorem (proved in fact before the appearance of either obstruction theory or cohomology theory) is, from a modern point of view, a corollary of the preceding theorem. In fact if the construction of $K(\pi, n)$ is examined, it can be seen that $K(\mathbb{Z}, n)$ is obtained from S^n by attaching cells of dimensions $\geq n + 2$ ($\mathbb{Z}$ has one generator and no relations). Hence $\pi(X, K(\mathbb{Z}, n)) = \pi(X, S^n)$ by the cellular approximation theorem (any map $X \to K(\mathbb{Z}, n)$ is homotopic to a map whose image lies in the n-skeleton, and a homotopy between two such maps can always be effected in the $(n + 1)$-skeleton.

The theorem can be proved directly by repeating the proof of the previous theorem more or less word for word. The main difference between the two proofs will be that the higher obstructions and difference cochains will vanish not because the higher homotopy groups of S^n are trivial, which is false, but because X has no cells of higher dimension.

B. Another theorem of Hopf can be proved similarly.

Theorem. *Let (X, A) be a cellular pair, in which X is n-dimensional, and A has the same integral cohomology as the $(n-1)$-sphere. Then A is a retract of X if and only if the homomorphism $H^{n-1}(X; \mathbb{Z}) \to H^{n-1}(A, \mathbb{Z})$ induced by the inclusion is an epimorphism.* □

9.5. Obstructions to the extension of sections.

A. Let (E, B, F, p) be a locally trivial bundle. We shall assume that the fibre F is homotopically simple (for example, simply connected) and that the base is simply connected. (The assumption that the base is simply connected is unnecessarily strong, and can be replaced by assuming that the bundle is *simple*. This means that the bundle over S^1 induced by any continuous map $S^1 \to B$ is trivial; we shall meet important examples of this situation in the next section.)

Suppose that the base B is cellular and that a section $s : B^n \to E$ is given over its n-skeleton. Let e be an $(n+1)$-cell of the base, and $h : D^{n+1} \to B$ its characteristic map. The bundle over D^{n+1} induced by h is trivial. The section s induces a section of the induced bundle over $S^n \subset D^{n+1}$, that is, a map $S^n \to D^{n+1} \times F$, which is an element of the group $\pi_n(D^{n+1} \times F) = \pi_n(F)$. (It would be more correct to say an element of the homotopy group of some fibre over a point of the cell e; however, since the base is simply connected or the bundle is simple we can identify the homotopy groups of all the fibres.) The function taking the cell e to the element just constructed in $\pi_n(F)$ is a cochain in $\mathscr{C}^{n+1}(B; \pi_n(F))$, which we denote by c_s. It is the *obstruction* (or *obstruction cochain*) to *extending s to the $(n + 1)$-skeleton*. The properties of this obstruction are completely analogous to those of the obstruction to extending a map, and we simply list them.

(1) The section s can be extended to a section over the $(n + 1)$-skeleton of the base if and only if $c_s = 0$.

(2) $\delta c_s = 0$.

(3) The cohomology class C_s (also called the obstruction) of the cocycle c_s vanishes if and only if the section can be extended to the $(n + 1)$-skeleton, possibly

by first changing it on the n-skeleton, but leaving it unchanged on the $(n-1)$-skeleton.

Difference cochains can also be introduced and are the analogues of those defined earlier.

Obstructions to extending maps are a special case of obstructions to extending sections: a map $X \to Y$ is equivalent to its graph $X \to X \times Y$, which can be regarded as a section of the trivial bundle. The obstructions to extending the map and to extending its graph are identical. At the same time the obstruction theory of sections is not included in that of maps; in particular, there is no analogue of the following construction in the latter.

B. Suppose that $\pi_0(F) = \ldots = \pi_{n-1}(F) = 0$, $\pi_n(F) \neq 0$. For $k \leq n$ the obstruction to extending a section to the k-skeleton of the base is 0 (as it takes values in the zero group). The first non-trivial obstruction arises in $H^{n+1}(B, \pi_n(F))$. It turns out that this obstruction does not depend on how the section was constructed on the n-skeleton. This fact is almost obvious and is proved by means of difference cochains and the analogue of the theorem stating that $\delta d_{f,g} = c_g - c_f$. Hence this class is determined by the bundle itself, and is called the *characteristic class of the bundle*.

Proposition. (i) (Basic property of characteristic classes.) *If* $\xi \in H^{n+1}(B; \pi_n(F))$ *is the characteristic class of the bundle* (E, B, F, p) *and* $f : B' \to B$ *is a continuous map, then the characteristic class of the bundle over* B' *induced by* f *is* $f^*\xi$.

(ii) *The characteristic class is homotopy invariant; in particular, it is independent of the cell decomposition of the base.*   □

We shall meet some concrete examples of characteristic classes in the next section.

Note. Let X be a cellular space with $\pi_0(X) = \ldots = \pi_{n-1}(X) = 0$, $\pi_n(X) \neq 0$. Then the characteristic class of the Serre fibration $EX \to X$ with fibre ΩX, which lies in $H^n(X; \pi_{n-1}(\Omega X)) = H^n(X; \pi_n(X))$ is just the fundamental class of the space X. (Hence the invariant meaning of the fundamental class is obvious.)

§10. Characteristic classes of vector bundles

10.1. Vector bundles.

In this section we shall give a short account of the necessary elements of the theory of vector bundles and the more general theory of Steenrod fibre bundles that were not included in Part I. For details see §§4.3–4.5 of Rokhlin and Fuchs (1977).

A. A *real (complex) vector bundle* is a locally trivial bundle (E, B, F, p) with additional structure: there are continuous operations $e_1, e_2 \to e_1 + e_2$ and $e, c \to ce$ defined for $e_1, e_2, e \in E$, $p(e_1) = p(e_2)$ and $c \in \mathbb{R}$ ($c \in \mathbb{C}$), satisfying the axioms for a vector space. With respect to these operations the fibres of the bundle are real (complex) vector spaces. It is assumed that these spaces are finite dimensional and have the same dimension, which is defined to be the dimension of the bundle.

A real vector bundle is called *oriented* if its fibres are endowed with compatible orientations.

Henceforth when we speak of a vector bundle we shall always mean a real, or oriented real, or complex vector bundle.

B. As we have already said, the theory of vector bundles is included in the theory of Steenrod bundles. In particular, an n-dimensional real vector bundle is a Steenrod bundle with standard fibre $\mathbb{R}^n$ and structure group $GL(n, \mathbb{R})$, an n-dimensional oriented real vector bundle is a Steenrod bundle with standard fibre $\mathbb{R}^n$ and structure group $GL_+(n, \mathbb{R})$, and an n-dimensional complex vector bundle is a Steenrod bundle with standard fibre $\mathbb{C}^n$ and structure group $GL(n, \mathbb{C})$.

C. A vector bundle of any of the above types with a cellular base has a homotopically unique Euclidean or Hermitian structure (that is, its fibres can be endowed with such a structure depending continuously on the point in the base). When necessary we shall assume that such a structure exists, that is, that the structure groups of our bundles have been contracted to $O(n)$, $SO(n)$, $U(n)$.

D. From the general theory of Steenrod bundles we also need the operation of forming the *associated bundle*, which, in the case of vector bundles, means the following. Let ξ be a vector bundle with structure group G, and let F be the space on which G acts. Let E^* denote the collection of all n-frames in all the fibres of ξ. The set E^* has a natural topology, a natural projection onto B, which we denote by p^* and a natural (right) action of G. Let E' denote the space $(E^* \times F)/G$, that is, the space obtained from $E^* \times F$ by the identification $(e^*g, f) = (e^*, gf)$, and consider the bundle (E', B, F, p'), where p' is the natural projection of E' onto B (the class of the pair (e^*, f) is mapped to $p^*(e^*)$). This is then the *associated bundle*.

In many important specific cases the associated bundle has a clear geometric interpretation. For example, if $G = O(n)$ or $SO(n)$, and $F = S^{n-1}$, then E' can be identified with the subspace of E consisting of vectors of unit length. If F is a Stiefel manifold, then E' consists of the frames of the corresponding type in the fibres of the original bundle, and so on.

10.2. Associated bundles and characteristic classes.

A. Let us fix one of the three types of vector bundles named above. We say that we are given a q-dimensional *characteristic class* α of an n-dimensional vector bundle (of our type) with coefficients in an Abelian group A, if to each n-dimensional vector bundle ξ there corresponds a q-dimensional cohomology class $\alpha(\xi)$ of the base with coefficients in A, in such a way that if ξ is an n-dimensional vector bundle with base B and $f : B' \to B$ is a continuous map of a cellular space B' into B, then $\alpha(f^*\xi) = f^*\alpha(\xi)$ (on the left-hand side f^* denotes passage to the induced bundle, and on the right-hand side the induced cohomology homomorphism).

We have already met the term "characteristic class", which was used in §9 for the first obstruction to constructing a section, and an equation of the type $\alpha(f^*\xi) = f^*\alpha(\xi)$ was satisfied by these characteristic classes. However, such a construction is not directly applicable to vector bundles because their fibres are contractible. (Recall that the coefficients for the characteristic classes in §9 were in the first non-trivial homotopy group of the fibre.) The way out of this situation is to apply the construction of §9 not to the vector bundle but to some derived bundle that can be constructed naturally in terms of the vector bundle. Associated bundles provide a supply of derived bundles of this kind.

In more detail our method for constructing characteristic classes is as follows. We fix a space F on which the structure group of our bundle acts, and take the bundle with fibre F associated with our vector bundle. Then the characteristic class of the associated bundle (in the sense of §9) is called the characteristic class of our vector bundle. This class lies in $H^{m+1}(B; \pi_m(F))$, where B is the base of the bundle, m the dimension of the first non-trivial homotopy group of F (we assume that the conditions imposed on F and the bundle in §9 are fulfilled).

F may be taken to be any space on which the structure group of our bundle acts. However, as a rule we shall take F to be a real or complex Stiefel manifold (the space of orthonormal or unitary frames), in particular, a sphere.

B. Let ξ be an oriented n-dimensional vector bundle with cellular base B. Consider the associated bundle ξ' with fibre S^{n-1}. It is easy to see that the orientability of ξ implies that ξ' is simple (the reader who wishes to examine this topic more deeply can easily convince himself that a Steenrod bundle with connected structure group is always simple), and the choice of a definite orientation determines an isomorphism between the $(n-1)$th homotopy groups of the fibres of ξ' and the group $\mathbb{Z}$. Hence the first obstruction to the construction of a section to ξ' is defined, and is an element of $H^n(B; \mathbb{Z})$. This element – the characteristic class of ξ – is called the *Euler class* of ξ and denoted by $e(\xi)$.

C. Lemma. *Let $1 \leq k < n$. Then:*
(i) $\pi_i(V(n, k)) = 0$ *for* $i < n - k$;
(ii) $\pi_{n-k}(V(n, k)) = \begin{cases} \mathbb{Z} & \text{if } k = 1 \text{ or } n - k \text{ is even,} \\ 0 & \text{otherwise.} \end{cases}$
(See Part I, §10.7.)   □

Again let ξ be an oriented n-dimensional vector bundle. Let ξ_k denote the associated bundle with fibre $V(n, k)$, the manifold of orthogonal k-frames in n-dimensional space. The first obstruction to the construction of a section of ξ_k takes values in $H^{n-k+1}(B; \pi_{n-k}(V(n, k)) = \mathbb{Z}$ or $\mathbb{Z}_2$. This obstruction is reduced mod 2 and the resulting class $w_j(\xi) \in H^j(B; \mathbb{Z}_2)$, $j = n - k + 1$, is called the *jth Stiefel-Whitney class* of the bundle. Note that the orientability of ξ which is necessary, as in the construction of the Euler class, for the bundle ξ_k to be simple, becomes superfluous after the reduction of the coefficients mod 2, so that the Stiefel-Whitney classes are defined for any real vector bundles.

Note also that $w_1(\xi) = 0$ if and only if ξ is orientable.

For oriented n-dimensional vector bundles we have the equation $w_n(\xi) = \rho_2 e(\xi)$, where ρ_2 denotes reduction mod 2.

D. Now let ξ be a complex n-dimensional vector bundle.

Lemma. *Let $1 \leq k < n$. Then*

$$\pi_i(CV(n, k)) = \begin{cases} 0 & \text{for } i < 2(n - k) + 1, \\ \mathbb{Z} & \text{for } i = 2(n - k) + 1. \end{cases} \quad \square$$

In view of this lemma, the first obstruction to the construction of a section of the bundle ξ_k is a class $c_j(\xi) \in H^{2j}(B; \mathbb{Z})$, $j = n - k + 1$. This class is called the *jth Chern class of the bundle ξ*.

It can be shown that $e(\mathbb{R}\xi) = c_n(\xi)$, $w_{2j}(\mathbb{R}\xi) = \rho_2 c_j(\xi)$, and $w_{2j+1}(\mathbb{R}\xi) = 0$.

E. Finally, if ξ is again an n-dimensional real vector bundle, then we put $(-1)^j c_{2j}(\mathbb{C}\xi) = p_j(\xi) \in H^{4j}(B; \mathbb{Z})$, and call the classes $p_j(\xi)$ the *Pontryagin classes* of ξ. (The sign $(-1)^j$ has a historical origin; the reason why we restrict ourselves to even Chern classes will be partially explained below.) The Pontryagin classes can also be defined directly: we have to form the bundle associated with ξ whose fibre is the space of systems of $n - 2j + 2$ vectors in the fibres having rank $> n - 2j$, and take the first obstruction to the construction of a section. This is again the jth Pontryagin class $p_j(\xi)$.

10.3. Characteristic classes and classifying spaces.

A. There is a general classification theory of Steenrod bundles in topology (see §4.5 of Rokhlin and Fuchs (1977)). Here we shall briefly describe the results of this theory that concern vector bundles.

We again fix on one of the three types of vector bundle, and call a bundle of this type simply a vector bundle.

Let G_n denote $G(\infty, n)$, $G_+(\infty, n)$, or $\mathbb{C}G(\infty, n)$, depending on the type of the bundle, and let V denote $\mathbb{R}^\infty$, $\mathbb{R}^\infty$, or $\mathbb{C}^\infty$. Consider the n-dimensional vector bundle η_n over G_n, with total space the subspace of $G_n \times V$ consisting of all pairs (π, x) with $x \in \pi$; the projection is $(\pi, x) \to \pi$.

Theorem. *Let X be a finite cellular space. Then: (i) for any n-dimensional vector bundle ξ over X there is a continuous map $f : X \to G_n$, such that $\xi \sim f^*\eta$ ($\sim$ denotes bundle equivalence); (ii) the map f is uniquely determined up to homotopy, that is, if $f_1^*\eta \sim f_2^*\eta$, then $f_1 \sim f_2$ (here $\sim$ denotes homotopy).* $\square$

This theorem shows that the set of equivalence classes of n-dimensional real, oriented, or complex bundles over X is in one-one correspondence with the set $\pi(X, G(\infty, n))$, $\pi(X, G_+(\infty, n))$, or $\pi(X, \mathbb{C}G(\infty, n))$.

B. **Theorem.** *The group of q-dimensional characteristic classes of n-dimensional real (n-dimensional oriented, n-dimensional complex) vector bundles with coefficients in A is isomorphic to the group $H^q(G(\infty, n); A)$ (the group $H^q(G_+(\infty, n); A)$, $H^q(\mathbb{C}G(\infty, n); A)$).* $\square$

The proof is the same in the three cases; we confine ourselves to the real case. A given characteristic class can be computed for the bundle η, resulting in an element of $H^q(G(\infty, n); A)$. We have to show that: (i) a nonzero characteristic class takes a nonzero value on η; (ii) every element of $H^q(G(\infty, n); A)$ is the value of some characteristic class on η.

(i) If a characteristic class vanishes on η, then it vanishes on any bundle with a finite cellular base, since such a bundle can always be represented in the form $f^*\eta$. If our characteristic class takes a nonzero value $\alpha \in h^q(X; A)$ for some bundle ξ with an arbitrary cellular base X, then we find a finite cellular subspace $Y \subset X$ such that the inclusion homomorphism $H^q(X; A) \to H^q(Y; A)$ maps α to a nonzero element. Then our characteristic class takes a nonzero value on the bundle $\xi \mid Y$ which contradicts what has been proved.

(ii) Let $\gamma \in H^q(G(\infty, n); A)$. The value of the characteristic class on the bundle $f^*\eta$ with base X is defined as $f^*\gamma \in H^q(X; A)$; this defines a class for all bundles with finite cellular bases. If ξ is a bundle with an infinite cellular base X, then for any finite subspace $Y \subset X$ the value $\alpha_Y \in H^q(Y; A)$ of the characteristic class on the bundle $\xi \mid Y$ is defined, and there clearly exists a unique class $\alpha \in H^q(X; A)$ that is mapped to α_Y by the homomorphism induced by each inclusion $Y \to X$; this is the value of the characteristic class on ξ. $\qquad\square$

C. Theorem. (i) *Every characteristic class with coefficients in $\mathbb{Z}_2$ of a real n-dimensional vector bundle is a polynomial in the Stiefel-Whitney classes $w_1, \ldots, w_n$, and different classes correspond to different polynomials.*

(ii) *Every characteristic class with coefficients in $\mathbb{Z}$ of a complex n-dimensional vector bundle is a polynomial in the Chern classes $c_1, \ldots, c_n$, and different classes correspond to different polynomials.*

(iii) *Each characteristic class with coefficients in $\mathbb{Q}$ or $\mathbb{R}$ or $\mathbb{C}$ of a real n-dimensional vector bundle is a polynomial in the (images under the homomorphism determined by the inclusion of $\mathbb{Z}$ in the coefficient domain of) the Pontryagin classes $p_1, p_2, \ldots, p_{[n/2]}$, and different classes correspond to different polynomials.* $\qquad\square$

The proof (in each case) consists of two parts. Firstly, it is necessary to verify that the qth cohomology group of $G(\infty, n)$, $G_+(\infty, n)$, or $\mathbb{C}G(\infty, n)$ with coefficients in $\mathbb{Z}_2$, $\mathbb{Z}$, $\mathbb{Q}$, $\mathbb{R}$, or $\mathbb{C}$ has precisely the same dimension as the group of polynomials of degree q of the indicated form. It is easy to extract this result from computations of the cohomology of Grassmannians (see Fuchs (1986) for example). Secondly, it is necessary to establish that none of these polynomials is identically zero as a characteristic class. (In the complex case (ii) slightly more is required – the details are left to the reader.) For this we need a supply of explicitly computed characteristic classes, which we do not yet have, so we postpone the second part of the proof until 10.4 and 10.5.

10.4. The most important properties of Stiefel-Whitney classes.

A. Theorem. *The classes w_i have the following properties:*

(i) *If ζ is the canonical line bundle over $\mathbb{R}P^n$, then $0 \neq w_1(\zeta) \in H^1(\mathbb{R}P^n; \mathbb{Z}_2) = \mathbb{Z}_2$ and $w_i(\zeta) = 0$ for $i > 1$.*

(ii) *For any vector bundles ξ, η with the same (cellular) base*

$$w_i(\xi \oplus \eta) = \sum_{\alpha+\beta=i} w_\alpha(\xi) w_\beta(\eta)$$

(we assume that for any bundle $w_0 = 1$). $\qquad\square$

Assertion (ii) can be written in the equivalent form:

(ii')
$$w_i(\xi \times \eta) = \sum_{\alpha+\beta=i} w_\alpha(\xi) \times w_\beta(\eta),$$

where ξ, η are bundles with cellular bases X and Y (not necessarily the same), $\xi \times \eta$ is the direct product of these bundles (with base $X \times Y$), and $\times$ on the right-hand side denotes the cohomology $\times$-product.

Note. The properties of Stiefel-Whitney classes in the above theorem are often used as the basis of an axiomatic definition of Stiefel-Whitney classes (see Milnor and Stasheff (1974) for example).

Formulae (ii) and (ii′) can be written in the form $w(\xi \oplus \eta) = w(\xi)w(\eta)$ and $w(\xi \times \eta) = w(\xi) \times w(\eta)$, where w denotes the formal sum $1 + w_1 + w_2 + \ldots$.

B. Theorem. *The Stiefel-Whitney classes of stably equivalent bundles are the same.*   □

In fact a section of the bundle ξ can obviously be extended to a section of the bundle $\xi + n$, where n denotes the trivial n-dimensional bundle.

C. *Example.* We shall find $w_i(\zeta \times \cdots \times \zeta)$, where $\zeta \times \cdots \times \zeta$ is the product of n copies of the Hopf bundle over $\mathbb{R}P^n$, $N > n$. Recall that $H^*(\mathbb{R}P^N; \mathbb{Z}_2) = \mathbb{Z}_2[x]/x^{N+1}$, $\dim x = 1$. By (ii′) and (i),

$$w_i(\zeta \times \cdots \times \zeta) = w(\zeta) \times \cdots \times w(\zeta) = (1 + x) \times \cdots \times (1 + x),$$

and hence $w_i(\zeta \times \cdots \times \zeta) = e_i(p_1^* x, \ldots, p_n^* x)$, where e_i is the elementary symmetric polynomial, and $p_i : \mathbb{R}P^N \times \cdots \times \mathbb{R}P^N \to \mathbb{R}P^N$ is the ith projection.

Corollary. *No polynomial in Stiefel-Whitney classes is identically zero as a characteristic class.*   □

In fact, this polynomial is already nontrivial on the product $\zeta \times \cdots \times \zeta$ with a large enough number of factors, but no polynomial in the elementary symmetric polynomials can vanish.

Recall that the last assertion was necessary for the proof of the basic theorem in 10.3.C.

D. There are formulae expressing the Stiefel-Whitney classes of the bundles $\xi \otimes \eta$, $S^k\xi$, and $\Lambda^k\xi$ in terms of the Stiefel-Whitney classes of ξ and η and the dimensions of ξ and η, but they are more complicated. Here are some of them (we put $\dim \xi = \eta$, $\dim \eta = m$):

$$w_1(\xi \otimes \eta) = mw_1(\xi) + nw_1(\eta),$$

$$w_2(\xi \otimes \eta) = m^2 w_2(\xi) + (mn - 1)w_1(\xi)w_1(\eta) +$$
$$+ n^2 w_2(\eta) + \frac{m(m-1)}{2} w_1(\xi)^2 + \frac{n(n-1)}{2} w_1(\eta)^2,$$
$$\cdots$$

$$w_1(S^2\xi) = (n + 1)w_1(\xi),$$

$$w_2(S^2\xi) = (n + 2)w_2(\xi) + \frac{(n-1)(n+2)}{2} w_1(\xi)^2,$$
$$\cdots$$

$$w_1(\Lambda^2\xi) = (n - 1)w_1(\xi),$$

$$w_2(\Lambda^2(\xi)) = (n - 2)w_2(\xi) + \frac{(n-1)(n-2)}{2} w_1(\xi)^2,$$

General formulae for $w_i(\xi \otimes \eta)$, $w_i(S^k \xi)$, $w_i(\Lambda^k \xi)$ can be obtained as follows. We take formal variables $x_1, \dots, x_n, y_1, \dots, y_m$, form the ith symmetric polynomial in the mn variables $x_s + y_t$, and represent it as a polynomial in the elementary symmetric polynomials in the x's and y's:

$$e_i(x_s + y_t) = P_i(e_u(x), e_v(y)).$$

Then

$$w_i(\xi \otimes \eta) = P_i(w_u(\xi), w_v(\eta)).$$

In addition we form the ith elementary symmetric polynomial in the $\binom{n+r}{r}$ variables $x_{i_1} + \dots + x_{i_r}$, $i_1 \leq \dots \leq i_r$, and express it in terms of the elementary symmetric polynomials in the x's:

$$e_i(\{x_{i_1} + \dots + x_{i_r} \mid i_i \leq \dots \leq i_r\}) = P_i(e_s(x)).$$

Then

$$w_i(S^r \xi) = P_i(w_s(\xi)).$$

Similarly

$$e_i(\{x_{i_1} + \dots + x_{i_r} \mid i_1 < \dots < i_r\}) = P_i(e_s(x)),$$
$$w_i(\Lambda^r \xi) = P_i(w_s(\xi)).$$

An outline of the proof of all these formulae is as follows. For bundles which split into the sum of line bundles, all the formulae can easily be derived from the product formula (part (ii) of the theorem of 10.4.A). They can be carried over to the general case on the basis of a general splitting principle, which says, roughly speaking, that a relation between characteristic classes that is valid for bundles that split into the sum of line bundles is also valid for arbitrary bundles. The formal basis of this principle is a theorem of Borel stating that the natural map $\mathbb{R}P^\infty \times \dots \times \mathbb{R}P^\infty \to G(\infty, n)$ induces a monomorphism in mod 2 cohomology.

10.5. The most important properties of Euler, Chern, and Pontryagin classes.

A. For Euler classes we have the product formula

$$\mathrm{eu}(\xi \oplus \eta) = \mathrm{eu}(\xi)\mathrm{eu}(\eta).$$

B. Everything that has been said for the Stiefel-Whitney classes can be carried over word for word to Chern classes; in particular, for the canonical line bundle ζ_C over $\mathbb{C}P^\infty$, the class c_1 is the standard generator of $H^2(\mathbb{C}P^\infty; \mathbb{Z}) = \mathbb{Z}$, and the remaining Chern classes are zero; there is the product formula

$$c_i(\xi \oplus \eta) = \sum_{\alpha+\beta=i} c_\alpha(\xi)c_\beta(\eta)$$

and the formulae for the Chern classes of tensor products, symmetric and exterior powers are precisely the same as the corresponding formulae for Stiefel-Whitney classes. Chern classes, like Stiefel-Whitney classes, are the same for stably equivalent bundles.

C. Proposition. $c_i(\bar{\xi}) = (-1)^i c_i(\xi)$. ☐

It follows that $2c_i(\mathbb{C}\xi) = 0$ for any *real* vector bundle ξ and any odd i. (In fact $\mathbb{C}\bar{\xi} = \mathbb{C}\xi$.) (Cf. the comment on the definition of the Pontryagin classes.)

D. We define the polynomial Q in r variables by

$$N_r = Q_r(e_1, \ldots, e_r),$$

where the e_i are the elementary symmetric polynomials, and N_r is the sum of the rth powers. Put

$$\mathrm{ch}_r(\xi) = \frac{1}{r!} Q_r(c_1(\xi), \ldots, c_r(\xi)) \in H^{2r}(-; \mathbb{Q}).$$

The (dimensionally non-homogeneous) characteristic class

$$\mathrm{ch} = \mathrm{ch}_0 + \mathrm{ch}_1 + \mathrm{ch}_2 + \ldots$$

with coefficients in $\mathbb{Q}$ is called the *Chern character*. (Note that $\mathrm{ch}_0(\xi) \in H^0(-; \mathbb{Q})$ is just the dimension of ξ.) The basic property of the Chern character is:

Theorem. $\mathrm{ch}(\xi \otimes \eta) = \mathrm{ch}\,\xi\,\mathrm{ch}\,\eta$. ☐

E. The product formula and all the other formulae for the Pontryagin classes can be derived from the corresponding formulae for Chern classes, and hold "modulo 2-torsion". For example,

$$2\left[p_i(\xi \oplus \eta) - \sum_{\alpha+\beta=i} p_\alpha(\xi)p_\beta(\eta) \right] = 0.$$

Nevertheless, it is true that the Pontryagin classes of stably equivalent bundles are the same.

F. In conclusion we give two formulae by means of which the Stiefel-Whitney classes can be expressed in terms of the Euler class. Let ξ be a real vector bundle of dimension n over a cellular space X, and let ζ be the canonical line bundle over $\mathbb{R}P^\infty$. We form the bundle $\xi \otimes \zeta$ over $X \times \mathbb{R}P^\infty$ (more precisely this is the tensor product of the bundles over $X \times \mathbb{R}P^\infty$ induced from the bundles ξ and η by the projections onto the factors). Then

$$\rho_2\mathrm{eu}(\xi \oplus \zeta) = \sum_{i=0}^{n}[w_i(\xi) \times x^{n-i}] \in H^n(X \times \mathbb{R}P^\infty; \mathbb{Z}_2),$$

where $x \in H^1(\mathbb{R}P^\infty; \mathbb{Z}_2)$ is the generator (see 8.7.**B**). Similarly if ξ is a complex vector bundle of dimension n over X and $\zeta_\mathbb{C}$ is the canonical (complex) line bundle over $\mathbb{C}P^\infty$, then

$$e(\mathbb{R}(\xi \otimes \zeta_\mathbb{C})) = c_n(\xi \otimes \zeta_\mathbb{C}) =$$

$$= \sum_{i=0}^{n}(c_i(\xi) \times x^{n-1}) \in H^{2n}(X \times \mathbb{C}P^\infty; \mathbb{Z}).$$

10.6. Characteristic classes in the topology of smooth manifolds.

We can only touch on this vast subject.

A. We begin with a geometrical interpretion of the first obstruction. Let (E, B, F, p) be a homotopically simple locally trivial bundle in which E and B are smooth manifolds with B closed, n-dimensional and oriented, and p is a smooth map with Jacobian matrix of maximal rank at each point of E (a submersion). We assume that $\pi_0(F) = \ldots = \pi_{k-2}(F) = 0$ and $\pi_{k-1}(F) = \pi$. Then the first obstruction to the construction of a section of our bundle lies in $H^k(B; \pi)$. We assume in addition that we have succeeded in constructing a section over $B - X$, where X is a submanifold of B (perhaps with singularities of codimension ≥ 2) of dimension $n - k$ (simple considerations of the type of reduction to general position show that this can always be done) or the union of several such manifolds intersecting transversally: $X = \bigcup X_i$. We choose a nonsingular point of one of these manifolds X_i (which we assume to be connected) and consider a small $(k - 1)$-sphere containing the manifold near this point. Since there is a section over this sphere, and the bundle can be regarded as trivialized near this point, we obtain a map $S^{k-1} \to$ the fibre, determining (since the bundle and the fibre are simple) an element α_i of $\pi_{k-1}(F) = \pi$.

Proposition. *Under the Poincaré isomorphism the class* $\sum \alpha_i[X_i] \in H_{n-k}(B; \pi)$ *corresponds to the first obstruction to the extension of the section to our bundle.*   □

To prove this it is sufficient to triangulate B in such way that X does not intersect any simplexes of dimension $< k$, and intersects each k-simplex transversally in not more than one point.

This proposition provides the technical basis for the proof of most of the following results. We shall not explain how it is applied every time, leaving this to the reader.

B. (Differential-topological interpretation of the Euler class.) Let $\xi = (E, B, \mathbb{R}^n, p)$ be a smooth vector bundle (that is, a vector bundle in which E and B are smooth manifolds, p is a submersion and the vector operations in E are smooth). We assume that B is a closed oriented manifold. Let $s : B \to E$ be a section of ξ in general position with respect to the zero section. Then $B \cap s(B)$ (we regard B as embedded in E as the image of the zero section) represents a homology class of B corresponding to the Euler class $\mathrm{eu}(\xi)$ of ξ under the Poincaré isomorphism.

Further, let Y be a closed oriented submanifold of a smooth manifold X and let $\nu_X(Y)$ be the corresponding normal bundle. Then

$$D[\mathrm{eu}(\nu_X(Y))] = i[Y],$$

where D is the Poincaré isomorphism (in Y), $i : Y \to X$ is the inclusion, and $[Y]$ is the homology class of X represented by Y.

Corollary. If $[Y] = 0$, then $\mathrm{eu}(\nu_X Y) = 0$; for example, the Euler class of the normal bundle of a manifold Y embedded in Euclidean space or a sphere is equal to 0.   □

The last assertion is false for immersions. In particular, if f is an immersion of a closed oriented manifold X of even dimension n in $\mathbb{R}^{2n}$ with transverse self-intersections, then the algebraic number of points of self-intersection (the definition of the sign corresponding to a point of transverse self-intersection is left to the reader) is equal to half the *normal Euler number*, that is, the value of the Euler class of the normal bundle on the fundamental homology class of X.

On the other hand, the *tangential Euler number* of a closed oriented manifold, that is, the value of the Euler class of the tangent bundle on the fundamental class is equal to the Euler characteristic of the manifold. From this follows *Euler's theorem*: there is a non-vanishing tangent vector field on a closed manifold if and only if its Euler characteristic is zero. We emphasize that this theorem is also valid for non-orientable manifolds (it can be extended to this case, for example, by means of a two-sheeted orientable covering). Another version of this theorem: there is a continuous field of tangent lines on a closed manifold if and only if its Euler characteristic is zero.

C. (Differential topological interpretation of the Stiefel-Whitney classes.) Here we are concerned only with cohomology and homology mod 2, and the Poincaré isomorphism D is understood in the corresponding sense.

The Stiefel-Whitney classes of the tangent bundle of a smooth manifold X are called the *Stiefel-Whitney classes of the manifold*, and denoted by $w_i(X)$. Similarly we speak of the *Pontryagin classes* $p_i(X)$ *of a smooth manifold*, and the *Chern classes* $c_i(X)$ *of a complex manifold* X. Since the normal bundle of a smooth manifold in Euclidean space is independent, up to stable equivalence, of the method of embedding (proof: let ν, ν' be the normal bundles corresponding to embeddings of X in $\mathbb{R}^N$ and $\mathbb{R}^{N'}$, and τ the tangent bundle of X; then $\nu \oplus N' = \nu \oplus (\nu' \oplus \tau) = \nu' \oplus (\nu \oplus \tau) = \nu' \oplus N$), we can speak of the *normal Stiefel-Whitney classes* of X, namely the Stiefel-Whitney classes of the normal bundle in any embedding (or immersion – it makes no difference) of X in Euclidean space; these classes are denoted by $\bar{w}_i(X)$. From the product formula it follows that

$$\sum_{\alpha+\beta=i} w_\alpha(X)\bar{w}_\beta(X) \quad \text{for } i > 0,$$

and in particular the normal Stiefel-Whitney classes can be expressed in terms of the tangential classes and conversely.

Consider a smooth map in general position of an n-dimensional manifold X into $\mathbb{R}^q$, $q \leq n$; let $Y \subset X$ be the set of points at which this map is not a submersion (the rank of the Jacobian matrix is less than q). Then $\dim Y = q - 1$ and

$$D^{-1}[Y] = w_{n-q+1}(X).$$

Consider a smooth map in general position of a closed n-dimensional manifold X into $\mathbb{R}^q$, $q \geq n$; let $Y \subset X$ be the set of points at which this map is not an immersion (the rank of the Jacobian matrix is less than n). Then $\dim Y = 2n-q-1$ and

$$D^{-1}[Y] = \bar{w}_{q+1-n}(X).$$

If an n-dimensional manifold X admits an immersion in $\mathbb{R}^{n+q}$, then $\bar{w}_i(X) = 0$ for $i > q$. (In the closed case this assertion of course follows from the preceding ones, but it is much simpler and more natural to prove it directly.)

If a closed n-dimensional manifold X admits an embedding in $\mathbb{R}^{n+q}$, then $\bar{w}_i(X) = 0$ for $i \geq q$. (To prove this it is necessary to use both the previous assertion and the corresponding property of the Euler classes.) As an example, it is easy to show that if $2^k \leq n < 2^{k+1}$, then there is no submersion of $\mathbb{R}P^n$ in $\mathbb{R}^{2^{k+1}-2}$, and no embedding in $\mathbb{R}^{2^{k+1}-1}$.

Let X be a triangulated closed smooth manifold. Let C_i denote the classical i-chain of the barycentric triangulation with coefficients in $\mathbb{Z}_2$, that is, the sum of all the i-simplexes of the barycentric subdivision. Then C_i is a cycle, and

$$D^{-1}[C_i] = w_i(X).$$

The values of classes of the form $w_{i_1}(X) \dots w_{i_r}(X)$ with $i_1 + \dots i_r = n$ on the fundamental class of a closed n-dimensional manifold (these are residues mod 2) are called the *Stiefel-Whitney numbers* of X, and denoted by $w_{i_1 \dots i_r}[X]$. For example, two-dimensional manifolds have two Stiefel-Whitney numbers: $w_{11}[X]$ and $w_2[X]$ (we recommend the reader to compute them for the classical surfaces).

Proposition. *If a closed manifold is the boundary of a compact manifold, then all its Stiefel-Whitney numbers are zero.* □

Proof. If $X = \partial Y$ and $i : X \to Y$ is the inclusion, then $\tau_X = (i^*\tau_Y) \oplus 1$, where τ_X, τ_Y are the tangent bundles (the normal bundle to a boundary is always trivial). Hence $w_j(X) = i^*w_j(Y)$ for any j, and

$$\langle w_{j_1}(X) \dots w_{j_r}(X), [X] \rangle = \langle i^*(w_{j_1}(Y) \dots w_{j_r}(Y)), [X] \rangle =$$
$$= \langle w_{j_1}(Y) \dots w_{j_r}(Y), i_*[X] \rangle = 0,$$

because $i_*[X] = 0$ (the boundary of a compact manifold is its boundary in the topological sense also). □

This proposition provides a powerful necessary condition for a given closed manifold to be the boundary of a compact manifold. (Show that if $n + 1$ is not a power of 2, then neither $\mathbb{R}P^n$ nor $\mathbb{C}P^n$ is the boundary of a compact manifold.) However, it is most remarkable that this condition is also sufficient:

Theorem. *A closed manifold is the boundary of a compact manifold if and only if all its Stiefel-Whitney numbers are zero* (Thom (1954)).

We add that the Stiefel-Whitney numbers are not linearly independent. For example, if X is a closed one-dimensional manifold (that is, a union of circles), then $w_i[X] = 0$; if it is a closed two-dimensional manifold, then $w_{11}[X] = w_2[X]$ (the reader will know this if he has followed our advice and computed the Stiefel-Whitney numbers of the classical surfaces). In fact the Stiefel-Whitney numbers $w_{j_1 \dots j_r}[X]$ of a closed n-dimensional manifold form a linearly independent system when $j_1 + \dots + j_r = n$, $j_1 \leq \dots \leq j_r$, and no number of the form $j_s + 1$ is a power of 2.

D. (Differential-topological interpretation of the Pontryagin classes.) Let X be a closed oriented n-dimensional manifold, and $f : X \to \mathbb{R}^{n-2q+2}$ a smooth map in general position. Let Y denote the set of points of X at which the rank of the Jacobian of f is not greater than $n - 2q$ (that is, at least 2 less than its maximum possible value). Then Y is an oriented $(n-4q)$-dimensional manifold (possibly with singularities), and the class $[Y] \in H_{n-4q}(X)$ corresponds to the Pontryagin class $p_q(X) \in H^{4q}(X)$ (of the tangent bundle) of X under the Poincaré isomorphism. There is a similar description for the normal Pontryagin classes.

(The orientation, and even the orientability, of X is not really necessary here; however, to attach meaning to this expression it is necessary to use the version of Poincaré duality explained in 8.8.D.)

If X is a closed oriented manifold of dimension $4m$, then the value of the class $p_{j_1}(X) \ldots p_{j_r}(X)$, $j_1 + \ldots + j_r = m$, on the fundamental homology class of X is called the *Pontryagin number*, denoted by $p_{j_1 \ldots j_r}[X]$. (It is convenient to assume that X is not necessarily connected; the fundamental class of a non-connected manifold is the sum of the fundamental classes of its components.) If X is the boundary of a compact oriented manifold, then all the Pontryagin numbers are zero (this is proved in exactly the same way as for the Stiefel-Whitney classes). There is also *Thom's theorem*, which states that if all the Pontryagin numbers of a closed oriented manifold are zero (for example, if its dimension is not divisible by 4), then the union of a number of copies of X (all taken with the same orientation) is the boundary of a closed oriented manifold. Furthermore, every set of integers $\{p_{j_1 \ldots j_r} \mid j_1 + \ldots + j_r = m\}$ becomes, after multiplication by a certain positive integer (the same for all the numbers in the set), the set of Pontryagin numbers of some closed oriented manifold. (This theorem is in fact far simpler than the corresponding assertion for Stiefel-Whitney numbers.)

A useful corollary of Thom's theorem (and of the fact that if $Y = X_1 \sqcup X_2$ is the sum of two closed oriented $4m$-dimensional manifolds, then

$$p_{j_1 \ldots j_r}[Y] = p_{j_1 \ldots j_r}[X_1] + p_{j_1 \ldots j_r}[X_2]$$

for any $j_1, \ldots, j_r$ with $j_1 + \ldots + j_r = m$) is the following:

Theorem. *Suppose that with each closed oriented n-dimensional manifold X there is associated an integer $\sigma(X)$ with the following properties: (i) if X is the boundary of a compact oriented manifold, then $\sigma(X) = 0$; (ii) $\sigma(X_1 \sqcup X_2) = \sigma(X_1) + \sigma(X_2)$. Then*

$$\sigma(X) = \sum_{j_1 + \ldots + j_r = n/4} a_{j_1 \ldots j_r} p_{j_1 \ldots j_r}[X],$$

where $a_{j_1 \ldots j_r}$ are rational numbers independent of X. In particular $\sigma(X) \equiv 0$ if n is not divisible by 4. $\square$

This result has a widely known application. Let $\sigma(X)$ denote the signature of the "intersection number" form in the $2m$-dimensional homology of the $4m$-dimensional manifold X. The theorem of 8.8.B shows that σ satisfies condition (i) of the preceding theorem; condition (ii) is obvious for σ. Hence the signature is a linear combination of the Pontryagin numbers with rational coefficients. In particular, $\sigma(X) = a p_1[X]$ for $\dim X = 4$, $\sigma(X) = b p_2[X] + c p_{11}[X]$ for $\dim X = 8$, and so on. To find $a, b, c, \ldots$, we need a large enough supply of computations of specific examples. For example, $H_{2m}(\mathbb{C}P^{2m}) = \mathbb{Z}$, the matrix of the intersection form is (1), so that $\sigma(\mathbb{C}P^{2m}) = 1$. Furthermore it is easy to show that

$$\tau_{\mathbb{C}P^{2m}} \oplus 1 = (2m+1)\zeta_{\mathbb{C}},$$

and hence

$$\mathbb{C}\tau_{\mathbb{C}P^{2m}} \oplus 1 = (2m+1)(\zeta_{\mathbb{C}} \oplus \bar{\zeta}_{\mathbb{C}})$$

(the identity has also been complexified!) and

$$(p_0 - p_1 + p_2 - \ldots + (-1)^m p_m)(\mathbb{C}P^{2m}) = [(1+x)(1-x)]^{2m+1} = (1-x^2)^{2m+1},$$

where $x \in H^2(\mathbb{C}P^{2m}) = \mathbb{Z}$ is the canonical generator. Consequently

$$p_i(\mathbb{C}P^{2m}) = \begin{cases} \binom{2m+1}{i}x^{2i} & \text{for } i \leq m, \\ 0 & \text{for } i > m. \end{cases}$$

In particular, $p_1(\mathbb{C}P^2) = 3x^2$, $p_1[\mathbb{C}P^2] = 3$, and since $\sigma(\mathbb{C}P^2) = 1$, for any four-dimensional manifold X we have

$$\sigma(X) = \frac{1}{3}p_1[X]. \tag{14}$$

(In particular, the first Pontryagin number of any closed oriented four-dimensional manifold is divisible by 3.) Furthermore, $p_{11}[\mathbb{C}P^4] = 25$, $p_2[\mathbb{C}P^4] = 10$, $\sigma(\mathbb{C}P^4) = 1$. To this we add that

$$(p_0 + p_1 + p_2)(\mathbb{C}P^2 \times \mathbb{C}P^2) = (p_0 + p_1)(\mathbb{C}P^2) \times (p_0 + p_1)(\mathbb{C}P^2)$$

(the multiplication formula for Pontryagin classes is valid modulo 2-torsion in general, but there is no torsion in the cohomology of complex projective spaces), and hence, $p_1(\mathbb{C}P^2 \times \mathbb{C}P^2) = (1 \times 3x^2) + (3x^2 \times 1)$, $p_1(\mathbb{C}P^2 \times \mathbb{C}P^2) = 18(x^2 \times x^2)$, $p_2(\mathbb{C}P^2 \times \mathbb{C}P^2) = p_1(\mathbb{C}P^2) \times p_1(\mathbb{C}P^2) = 3x^2 \times 3x^2$, $p_{11}[\mathbb{C}P^2 \times \mathbb{C}P^2] = 18$, $p_2[\mathbb{C}P^2 \times \mathbb{C}P^2] = 9$; also $\sigma(\mathbb{C}P^2 \times \mathbb{C}P^2) = 1$ (the signature is easily seen to be multiplicative). Thus $1 = 10b + 25c$, $1 = 9b + 18c$, from which $b = 7/45$, $c = -1/45$. Thus for dim $X = 8$

$$\sigma(X) = \frac{7p_2[X] - p_{11}[X]}{45} \tag{15}$$

(Hence $7p_2[X] - p_{11}[X]$ is divisible by 45, and if an eight-dimensional manifold has $p_1 = 0$ then its signature is divisible by 7.) Formulae (14) and (15) are the beginning of an infinite chain of formulae. Hirzebruch worked on their description in the 1950s. He computed the Pontryagin numbers of manifolds of the form $\mathbb{C}P^{2m_1} \times \cdots \times \mathbb{C}P^{2m_k}$ (we have essentially done this), and using the fact that their signature is 1, he found the coefficients of the Pontryagin numbers in the formulae for the signature. The resulting formulae can be found in Hirzebruch (1966).

E. As we know, the Euler class of a manifold can be computed knowing only the ranks of its homology groups. It turns out that although the Stiefel-Whitney classes are not determined by the cohomology groups of the manifold or even by its cohomology ring, they are none the less homotopy invariant (see details in 11.3). It has long been known that there is no similar homotopy invariance theorem for the Pontryagin classes: the signature is the only homotopy invariant polynomial in the Pontryagin classes. In the 1960s S.P. Novikov proved a difficult theorem on the topological invariance of rational Pontryagin classes (a homeomorphism between smooth oriented closed manifolds takes Pontryagin classes into classes differing from Pontryagin classes by elements of finite order; there are examples showing that these elements of finite order may be non-zero). V.A. Rokhlin, A.S. Schwarz and Thom had proved this for homeomorphisms between certain smooth triangulated two-dimensional manifolds (see Milnor and Stasheff (1974)). The latter result naturally led to the problem of the "combinatorial computation of Pontryagin classes", that is, computation by means of triangulations (see the end of C). At present this problem has been solved for the first Pontryagin class (see Gabrielov, Gel'fand and Losik (1975)).

§11. Steenrod squares

The subject of this section is a fragment of the theory of cohomology operations, which will be studied in detail in other surveys in this series. The *Steenrod squares* are the best-known cohomology operations. They act on cohomology with coefficients in $\mathbb{Z}^2$. Their chief application, which we shall not touch on here, is concerned with problems of homotopy classification. We shall also not deal here with the numerous generalizations of Steenrod squares, among which the *Steenrod powers* deserve first place, being the $\mathbb{Z}_p$ analogue of the Steenrod squares. We shall mainly be concerned with the application of Steenrod squares to the problem of obstructions to Stiefel-Whitney classes which we have already studied.

All the same we begin with a brief account of the general theory of cohomology operations.

11.1. General theory of cohomology operations.

A. Suppose we are given two integers n and q and two Abelian groups Π and G. We say that we are given a *cohomology operation* ϕ *of type* (n, q, Π, G) if for any cellular space X, there is defined a map $\phi = \phi_X : H^n(X; \Pi) \to H^q(X; G)$ (not necessarily a homomorphism), which is natural with respect to X in the sense that the diagram

$$
\begin{array}{ccc}
H^n(X; \Pi) & \xrightarrow{\phi_X} & H^q(X; G) \\
f^* \uparrow & & \uparrow f^* \\
H^n(Y; \Pi) & \xrightarrow{\phi_Y} & H^n(Y; G)
\end{array}
$$

is commutative for any continuous map $f : X \to Y$.

B. The collection of all cohomology operations of type (n, q, Π, G) is an Abelian group, which we denote by $\mathcal{O}(n, q, \Pi, G)$.

Theorem. $\mathcal{O}(n, q, \Pi, G) \cong H^q(K(\Pi, n); G)$. $\qquad\qquad\qquad\qquad\square$

In fact, $H^n(X; \Pi) = \pi(X, K(\Pi, n))$ (see 9.3), the correspondence being established by the formula $f \mapsto f^*e$, where $e \in H^n(K(\Pi, n); \Pi)$ is the fundamental class. A cohomology operation ϕ determines an element $\phi(e)$ of $H^q(K(\Pi, n); G)$, and a simple verification shows that the correspondence $\phi \mapsto \phi(e)$ establishes an isomorphism $\mathcal{O}(n, q, \Pi, G) \cong H^q(K(\Pi, n); G)$ (compare the proof of the theorem in 9.3). □

Corollary. *Nontrivial cohomology operations cannot lower dimension (that is, if $\mathcal{O}(n, q, \Pi, G) \neq 0$, then $q \geq n$).* □

For $H^q(K(\Pi, n); G) = 0$ for $q < n$. □

C. *Example.* The connecting homomorphism (Bockstein homomorphism)

$$H^n(X; G'') \to H^{n+1}(X; G'),$$

corresponding to a short exact sequence

$$0 \to G' \to G \to G'' \to 0,$$

is a cohomology operation of type $(n, n+1, G'', G')$. In particular, there are cohomology operations $\beta \in \mathcal{O}(n, n+1, \mathbb{Z}_m, \mathbb{Z}_m)$, $\tilde{\beta} \in \mathcal{O}(n, n+1, \mathbb{Z}_m, \mathbb{Z})$; see 2.6.**B**.

Note. It is easy to show (by means of the theorem in **B**) that the Bockstein homomorphisms are the only cohomology operations raising dimension by 1. Also cohomology operations that preserve dimension are just the homomorphisms induced by homomorphisms of the coefficient groups.

D. A *stable cohomology operation of type* (r, Π, G) is a sequence of cohomology operations $\phi_n \in \mathcal{O}(n, n+r, \Pi, G)$, $n = 1, 2, 3, \ldots$, such that for any space X and any $n \geq 1$, the diagram

$$
\begin{array}{ccc}
H^n(X; \Pi) & \xrightarrow{\Sigma} & H^{n+1}(\Sigma X; \Pi) \\
\downarrow{\phi_n} & & \downarrow{\phi_{n+1}} \\
H^{n+r}(X; G) & \longrightarrow & H^{n+r+1}(\Sigma X; G)
\end{array}
$$

(where Σ is the suspension isomorphism) is commutative.

For example, the Bockstein homomorphism is a stable cohomology operation.

11.2. Steenrod squares and their properties.

(Proofs of the assertions in this section can be found in §29 of Fomenko and Fuchs (1989).)

A. The Steenrod squares Sq^i are stable cohomology operations which are homomorphisms

$$Sq^i : H^n(X; \mathbb{Z}_2) \to H^{n+i}(X; \mathbb{Z}_2).$$

They are defined for all $i \geq 0$, and have the following properties.

1°.

$$Sq^i \alpha = \begin{cases} 0 & \text{if } i > \dim \alpha, \\ \alpha^2 & \text{if } i = \dim \alpha, \\ \alpha & \text{if } i = 0. \end{cases}$$

2°. (H. Cartan's formula)

$$Sq^i(\alpha\beta) = \sum_{p+q=i} Sq^p\alpha Sq^q\beta.$$

Note. Let us introduce the homomorphism $Sq = Sq^0 + Sq^1 + \ldots : H^*(X; \mathbb{Z}_2) \to H^*(X; \mathbb{Z}_2)$. Then property $2°$ can be written in the form $Sq(\alpha\beta) = Sq\,\alpha\,Sq\,\beta$, that is, Sq is a ring homomorphism.

$3°$. Sq^1 is equal to the Bockstein homomorphism β.

$4°$. (The Adem relations) If $a < 2b$, then

$$Sq^a Sq^b = \sum_{c=\max(a-b-1,0)}^{[a/2]} \binom{b-c-1}{a-2c} Sq^{a+b-c} Sq^c.$$

Corollary. *Every Steenrod square can be represented as a (noncommutative) polynomial in* $Sq^1, Sq^2, Sq^4, Sq^8, Sq^{16}, \ldots$. $\qquad\square$

Note. Bullett and Macdonald (1982) have found the following elegant method of writing the Adem relations:

$$P(s^2 + st)P(t^2) = P(s^2)P(st + t^2),$$

where $P(u) = \sum u^i Sq^i$.

B. Theorem. *There is a unique sequence of stable cohomology operations* $\{Sq^i\}$ *having property* $1°$. $\qquad\square$

In particular, this means that properties $2°$–$4°$ are consequences of stability and property $1°$.

C. Theorem. (i) *All cohomology operations from* $\mathbb{Z}_2$-*cohomology to* $\mathbb{Z}_2$-*cohomology are iterations of Steenrod squares and the cohomology product.*

(ii) *All stable cohomology operations from* $\mathbb{Z}_2$-*cohomology to* $\mathbb{Z}_2$-*cohomology are polynomials in Steenrod squares. Moreover, monomials of the form*

$$Sq^{i_1} Sq^{i_2} \ldots Sq^{i_k}$$

with $i_1 \geq 2i_2$, $i_2 \geq 2i_3$, $\ldots$, $i_{k-1} \geq 2i_k$, *constitute a basis for the space of all such stable operations.* $\qquad\square$

11.3. Steenrod squares and Stiefel-Whitney classes.

A. If we associate with a real vector bundle ξ the cohomology class $Sq^k w_m(\xi)$ of its base, we obtain a new characteristic class for real vector bundles. But there are no "new characteristic classes": in 10.3.C we saw that every characteristic class of a real vector bundle with coefficients in $\mathbb{Z}_2$ was a polynomial in the Stiefel-Whitney classes. What is this polynomial? It turns out that

$$Sq^k w_m = \sum_{j=0}^{k} \binom{k-m}{j} w_{k-j} w_{m+j}.$$

To prove this it is sufficient to verify that the left and right sides take the same value on the bundles $\zeta \times \cdots \times \zeta$ with base $\mathbb{R}P^\infty \times \cdots \times \mathbb{R}P^\infty$, where ζ is the canonical line bundle over $\mathbb{R}P^\infty$, and the number of factors is sufficiently large (cf. 10.3.C).

B. (Digression: Thom spaces and isomorphisms.) Let ξ be a real vector bundle of dimension n. We fix a Euclidean structure on the fibres and let $D(\xi)$ and $S(\xi)$ denote the spaces of the ball and sphere bundles (with fibres D^n and S^{n-1}) associated with ξ. The space $T(\xi) = D(\xi)/S(\xi)$ is called the *Thom space* of the bundle ξ. It can be constructed as follows: the base B of ξ lies in it (as the zero section), and through each of its points passes an n-sphere. These spheres are not canonically homeomorphic to S^n, since the fibres of ξ are not canonically homeomorphic to R^n, but they cover the whole space $T(\xi)$ and they all have a point in common, the point antipodal to the intersection with B, and otherwise do not intersect (see Fig. 8).

Fig. 8

Example. The Thom space of the canonical line bundle over $\mathbb{R}P^n$ is homeomorphic to $\mathbb{R}P^{n+1}$ (prove this). There is a complex analogue of this statement.

If the base B of ξ is cellular, then $T(\xi)$ has a natural cellular structure: the cells are the inverse images in $D(\xi) - S(\xi)$ of the cells of B, and the point obtained from $S(\xi)$. Thus if B has n_0 0-cells, n_1 1-cells, ..., then $T(\xi)$ has one 0-cell, n_0 n-cells, n_1 $(n+1)$-cells, n_2 $(n+2)$-cells, ... Moreover, if ξ is oriented, then oriented cells of $T(\xi)$ correspond to oriented cells of B, and this correspondence is compatible with the incidence numbers. Hence for an oriented bundle ξ we have the *Thom isomorphisms*

$$t : H_q(B) \cong \tilde{H}_{q+n}(T(\xi)), \quad t : H^q(B) \to \tilde{H}^{q+n}(T(\xi)).$$

Similar isomorphisms can be constructed mod 2 independently of orientation. The class $u = t(1) \in H^n(T(\xi); \mathbb{Z}$ or $\mathbb{Z}_2)$ is called the *Thom class* of the bundle ξ.

C. Theorem. *For any bundle ξ*

$$w_m(\xi) = t^{-1}\mathrm{Sq}^m t(1),$$

where 1 is the 0-dimensional cohomology class of the base. $\qquad\qquad\square$

The proof, of which we leave the details to the reader (they can be found in Milnor and Stasheff (1974)), consists of the successive verification of the axioms for the Stiefel-Whitney classes (see 10.3.A) for the classes $t^{-1}\mathrm{Sq}^m t(1)$.

D. (Wu's formula) Put $w = 1 + w_1 + w_2 + \ldots$, and $\mathrm{Sq} = 1 + \mathrm{Sq}^1 + \mathrm{Sq}^2 + \ldots$. Note that the operation $\mathrm{Sq} : H^*(X; \mathbb{Z}_2) \to H^*(X; \mathbb{Z}_2)$ is invertible: $(\mathrm{Sq})^{-1} = 1 + \mathrm{Sq}^1 + \mathrm{Sq}^2 + \mathrm{Sq}^2\mathrm{Sq}^1 + \ldots$.

Theorem. *Let X be a closed manifold. Then for any $\alpha \in H^*(X; \mathbb{Z}_2)$*

$$\langle \mathrm{Sq}^{-1}w(X), D(\alpha)\rangle = \langle \mathrm{Sq}\,\alpha, [X]\rangle. \quad\square$$

This is "Wu's formula" and it completely determines the Stiefel-Whitney classes of the manifold. Its proof can be found in Fomenko and Fuchs (1989), Milnor and Stasheff (1974), or Spanier (1966).

Corollary 1. *The Stiefel-Whitney classes are homotopy invariants.* □

Corollary 2. (Stiefel's theorem.) *Every oriented closed three-dimensional manifold is parallelizable.* □

Proof. To prove that an oriented three-dimensional manifold is parallelizable, it is sufficient to construct two linearly independent tangent vector fields on X, that is, to construct a section of the bundle $E \to X$ with fibre $V(3,2)$ associated with the tangent bundle. Since $V(3,2)$ is homeomorphic to $\mathbb{R}P^3$, $\pi_1(V(3,2)) = \mathbb{Z}_2$, $\pi_2(V(3,2)) = 0$. The primary obstruction to the construction of a section lies in $H^2(X; \pi_1 V(3,2))) = H^2(X; \mathbb{Z}_2)$ and is equal to the class w_2. It is enough for us to prove that $w_2 = 0$, since the next obstruction lies in the zero group $H^3(X; \pi_2 V(3,2)))$. To apply Wu's formula, we see how the Steenrod squares act on $H^*(X; \mathbb{Z}_2)$ taking values in H^3. Since the manifold is orientable, $\mathrm{Sq}^1 : H^2(X; \mathbb{Z}_2) \to H^3(X; \mathbb{Z}_2)$ is trivial (the Bockstein homomorphism $\beta : H^2(X; \mathbb{Z}_2) \to H^3(X; \mathbb{Z}_2)$ is by definition the composition $H^2(X; \mathbb{Z}_2) \to H^3(X; \mathbb{Z}) \to H^3(X; \mathbb{Z}_2)$ and the first arrow, being a map of a finite group into $\mathbb{Z}$, is trivial). Further, $\mathrm{Sq}^2 : H^1(X; \mathbb{Z}_2) \to H^3(X; \mathbb{Z}_2)$ is zero (since $\mathrm{Sq}^i x = 0$ for $i > \dim x$), and a fortiori $\mathrm{Sq}^3 : H^0(X; \mathbb{Z}_2) \to H^3(X; \mathbb{Z}_2)$ is zero. Consequently, $\langle \mathrm{Sq}\,\alpha, [X] \rangle = 0$ for dim $\neq 3$, and hence $\mathrm{Sq}^{-1}(X)$ can take nonzero values only on 0-dimensional cohomology classes, so that $\mathrm{Sq}^{-1} w(X) \in H^0(X; \mathbb{Z}_2)$, $w(X) \in H^0(X; \mathbb{Z}_2)$ and $w_2(X) = 0$. □

11.4. Secondary obstructions.

It goes without saying that Steenrod squares were discovered by Steenrod. They first appeared in Steenrod (1947). Not everyone knows, however, that Steenrod did not introduce them just like that, but in pursuit of a definite goal, the generalization of Hopf's classification theorem. We proved this theorem of Hopf's in 9.4.A; it says that if X is an n-dimensional cellular space then

$$\pi(X, S^n) = H^n(X; \mathbb{Z}),$$

where the correspondence takes the class of a map $f : X \to S^n$ to the cohomology class $f^*(s)$, where $s \in H^n(S^n; \mathbb{Z})$ is a generator. Steenrod solved the following problem completely: find $\pi(X, S^n)$ when dim $X = n + 1$. Clearly there is the map $\pi(X, S^n) \to H^n(X; \mathbb{Z})$, $f \to f^*(s)$, in this case too, but it is not necessarily one-one. It is necessary firstly to find its image, and secondly for a given $\alpha \in H^n(X; \mathbb{Z})$ to describe the set of classes of maps $f : X \to S^n$ for which $f^*(s) = \alpha$. The following theorem of Steenrod's contains the answers to these questions.

Theorem. *Let $n > 2$, and let X be a cellular space. Then:*

(i) *if* dim $X \leq n + 2$, *the image of the map* $\psi : \pi(X, S^n) \to H^n(X; \mathbb{Z})$, $f \to f^*(s)$, *is the kernel of the composite map*

$$H^n(X; \mathbb{Z}) \xrightarrow{\rho_2} H^n(X; \mathbb{Z}_2) \xrightarrow{\mathrm{Sq}^2} H^{n+2}(X; \mathbb{Z}_2);$$

(ii) *if* $\dim X \leq n+1$, *then the set* $\psi^{-1}(u)$ *is in one-one correspondence with the cokernel of the composite map*

$$H^{n-1}(X; \mathbb{Z}) \xrightarrow{\rho_2} H^{n-1}(X; \mathbb{Z}_2) \xrightarrow{Sq^2} H^{n+1}(X; \mathbb{Z}_2). \quad \Box$$

For the proof see Spanier (1966).

11.5. The non-existence of spheroids with odd Hopf invariant.

Theorem. *If* n *is not a power of* 2, *then* $\pi_{2n-1}(S^n)$ *contains no elements with odd Hopf invariant* (see 7.3).

Proof. Let $\alpha \in \pi_{2n-1}(S^n)$ be an element with odd Hopf invariant, and let $Y = S^n \cup_f D^{2n}$, where f is a spheroid in the class α. Clearly

$$H^q(Y; \mathbb{Z}_2) = \begin{cases} \mathbb{Z}_2 & \text{for } q = 0, n, 2n, \\ 0 & \text{otherwise,} \end{cases}$$

and, by definition, the Hopf invariant, the operation of squaring $H^n(Y; \mathbb{Z}_2) \to H^{2n}(Y; \mathbb{Z}_2)$ is non-trivial. But this operation is Sq^n, and if n is not a power of 2, Sq^n can be represented as a polynomial in Sq^i with $i < n$ (for example, $Sq^{2n+1} = Sq^1 Sq^{2n}$, $Sq^{4n+2} = Sq^2 Sq^{4n}$, and so on; see the corollary in 11.2.A). Since $H^q(Y; \mathbb{Z}_2) = 0$ for $n < q < 2n$, any combination of Steenrod squares with indices less than n is trivial on $H^n(Y; \mathbb{Z}_2)$ which is a contradiction.

Note. The problem of describing the set of values of the Hopf invariant is one of the oldest in algebraic topology. Since $h([\iota_{2n}, \iota_{2n}]) = 2$ (see lemma 2 in 7.3), it reduces to the problem of the existence of elements in $\pi_{2n-1}(S^n)$ with odd Hopf invariant. This problem has many equivalent formulations; in particular, the existence of an element with this property is equivalent to the existence of a bilinear multiplication on $\mathbb{R}^n$ with unique division, and to the parallelizability of the $(n-1)$-sphere. In fact, there are elements with odd Hopf invariant in $\pi_{2n-1}(S^n)$ only when $n = 2, 4, 8$. One possible proof of this fact (historically the first – see Adams (1960)) depends on the fact that the operations $Sq^{16}, Sq^{32}, \ldots$, are indecomposable in the class of the usual (primary) cohomology operations, but decompose into non-trivial compositions of the so-called secondary cohomology operations (see Fomenko and Fuchs (1989)). A simpler proof using K-theory is known (see Adams and Atiyah (1966), Fomenko and Fuchs (1989)).

References*

Adams, J.F. (1960): On the non-existence of elements of Hopf invariant one. Ann. of Math. (2), **72**, No. 1, 20–104. Zbl. 096.17404

Adams, J.F., Atiyah, M.F. (1966): K-theory and the Hopf invariant. Quart. J. Math., Oxford Ser. (2), **17**, 31–38. Zbl. 136.43903

Atiyah, M.F. (1965): K-Theory. Cambridge, MA: Harvard University. 160 pp.

Borel, A. (1960): Seminar on Transformation Groups. Ann. Math. Stud. 46. Princeton, NJ: Princeton Univ. Press. Zbl. 091.37202

Bullett, S.R., Macdonald, I.G. (1982): On the Adem relations. Topology **21**, 329–332. Zbl. 506.55015

Dold, A. (1972): Lectures on Algebraic Topology. Berlin-Heidelberg-New York: Springer-Verlag. 377 pp. Reprint: Springer-Verlag 1995. Zbl. 872.55001

Eilenberg, S., Steenrod, N.E. (1952): Foundations of Algebraic Topology. Princeton Math. Ser. 15. Princeton, NJ: Princeton Univ. Press. 328 pp. Zbl. 047.41402

Fomenko, A.T., Fuks, D.B. (1989): A Course in Homotopic Topology. Moscow: Nauka. (Russian, English summary) Zbl. 675.55001

Fuks, D.B. (1971): Homotopy theory. Itogi Nauki Tekh., Algebra. Topology. Geometry. 1969, 71–122. Moscow.

Fuks, D.B. (1986): Classical manifolds. Itogi Nauki Tekh. VINITI. Sovrem. Probl. Matem. Fundamen. Napravleniya, **12**, 253–314. English transl. in: Topology II, Encyl. Math. Sci. **24**. Berlin-Heidelberg-New York: Springer-Verlag (2003).

Gabrielov, A.M., Gel'fand, I.M., Losik, M.V. (1975): Combinatorial computation of characteristic classes Funkts. Anal. Prilozh. **9**, No. 2, 12–28. English transl.: Functional. Anal. Appl. 9, 103–115 (1975). Zbl. 312.57016

Godement, R. (1958): Topologie algébrique et théorie des faisceaux. Paris: Hermann, Actual. Sci. Ind. No. 1252. 283 pp. Zbl. 080.16201. 3rd edn.: Paris: Hermann (1973). Zbl. 275.55010

Goresky, M., MacPherson, R. (1980): Intersection homology theory. Topology **19**, 135–165. Zbl. 448.55004

Hirzebruch, F. (1966): Topological Methods in Algebraic Geometry. Berlin-Heidelberg-New York: Springer-Verlag. 232 pp. Zbl. 138.42001. Reprint of the 2nd corr. print. of the 3rd edn. 1978: Springer-Verlag 1995. Zbl. 843.14009

Homotopy theory of differential forms. Collection of articles. (1981) Moscow: Mir. 191 pp.

Karoubi, M. (1978): K-Theory. An Introduction. Grundl. Math. Wiss. 226, Berlin-Heidelberg-New York: Springer-Verlag. 308 pp. Zbl. 382.55002

Milnor, J. (1959): On spaces having the homotopy of a CW complex. Trans. Amer. Math. Soc. **90**, 272–280. Zbl. 084.39002

Milnor, J. (1963): Morse theory. Ann. Math. Stud. No. 51. Princeton, NJ: Princeton Univ. Press. 153 pp. Zbl. 108.10401

Milnor, J., Stasheff, J.D. (1974): Characteristic classes. Ann. Math. Stud. No. 76. Princeton, NJ: Princeton Univ. Press. 331 pp. Zbl. 298.57008

Mishchenko, A.S. (1984): Vector Bundles and Their Applications. Moscow: Nauka. 208 pp. Zbl. 569.55001. English transl., revised version: Luke, G., Mischenko, A.S., Dordrecht Kluwer Acad. Publ. 254 pp. (1998). Zbl. 907.55002

---

* For the convenience of the reader, references to reviews in Zentralblatt für Mathematik (Zbl.), compiled by means of the MATH database, and Jahrbuch über die Fortschritte der Mathematik (Jbuch) have, as far as possible, been included in this bibliography.

Munkres, J.R. (1963): Elementary differential topology. Ann. Math. Stud. No. 54. Princeton, NJ: Princeton Univ. Press. 107 pp. Zbl. 107.17201

Postnikov, M.M. (1985): Lectures on Algebraic Topology. Homotopy Theory of Cellular Spaces. Moscow: Nauka. 416 pp. (Russian) Zbl. 578.55001

Rokhlin, V.A., Fuks, D.B. (1977): Beginner's Course in Topology; Geometric chapters. Moscow: Nauka. 488 pp. English transl.: Springer-Verlag 1984. Zbl. 562.54003

Snaith, V.P. (1979): Algebraic cobordism and K-theory. Mem. Am. Math. Soc. 221. 152 pp. Zbl. 413.55004

Spanier, E.H. (1966): Algebraic Topology. New York: McGraw-Hill. 528 pp. Second printing: Springer-Verlag 1995. Zbl. 810.55001

Steenrod, N.E. (1947): Products of cocycles and extensions of mappings. Ann. of Math. (2) **48**, 290–320. Zbl. 030.41602

Stong, R.E. (1968): Notes on cobordism theory. Princeton, NJ: Princeton Univ. Press and Univ. Tokyo Press. 387 pp. Zbl. 181.26604

Switzer, R.M. (1975): Algebraic Topology – Homotopy and Homology. Grundl. Math. Wiss. 212, Berlin-Heidelberg-New York: Springer-Verlag. 526 pp. Zbl. 305.55001

Thom, R. (1954): Quelques propriétés globales des variétés différentiables. Comment. Math. Helv. **28**, 17–86. Zbl. 057.15502

Whitney, H. (1957): Geometric Integration Theory. Princeton, NJ: Princeton Univ. Press. Zbl. 083.28204

III. Classical Manifolds

D.B. Fuchs

Translated from the Russian
by the author

Contents

Introduction

This article contains a variety of information (mainly topological, but also geometric and analytic) about classical manifolds, such as spheres, Stiefel and Grassmann manifolds, Lie groups, and lens spaces (a fuller list of the manifolds considered can be found in the table of contents). A considerable part of the material here will undoubtedly appear in other volumes on topology in the *Encyclopaedia of Mathematical Sciences*, but a systematic exposition in a single article seems useful.

The text is mainly conceived as a reference manual, but in my opinion, students of topology would profit by reading it straight through from beginning to end. They will have the opportunity of acquanting themselves with a range of problems that have stimulated the creation of the basic methods of topology, and this will help them to find their way around the subject.

Chapter 1
Spheres

This chapter consists of two main sections, dealing with the homotopy groups of spheres and differential structures on spheres, with an appendix containing some unrelated facts that deserve to be mentioned. Note that the homology and cohomology spheres are too simple to be discussed here; on the other hand, the extraordinary homology and cohomology of spheres is of great significance, but these groups are simply the homotopy groups of appropriate classifying spaces and have only an indirect relation to spheres.

§1. Homotopy Groups

1.1. Generalities. A. If $i < n$, then $\pi_i(S^n) = 0$ (see, Fuks, Rokhlin (1977) 5.2.2.1). If $i > 1$, then $\pi_i(S^1) = 0$ (see Fuks, Rokhlin (1977) 5.2.2.2).

B. The Brouwer-Hopf Theorem. *The group $\pi_n(S^n)$ is isomorphic to $\mathbb{Z}$ (for all $n \geqslant 1$). The standard isomorphism $\pi_n(S^n) \to \mathbb{Z}$ takes the class of a map of spheres $S^n \to S^n$ into its degree* (see Fuks, Rokhlin (1977) 5.2.2.7). □

C. The *suspension homomorphism*

$$\sum \colon \pi_q(S^n) \to \pi_{q+1}(S^{n+1}) \tag{1}$$

is defined at the level of maps of spheres by the formula

$$\sum \varphi(tx_1, \ldots, tx_{q+1}, u) = (t\varphi(x_1, \ldots, x_{q+1}), u),$$

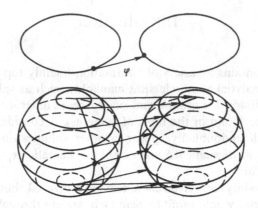

Fig. 1

where $\sum x_i^2 = 1$, $t^2 + u^2 = 1$ (see Fig. 1). The homomorphism (1) is a special case of the homomorphism $\sum: \pi_q(X) \to \pi_{q+1}(\sum x)$, where X is an arbitrary space; the definition in the general case is the same as in the case of spheres.

The Pontryagin-Freudenthal Theorem. "The easy part". *If $q < 2n - 1$, then (1) is an isomorphism; if $q = 2n - 1$, then (1) is an epimorphism.* (See Fuks, Rokhlin (1977) 5.2.1.4.)

"The difficult part". *The kernel of the epimorphism (1) with $q = 2n - 1$ is a cyclic group generated by the Whitehead square*[1] *$[\iota_n, \iota_n]$ of the homotopy class $\iota_n \in \pi_n(S^n)$ of the identity map* id: $S^n \to S^n$ *(see Postnikov Lecture 8).* □

(Further information on the kernels and images of the homomorphisms $\sum$ may be found in Postnikov (1985)).

Thus for any k the sequence

$$\pi_{1+k}(S^1) \xrightarrow{\Sigma} \pi_{2+k}(S^2) \xrightarrow{\Sigma} \pi_{3+k}(S^3) \xrightarrow{\Sigma} \cdots$$

stabilizes at the $(k + 2)$nd term:

$$\cdots \xrightarrow{\Sigma} \pi_{2k+1}(S^{k+1}) \underset{\text{epi}}{\xrightarrow{\Sigma}} \pi_{2k+2}(S^{k+2}) \xrightarrow{\Sigma} \cong \pi_{2k+3}(S^{k+3}) \xrightarrow{\Sigma} \cong \cdots$$

The stable group (i.e. the group $\pi_{N+k}(S^N)$ with $N \geqslant k + 2$) is denoted by π_k^S.

---

[1] Recall the definition of the *Whitehead product*. The sphere S^{m+n-1} is cut into two solid tori:

$$\{(x_1, \ldots, x_{m+n}) \in S^{m+n-1} | x_1^2 + \cdots + x_m^2 \geqslant x_{m+1}^2 + \cdots + x_{m+n}^2\} \approx S^{m-1} \times D^n,$$

$$\{(x_1, \ldots, x_{m+n}) \in S^{m+n-1} | x_1^2 + \cdots + x_m^2 \leqslant x_{m+1}^2 + \cdots + x_{m+n}^2\} \approx D^m \times S^{n-1}.$$

At the level of maps of the Whitehead product $\pi_m(X) \times \pi_n(X) \xrightarrow{[\cdot,\cdot]} \pi_{m+n-1}(X)$ is then defined by the formula

$$[f, g](x) = \begin{cases} f(\pi z), & \text{if } x = (y, z) \in S^{m-1} \times D^n \subset S^{m+n-1}, \\ g(\pi y), & \text{if } x = (y, z) \in D^m \times S^{n-1} \subset S^{m+n-1}, \end{cases}$$

where π denotes the canonical projection of the ball onto the sphere (this projection is homeomorphic in the interior of the ball and sends the boundary of the ball into a single point).

D. A k-dimensional *framed manifold* in $\mathbb{R}^{n+k}$ is a smooth closed (that is, compact and having empty boundary) k-dimensional submanifold of $\mathbb{R}^{n+k}$ furnished with a family of n pairwise orthogonal unit normal vector fields. Two such framed manifolds X_0, X_1 are called (framed) *cobordant* if there exists a compact smooth $(k + 1)$-dimensional submanifold of the product $\mathbb{R}^{n+k} \times I = \{(x_1, \dots, x_{n+k+1}) \in \mathbb{R}^{n+k+1} | 0 \leqslant x_{n+k+1} \leqslant 1\}$ with the following three properties: (i) its boundary coincides with the union of the manifolds $X_i \subset \mathbb{R}^{n+k} = \mathbb{R}^{n+k} \times i$ ($i = 1, 2$); (ii) it meets the hyperplanes $x_{n+k+1} = 0$ and $x_{n+k+1} = 1$ orthogonally; (iii) it posesses n pairwise orthogonal unit normal vector fields extending the framings of X_0 and X_1. The set of classes of cobordant k-dimensional framed manifolds in $\mathbb{R}^{n+k}$ is denoted by $\Omega_{\mathrm{fr}}^k(n + k)$. This set has a natural Abelian group structure ($\gamma \in \Omega_{\mathrm{fr}}^k(n + k)$ is the sum of α and β if γ is represented by a framed manifold, not intersecting $\mathbb{R}^{n+k-1} \subset \mathbb{R}^{n+k}$, whose intersections with the half-spaces $x_{n+k} > 0$ and $x_{n+k} < 0$ represent α and β respectively).

The Pontryagin Theorem. $\Omega_{\mathrm{fr}}^k(n + k) \cong \pi_{n+k}(S^n)$ (see Pontryagin (1976)). $\square$

Here we give the constructions of the standard mutually inverse isomorphisms $\Omega_{\mathrm{fr}}^k(n + k) \leftrightarrows \pi_{n+k}(S^n)$. The isomorphism $\Omega_{\mathrm{fr}}^k(n + k) \to \pi_{n+k}(S^n)$ assigns to the class of a framed manifold $X \subset \mathbb{R}^{n+k}$ the class of the following map of spheres. Identify $\mathbb{R}^{n+k}$ with $S^{n+k} - \mathrm{pt}$ and S^n with D^n/S^{n-1}; choose $\varepsilon > 0$ such that metric ε-neighbourhood U of X is the disjoint sum of ε-balls normal to X; the complement of U in S^{n+k} maps to the point of S^n that is the image of S^{n-1} under the projection $D^n \to D^n/S^{n-1} = S^n$; each of the ε-balls which constitue U maps linearly onto D^n according to the framing (the vectors of the framing are taken by the differential into the vectors on the coordinate axes) and then projects onto S^n. The isomorphism $\pi_{n+k}(S^n) \to \Omega_{\mathrm{fr}}^k(n + k)$ assigns to the class of a smooth map $f: S^{n+k} \to S^n$ the class of the following framed manifold. Fix a regular value $y \in S^n$ of f not equal to $f(\infty)$ (we put $S^{n+k} - \mathbb{R}^{n+k} = \infty$); choose a tangent n-frame v_1, $\dots, v_n$ to S^n at the point y; set $X = f^{-1}(Y)$ and define normal vector fields η_1, $\dots, \eta_n$ to X by the condition: the differential of f at the point $x \in X$ takes the vector $\eta_i(x)$ into v_i; orthogonalize the fields $\eta_1, \dots, \eta_n$. The manifold X with the resulting framing is our framed manifold. $\square$

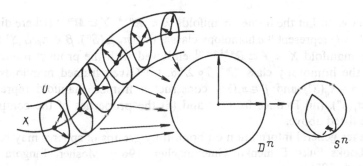

Fig. 2

Exercise. Deduce all the previous results from the Pontryagin Theorem. [How far the reader will get in this task, depends on his background. To simplify the exercise we give here the description of the suspension homomorphism and the Whitehead product in the language of framed cobordisms. The homomorphism $\sum$ assigns to the class of a framed manifold $X \subset \mathbb{R}^q$ the class of the same manifold placed in $\mathbb{R}^{q+1} \supset \mathbb{R}^q$ with the former framing supplemented with the vector parallel to the $(q+1)$st coordinate axis. The Whitehead product of the classes of framed manifolds $X \subset \mathbb{R}^m$ and $Y \subset \mathbb{R}^n$ is represented by the class of the framed manifold $Z \subset \mathbb{R}^{m+n-1}$ defined as follows. Without loss of generality we suppose that X and Y are contained in the interiors of the unit balls $D^m \subset \mathbb{R}^m$ and $D^n \subset \mathbb{R}^n$. In $\mathbb{R}^{m+n-1}$ consider the standard pair of linked spheres of dimensions $n-1$ and $m-1$; the sphere of radius 3 in the plane of the 1st 2nd, ..., $(n-1)$st and $(m+n-1)$st coordinate axes with center $(0, \ldots, 0)$, and the sphere of radius 3 in the plane of the nth, $(n+1)$st, ..., $(m+n-1)$st coordinate axes with center $(0, \ldots, 0, 3)$. The closed metric 1-neighbourhood of the union of these spheres is naturally identified with $(S^{n-1} \times D^m) \cup (S^{m-1} \times D^n)$. We set

$$Z = (S^{n-1} \times X) \cup (S^{m-1} \times Y) \subset (S^{n-1} \times D^m) \cup (S^{m-1} \times D^n) \subset \mathbb{R}^{m+n-1}$$

and endow Z with the framing which coincides on $s \times X$ ($s \in S^{n-1}$) with the given framing of $X = s \times X$ in $D^m = s \times D^m$ and coincides on $s \times Y$ ($s \in S^{m-1}$) with the given framing of $Y = s \times Y$ in $D^n = s \times D^n$.

E. A well as the Whitehead product, which is defined for the homotopy groups of arbitrary topological spaces, the homotopy groups of spheres posses a specific multiplicative structure, namely, the *composition product*.

$$\pi_q(S^n) \times \pi_r(S^q) \xrightarrow{\circ} \pi_r(S^n),$$

which is defined in terms of sphere maps by the formula $(f, g) \mapsto f \circ g$. Under stabilization this product takes the form

$$\pi_k^S \times \pi_l^S \xrightarrow{\circ} \pi_{k+l}^S;$$

this last product is associative, skew-commutative ($\beta \circ \gamma = (-1)^{kl}\alpha \circ \beta$) and distributive. (Warning: the unstable composition product is right distributive but in general not left distributive; see the details in Postnikov (1985), Lecture 6.)

Exercise. (i) Let the framed manifolds $X \subset \mathbb{R}^{n+k}$, $Y \subset \mathbb{R}^{m+1}$ (where dim $X = k$, dim $Y = l$) represent the homotopy classes $\alpha \in \pi_{n+k}(S^n)$, $\beta \in \pi_{m+l}(S^m)$. Show that the manifold $X \times Y \subset \mathbb{R}^{n+k+m+1}$ endowed with the product framing represents the homotopy class $\Sigma^{n+k}\beta \circ \Sigma^m\alpha$. (ii) Given framed manifolds representing $\alpha \in \pi_q(S^n)$ and $\beta \in \pi_r(S^q)$, construct a framed manifold representing $\alpha \circ \beta \in \pi_r(S^n)$. (iii) Deduce from (i) and (ii) the properties of the composition product listed above.

(Further general information on homotopy groups of spheres may be found in the books Fuks, Fomenko, Gutenmacher (1969), Mosher, Tangora (1968), Whitehead (1970).)

1.2. Tables and Related Information A. The groups $\pi_{n+k}(S^n)$ are only partially known. The groups $\pi_{n+k}(S^n)$ with $1 \leqslant k \leqslant 7$ are exhibited in Table 1. The groups π_k^S with $8 \leqslant k \leqslant 15$ are exhibited in Table 2.

Table 1

к \ П	2	3	4	5	6	7	8	$\geqslant 9$
1	$\mathbb{Z}$	$\mathbb{Z}_2$	$\mathbb{Z}_2$	$\mathbb{Z}_2$	$\mathbb{Z}_2$	$\mathbb{Z}_2$	$\mathbb{Z}_2$	$\mathbb{Z}_2$
2	$\mathbb{Z}_2$	$\mathbb{Z}_2$	$\mathbb{Z}_2$	$\mathbb{Z}_2$	$\mathbb{Z}_2$	$\mathbb{Z}_2$	$\mathbb{Z}_2$	$\mathbb{Z}_2$
3	$\mathbb{Z}_2$	$\mathbb{Z}_{12}$	$\mathbb{Z} \oplus \mathbb{Z}_{12}$	$\mathbb{Z}_{24}$	$\mathbb{Z}_{24}$	$\mathbb{Z}_{24}$	$\mathbb{Z}_{24}$	$\mathbb{Z}_{24}$
4	$\mathbb{Z}_{12}$	$\mathbb{Z}_2$	$\mathbb{Z}_2 \oplus \mathbb{Z}_2$	$\mathbb{Z}_2$	0	0	0	0
5	$\mathbb{Z}_2$	$\mathbb{Z}_2$	$\mathbb{Z}_2 \oplus \mathbb{Z}_2$	$\mathbb{Z}_2$	$\mathbb{Z}$	0	0	0
6	$\mathbb{Z}_2$	$\mathbb{Z}_3$	$\mathbb{Z}_2 \oplus \mathbb{Z}_{24}$	$\mathbb{Z}_2$	$\mathbb{Z}_2$	$\mathbb{Z}_2$	$\mathbb{Z}_2$	$\mathbb{Z}_2$
7	$\mathbb{Z}_3$	$\mathbb{Z}_{15}$	$\mathbb{Z}_{15}$	$\mathbb{Z}_{30}$	$\mathbb{Z}_{60}$	$\mathbb{Z}_{120}$	$\mathbb{Z} \oplus \mathbb{Z}_{120}$	$\mathbb{Z}_{240}$

Groups $\pi_{n+k}(S^n)$

Table 2

k	π_k^S	k	π_k^S
8	$\mathbb{Z}_2 \oplus \mathbb{Z}_2$	12	0
9	$\mathbb{Z}_2 \oplus \mathbb{Z}_2 \oplus \mathbb{Z}_2$	13	0
10	$\mathbb{Z}_2$	14	$\mathbb{Z}_6 \oplus \mathbb{Z}_2$
11	$\mathbb{Z}_{504}$	15	$\mathbb{Z}_{480} \oplus \mathbb{Z}_2$

(More complete tables may be found in Toda (1962).)

B. All the groups $\pi_{n+k}(S^n)$ are finitely generated. Their ranks are known:

$$\text{rank } \pi_q(S^n) = \begin{cases} 1, & \text{if } q = n \text{ or } n \text{ is even and } q = 2n - 1, \\ 0, & \text{in all other cases.} \end{cases}$$

(See, for example, Spanier (1966), §9.7.)

C. A non-trivial homomorphism $\pi_{4m-1}(S^{2m}) \to \mathbb{Z}$ may be obtained by means of the so-called *Hopf invariant*. We give here two equivalent definitions of it (prove the equivalence).

The **First Definition**: let $f: S^{4m-1} \to S^{2m}$ be a smooth map in the class $\alpha \in \pi_{4m-1}(S^{2m})$, and let y_1, y_2 be two regular values of f. The linking number of the inverse images $f^{-1}(y_1)$ and $f^{-1}(y_2)$ does not depend on the choice of f, y_1, and y_2 and is called the Hopf invariant of α; it is denoted by $\mathfrak{H}(\alpha)$.

Second Definition: let f be an arbitrary map in the class α. Attach D^{4m} to S^{2m} by means of f and denote the resulting space by X. The standard orientations of the sphere S^{2m} and the ball D^{4m} define canonical generators u and v of the groups $H^{2m}(X; \mathbb{Z}) = \mathbb{Z}$ and $H^{4m}(X; \mathbb{Z}) = \mathbb{Z}$. We set by definition $u^2 = \mathfrak{H}(\alpha)v$. (*Exercise*: check from each of the definitions that $\mathfrak{H}$ is a homomorphism.) (The Hopf invariant has various generalizations; see, for example Postnikov (1985) Lecture 6.) The non-triviality of the Hopf invariant is seen from the equality $\mathfrak{H}([\iota_n, \iota_n]) = 2$. (*Exercise*: deduce this equality from each of the two definitions.)

The Adams Theorem. $\mathfrak{H}$ *is an epimorophism only if* $m = 1, 2, 4$. (For the proof see Schwartz (1968).) □

D. Let p be an odd prime. If $k < 2p - 3$, then the p-primary component of the group $\pi_{n+k}(S^n)$ is trivial for any n. The p-primary components of the groups π_k^S with $k \leqslant 2p(p - 1) - 1$ are as follows.

$$(\pi_k^S)_p = \begin{cases} \mathbb{Z}_p, & \text{if } k = 2i(p - 1) - 1 \text{ with } i < p, \text{ or } k = 2p(p - 1) - 2, \\ \mathbb{Z}_{p^2}, & \text{if } k = 2p(p - 1) - 1, \\ 0, & \text{for other } k \leqslant 2p(p - 1) - 1. \end{cases}$$

(For further information on the groups $(\pi_k^S)_p$ see Toda (1962).)

1.3. The Groups $\pi_{n+1}(S^n)$.

A. The group $\pi_3(S^2) \cong \mathbb{Z}$ is generated by the class η_2 (the notation is standard) of the following *Hopf map* $h: S^3 \to S^2$: the sphere S^3 is regarded as the unit sphere of the complex plane $\mathbb{C}^2$, the sphere S^2 is identified with the complex projective line $\mathbb{C}P^1$, and the mapping h is defined by the formula $h(z, w) = (z; w)$. (*Exercise*: check from each of the two definitions of the Hopf invariant that $\mathfrak{G}(\eta_2) = 1$.) If $n > 2$ then the group $\pi_{n+1}(S^n) \cong \mathbb{Z}_2$ is generated by $\eta_n = \Sigma^{n-2} \eta_2$; the corresponding element of $\pi_1^S \cong \mathbb{Z}_2$ is denoted by η.)

(The above construction with complex numbers replaced by quaternions or Cayley numbers gives non-trivial elements of the groups $\pi_7(S^4)$ and $\pi_{15}(S^8)$. These elements have Hopf invariants equal to 1, and, like η_2, are called Hopf elements.)

B. In the language of framed cobordisms the k-fold class of the Hopf mapping $S^3 \to S^2$ is described as the class of the standard circle $S^1 \subset \mathbb{R}^2 \subset \mathbb{R}^3$ with the framing obtained from the standard framing (composed of the exterior normal to S^1 in $\mathbb{R}^2$ and the vector parallel to the axis OZ) by k-fold twisting in the positive direction (see Fig. 3) (*Exercise*. Consider in $\mathbb{R}^3$ the simplest non-trivial knot – the trefoil knot – see Fig. 4. Furnish it with the framing whose first vector is seen in Fig. 4, and whose second vector is normal to the plane of Fig. 4 and is directed upwards. Which element of $\pi_3(S^2)$ is represented by this framed manifold?)

C. For an arbitrary continuous mapping $f: S^{n+1} \to S^n$ $(n \geqslant 3)$ consider the space X obtained by attaching D^{n+1} to S^n by means of f. Clearly, $H^n(X; \mathbb{Z}_2) \cong H^{n+2}(X; \mathbb{Z}_2) \cong \mathbb{Z}_2$.

Theorem. *The mapping f represents a non-trivial element of the group* $\pi_{n+1}(S^n)$ *if and only if the operation* $Sq^2: H^n(X; \mathbb{Z}_2) \to H^{n+2}(x; z_2)$ *is non-trivial.* □

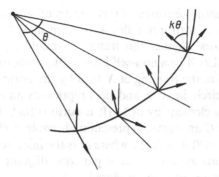

Fig. 3

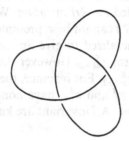

Fig. 4

(An analogous description by Steenrod squares is possible only for elements of the stable groups $\pi_{n+3}(S^n)$ ($n \geqslant 5$) and $\pi_{n+7}(S^n)$ ($n \geqslant 9$) corresponding to the Hopf elements of $\pi_7(S^4)$ and $\pi_{15}(S^8)$ – this is a consequence of the Adams Theorems from § 1.2.C.)

1.4. The Groups $\pi_{n+2}(S^n)$. A. If $n \geqslant 2$, then the non-trivial element η_n (the notation is standard) of the group $\pi_{n+2}(S^n) \cong \mathbb{Z}_2$ is represented by the composite sphere map

$$S^{n+2} \xrightarrow{\ \Sigma^{n-1}h\ } S^{n+1} \xrightarrow{\ \Sigma^{n-1}h\ } S^n;$$

thus $\zeta_n = \eta_{n+1} \circ \eta_n$ and $\zeta = \eta^2$ (where ζ is the element of π_2^S corresponding to ζ_n.)

B. The above description of the generator of the group $\pi_{n+2}(S^n)$, together with the discussion of §1.3.B, gives rise to an effective representation of this generator by a framed manifold. Namely, if $n = 2$, then it is represented by the standardly-embedded torus $S^1 \times S^1 \subset \mathbb{R}^3 \subset \mathbb{R}^4$ with the framing obtained from the standard framing (whose first vector is the exterior normal to $S^1 \times S^1$ and in $\mathbb{R}^3$, and whose second vector is parallel to the 4th basis vector) by rotation about the point $(\varphi, \psi) \in S^1 \times S^1$ through the angle $\varphi + \psi$. In the case $n > 2$, the whole picture is embedded in $\mathbb{R}^{n+2}$ and the framing is supplemented by the vectors parallel to the 5th, ..., basis vectors.

Note some interesting features of this construction. Suppose that a framed (2-dimensional) sphere with g handles X is given in $\mathbb{R}^{n+2}$; we shall exhibit an invariant which shows whether this framed manifold represents the non-zero element of $\pi_{n+2}(S^n)$. Let A be a non-self-intersecting smooth curve on X. Supplement the restriction of the framing of X to A by the normal to A in X. Then A becomes a framed circle in $\mathbb{R}^{n+2}$ and thus represents an element of the group $\pi_{n+1}(S^n)$; denote this element by $\sigma(A)$. (It is curious that the invariant σ is not additive: if A, B and C are curves representing homology classes α, β and $\alpha + \beta$, then $\sigma(C) = \sigma(A) + \sigma(B) + \phi_2(\alpha, \beta)$, where ϕ_2 is the intersection number modulo 2.) Now fix an arbitrary system of g pairwise disjoint non-self-intersection homologically independent smooth closed curves $A_1, \ldots, A_g$ and put $\sigma(X) = \sigma(A_1) \ldots \sigma(A_g)$. It turns out that σ is a well-defined invariant of the element of $\pi_{n+2}(S^n)$ determined by X, and that the last elements is non-trivial if and only if $\sigma \neq 0$. The invariant σ is called the *Arf invariant*. We see, in particular, that the non-trivial element of $\pi_{n+2}(S^n)$ cannot be represented by a framed sphere. (The above definition can be generalized to give an Arf invariant for an arbitrary element of a group of the form π^S_{4k-2}. However this invariant can be non-zero only in groups of the form $\pi^S_{2^m-2}$. For instance, the composition squares of the Hopf elements, which lie in π^S_6 and π^S_{14}, have non-trivial Arf invariants. Some other elements with non-trivial Arf invariant are known, for example, in π^S_{30}.)

1.5. The Whitehead J-Homomorphism. This is a homomorphism

$$J: \pi_k(SO(n)) \to \pi_{n+k}(S^n),$$

which is defined in two equivalent ways.

First Definition. The direct sum of the standard action of the group $SO(n)$ in $\mathbb{R}^n$ and its trivial action in $\mathbb{R}$ is an action of $SO(n)$ in $\mathbb{R}^{n+1}$. The latter defines an action of $SO(n)$ in S^n with (two) fixed points. A mapping $SO(n) \to \Omega^n S^n$ arises (where Ω is the loop operator). It induces, in turn, a mapping $\pi_k(SO(n)) \to \pi_k(\Omega^n S^n) = \pi_{n+k}(S^n)$, which is just our J.

Second Definition. Consider in $\mathbb{R}^{n+k}$ the standard k-sphere $S^k \subset \mathbb{R}^{k+1} \subset \mathbb{R}^{n+k}$ with the standard framing (the first vector is the exterior normal to S^k in $\mathbb{R}^{k+1}$, and the remaining vectors are parallel to the $(k + 2)$nd, $\ldots$, $(n + k)$th basis vectors). The homomorphism J assigns to the class of a map $f: S^k \to SO(n)$ the class of the framed manifold which is defined as the sphere S^k with the framing obtained from the standard one by applying the matrix $f(x)$ at the point $x \in S^k$. (*Exercise*: prove the equivalence of the definitions.)

Note that an element of the homotopy group of a sphere belongs to the image of the J-homomorphism if and only if it is represented by the standard sphere with some framing. For instance, the discussion in 1.4 shows that the non-trivial element of $\pi_{n+2}(S^n)$ does not belong to the image of the J-homomorphism (which is not surprising since $\pi_2(SO(n)) = 0$).

The stable J-homomorphism

$$J = J^S_k: \pi_k(SO) \to \pi^S_k \text{ with } k = 4m - 1,$$

where $SO = SO(\infty) = \bigcup_n SO(n)$, is particularly important. Indeed, the groups $\pi_{4m-1}(SO)$ are isomorphic to $\mathbb{Z}$ (cf. Chapter 2), and hence the image of the homomorphism J^S_{4m-1} is a cyclic group. The order of this group is known: it is equal to the denominator of the irreducible fraction $B_m/4m$, where B_m is the mth Bernoulli number. Recall that the Bernoulli numbers are defined by the formula

$$\frac{t}{e^t - 1} = 1 - \frac{t}{2} + \sum_{m=1}^{\infty} (-1)^{m-1} \frac{B_m}{(2m)!} t^{2m}.$$

In particular, $B_1 = \frac{1}{6}$, $B_2 = \frac{1}{30}$, $B_3 = \frac{1}{42}$, $B_4 = \frac{1}{30}$, $B_5 = \frac{5}{66}$. Thus the order of the image of the stable J-homomorphism in the groups π_3^S, π_7^S, π_{11}^S, π_{15}^S, π_{19}^S is equal to 24, 240, 504, 480, 264 respectively.

Note finally (this is important for §2) that the J-homomorphism in general is not onto. For example, it is seen from Table 2 that the homomorphism J_k^S is not onto for $k = 2, 6, 8, 9, 10, 13, 14, 15$.

§2. Differential Structures

It was mainly the immense popularity of this subject that compelled me to devote a separate section to it. Nevertheless, it has some applications both inside topology (e.g. in the theory of smoothings of combinatorial manifolds) and outside it (e.g. in algebraic geometry).

The theory of differential structures on spheres originates from the sensational work of J. Milnor who discovered (in 1956) that there exist smooth manifolds which are homeomorphic but not diffeomorphic to the seven-dimensional sphere. This phenomenon was commonly regarded as one of the most surprising results in topology till 1982, when M. Freedman and S. Donaldson showed that there exist smooth manifolds which are homeomorphic but not diffeomorphic to four-dimensional Euclidean space.

2.1. Generalities. (Cf. Milnor, Kervaire (1963), Browder (1972)) A. Consider the collection of all oriented smooth manifolds homomorphic to the n-sphere. Divide it into classes under the relation of oriented diffeomorphism, and denote the set of classes by θ_n. The connected sum operation (see Fig. 5) makes θ_n into an Abelian group.

The elements of the group θ_n are just *differential structures* on S^n. A differential structure on a sphere is called a $\partial\pi$-*structure* if the sphere with this structure is a

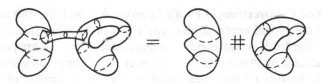

Fig. 5

boundary of a compact parallelizable smooth manifold. It may be shown that $\partial\pi$-structures form a subgroup of the group θ_n; denote this subgroup by $\theta_n(\partial\pi)$. In the theory of differential structures the group $\theta_n(\partial\pi)$ and the quotient-group $\theta_n/\theta_n(\partial\pi)$ are studied separately.

B. The First Milnor-Kervaire Theorem. *The group $\theta_n(\partial\pi)$ is trivial for even $n \neq 4$ (nothing is known for $n = 4$), and is a finite cyclic group for odd n.* □

For odd n, let ρ_n denote the order of the group $\theta_n(\partial\pi)$. The following is known about the numbers ρ_n. If $n \equiv 1 \bmod 4$, then $\rho_n \leqslant 2$. As a rule, $\rho_n = 2$ in this case; in particular, $\rho_n = 2$ if $n + 3$ is not a power of 2. But $\rho_1 = \rho_5 = \rho_{13} = \rho_{29} = 1$. (The problem of computing the numbers ρ_{4k+1} is closely related to the Arf invariant mentioned in § 1.4.B.) The numbers ρ_n with $n \equiv -1 \bmod 4$ may be very large; for example, $\rho_7 = 28$, $\rho_{11} = 992$, $\rho_{15} = 8128$. In general ρ_{4k-1} is equal to the denominator of the irreducible fraction $4B_k/k$ multiplied by $2^{2k-2}(2^{k-1} - 1)$. Alternatively the number ρ_{4k-1} may be described as one-eighth of the greatest common divisor of the signatures of closed almost parallelizable (that is, parallelizable in the complement of a point) $4k$-manifolds.

C. The second Milnor-Kervaire Theorem. *If $n \neq 2, 4, 6, 14$, then the quotient group $\theta_n/\theta_n(\partial\pi)$ is isomorphic to the quotient group $\pi_n^S/\mathrm{Im}\ J_n^S$: if $n = 2, 6, 14$, then the quotient group $\theta_n/\theta_n(\partial\pi)$ is embedded in $\pi_n^S/\mathrm{Im}\ J_n^S$ as a direct summand of index 2.* □

D. Thus the number of differential structures on the sphere S^n is finite for any $n \neq 4$ (nothing is known for $n = 4$). For $n \leqslant 15$ these numbers are shown in Table 3.

Table 3

n	$|\theta_n|$	n	$|\theta_n|$	n	$|\theta_n|$
1, 2, 3	1	8	2	12	1
4	?	9	8	13	3
5, 6	1	10	2	14	12
7	28	11	992	15	16256

Smooth manifolds which are homeomorphic but not necessary diffeomorphic to S^n are called n-dimensional *exotic spheres* or n-dimensional *Milnor spheres*.

2.2. Explicit Constructions of Exotic Spheres. A. The following construction of a seven-dimensional exotic sphere was given in Milnor's initial work. It is well-known that (real) vector bundles of dimension 4 over S^4 are parametrized, up to equivalence, by two integers (which may take arbitrary values); the Euler number and the (first) Pontryagin number. The *Euler number* of the bundle $p: E \to S^4$ is defined as the intersection number of a generic smooth section with the zero section. The (first) *Pontryagin number* of p is defined as follows. We take 4 generic sections $s_1, \ldots, s_4 \colon S^4 \to E$ and count the algebraic number of points $x \in S^4$ such that the vectors $s_1(x), \ldots, s_4(x)$ of the four-dimensional space $p^{-1}(x)$ form a system of rank 2 (for generic sections s_i the number of such points is finite; the definition of the *sign* of a point is left to the reader). For $m, n \in \mathbb{Z}$ we denote by $p(m, n)$ the bundle with Euler number m and Pontryagin number n. Denote then by $S(m, n)$ the manifold of unit vectors in the total space of the bundle $p(m, n)$ (with respect to an arbitrary Euclidean structure).

The Milnor Theorem. *The manifold $S(m, n)$ is homeomorphic to S^7 if and only if $m \neq \pm 1$. The manifolds $S(1, n_1)$, $S(1, n_2)$ are diffeomorphic if and only if $n_1 \equiv n_2 \bmod 28$.* □

Thus $S(1, 0)$, $S(1, 1)$, $\ldots$, $S(1, 27)$ is a complete list of all seven-dimensional exotic spheres.

B. The following ingenious construction of all $\partial\pi$-structures on spheres is due to E. Brieskorn and F. Hirzebruch.

Theorem. *If $n \geqslant 2$,[1] and $k \geqslant 0$ then the submanifold $B(n, k)$ of $\mathbb{C}^{2n+1}$ defined by the equations*

$$z_1^{5+8k} + z_2^3 + z_3^2 + \cdots + z_{2n+1}^2 = 0,$$

$$|z_1|^2 + |z_2|^2 + \cdots + |z_{2n+1}|^2 = 1,$$

is homeomorphic to S^{4n-1} and defines an element of the group $\theta_{4n-1}(\partial\pi)$. The manifolds $B(n, k_1)$ and $B(n, k_2)$ are diffeomorphic if and only if $k_1 \equiv k_2 \bmod \rho_{4n-1}$; moreover, the correspondence $B(n, k) \mapsto k + 1 \bmod \rho_{4n-1}$ induces an isomorphism $\theta_{4n-1}(\partial\pi) \to \mathbb{Z}_{\rho_{4n-1}}$. □

The author is not aware of any explicit constructions of exotic spheres from $\theta_m - \theta_m(\partial\pi)$.

§ 3. Appendix

3.1. Structures. A. An almost complex structure (i.e. a $GL(n, \mathbb{C})$-reduction of the tangent bundle) on the sphere S^{2n} exists if and only if $2n = 2$ or 6. (The construction of an almost complex structure on S^6 may be found in Steenrod (1951), 41.21.) The 2-sphere has the structure of a complex-analytic manifold (the Riemann sphere); it is not known if such a structure exists on the 6-sphere.

B. All the odd-dimensional spheres have contact structures. The standard contact structure on S^{2n-1} consists of the maximal complex subspaces of the tangent spaces to $S^{2n-1} \subset \mathbb{C}^n$. It is known that at least the 3-sphere S^3 has contact structures not diffeomorphic to the standard one (see Douady (1983)).

Obviously, among all the spheres, only the 2-sphere S^2 has a symplectic structure.

---

[1] For the case $n = 1$ see Chapter 5.

3.2. Vector Fields and Plane Fields. A. **The Adams Theorem.** *Let $n + 1 = a \cdot 2^{b+4c}$ where a, b, c are integers such that $a \equiv 1$ mod 2, $0 \leqslant b < 4$ and $c \geqslant 0$. Then the maximal number of linearly independent vector fields on the sphere S^n is equal to $2^b + 8c - 1$.* □

For the proof see Schwartz (1968), Chapter 10.

Corollary. *The sphere S^n is parallelizable if and only if $n = 1, 3, 7$.* □

B. **Theorem.** *If $q \leqslant n/2$, then the following statements are equivalent:*
(a) *there exists a continuous field of tangent q-planes on S^n;*
(b) *there exist q linearly independent vector fields on S^n.* □

Note that the existence of a continuous q-plane field on S^n is equivalent to the existence of a continuous $(n - q)$-plane field on S^n; thus for any q the problem of the existence of a continuous q-plane field on S^n is solved by the Adams theorem (see § 3.2.A).

3.3. Foliations (for details see Fuks (1981)). The Haefliger – Thurston classification theory of foliations reduces the problem of the existence and classification of foliations to pure homotopy problems. Sometimes (not very often) these homotopy problems can be completely solved. Here is one of the most significant results of this kind.

Theorem. *If $q \leqslant n/2$ then the statements (a) and (b) of the Theorem in §3.2.B are equivalent to the following:*
(c) *there exists a foliation of codimension q on S^n.* □

In particular, each odd-dimensional sphere has a foliation of codimension one. There are explicit construction of such foliations (Reeb, Lawson, Tamura, Durfee).

In certain cases the existence of foliations of codimension $> n/2$ on S^n has been proved. For example, foliations of all possible dimensions have been constructed on S^7.

(All the above results concern smooth foliations, e.g. of class C^∞. In the analytic case it is known that no sphere S^n with $n > 1$ has an analytic foliation of codimension 1.)

Chapter 2
Lie Groups and Stiefel Manifolds

§ 1. Lie Groups: Geometric Information

1.1. Generalities. A. In this chapter we shall consider only compact Lie groups since this is the only case of interest from the topological point of view: any (finite-dimensional) Lie group, no matter how complicated its algebraic structure, is topologically the direct product of its maximal compact subgroup and a Euclidean space. Our main consideration will be the classical groups: $SO(n)$ (the real group is meant), $U(n)$ and $Sp(n)$, the latter being the compact group of isometric linear automorphisms of (left, say) quaternionic space $\mathbb{H}^n$. The group $O(n)$ topologically is the direct product of the group $SO(n)$ and a two-element set, the group $U(n)$ topologically is the direct product $S^1 \times SU(n)$; the groups $C(n)$ and $SU(n)$ play an episodic rôle in our article. In the next section, devoted to cohomology, we shall also treat the spinor groups Spin(n) – that is, the two-fold covering of the group $SO(n)$ – and the simply-connected compact exceptional groups. The exceptional groups are not in fact mentioned anywhere else; their definition can be found in any text on Lie groups (see, for example, Adams (1969)), and there is extensive information on their homotopy groups in Fuks (1971) (though the case of F_4 is omitted there; I do not know any work where the homotopy groups of F_4, are computed, although it seems improbable that such work does not exist.

B. We shall regard the canonical embeddings

$$SO(1) \to SO(2) \to SO(3) \to \cdots,$$

$$U(1) \to U(2) \to U(3) \to \cdots,$$

$$Sp(1) \to Sp(2) \to Sp(3) \to \cdots,$$

as inclusions, and we put $SO = \bigcup_{n=1}^{\infty} SO(n)$, $U = \bigcup_{n=1}^{\infty} U(n)$, $Sp = \bigcup_{n=1}^{\infty} Sp(n)$. We also mention the important embeddings $O(n) \to U(n)$, $SO(n) \to SU(n)$ [a real orthogonal $n \times n$-matrix is regarded as a complex unitary $n \times n$-matrix], $U(n) \to SO(2n)$ [a unitary transformation of $\mathbb{C}^n$ is regarded as an orthogonal transformation of $\mathbb{R}^{2n}$] and the similar embeddings $U(n) \to Sp(n)$, $Sp(n) \to SU(2n)$. The homogeneous spaces arising are of great interest, due to their rôle in *Bott periodicity* (see § 1.3.B below), and some of them are also important for their own sake. For instance, $U(n)/O(n)$ is identified with the space of Lagrangian subspaces of the space $\mathbb{C}^n$ (i.e. real n-dimensional subspaces $\pi \subset \mathbb{C}^n$ such that $i\pi = \pi^\perp$), and $O(2n)/U(n)$ may be interpreted as the space of complex structures in the space $\mathbb{R}^{2n}$ compatible with the given Euclidean structure; see the details in Chapter 4.

C. The transitive actions of the groups $SO(n)$, $U(n)$, $Sp(n)$ on the spheres S^{n-1}, S^{2n-1}, S^{4n-1} correspondingly give rise to principal fibrations

$$SO(n) \xrightarrow{SO(n-1)} S^{n-1}, \qquad U(n) \xrightarrow{U(n-1)} S^{2n-1},$$

$$Sp(n) \xrightarrow{Sp(n-1)} S^{4n-1}, \tag{2}$$

where the fibres (over the point $(0, \ldots, 0, 1)$) are indicated above the arrows. (We can say that the projection maps of these fibrations assign to a matrix its last row.) We conclude inductively that the group $SO(n)$ is a "skew product" of the spheres $S^1, S^2, S^3, \ldots, S^{n-1}$; the group $U(n)$ is the skew product of the spheres $S^1, S^3, S^5, \ldots, S^{2n-1}$; and the group $Sp(n)$ is the skew product of the spheres $S^3, S^7,$

$S^{11}, \ldots, S^{4n-1}$ (for the sake of uniformity, we add that $O(n)$ is the skew product of the spheres $S^0, S^1, \ldots, S^{n-1}$, and $SU(n)$ is the skew product of the spheres $S^3, S^5, \ldots, S^{2n-1}$). We shall see below that, with the exception of some trivial cases, these products are not direct, though all the manifolds $SO(n)$, $U(n)$, $Sp(n)$ have much in common with the corresponding direct products of spheres.

1.2. Some Lie Groups of Low Dimension. A. Clearly, the manifold $SO(2)$ is the circle (the group $SO(2)$ is the group of rotations of the plane). The group $SO(3)$ is diffeomorphic to the real projective space $\mathbb{R}P^3$. A well-known diffeomorphism $\mathbb{R}P^3 \to SO(3)$ can be defined by means of the mapping $D^3 \to SO(3)$ which assigns to a point $x \in D^3$ the rotation of $\mathbb{R}^3$ around the oriented axis $0x$ through the angle $\pi \|x\|$ (the point 0 corresponds to the identity mapping). Obviously the same mapping is assigned to the points x and $-x \in S^2 \subset D^3$, namely the reflection in the line $0x$; otherwise our mapping $D^3 \to SO(3)$ is one-one, and it defines a diffeomorphism $\mathbb{R}P^3 = D^3(x \sim -x$ for $x \in S^2) \to SO(3)$. Another description of essentially the same diffeomorphism $\mathbb{R}P^3 \to SO(3)$ is given by the so-called *Rodrigues's formula*. This formula actually defines a mapping $S^3 \to SO(3)$, which sends a unit quaternion $\cos \varphi + \sin \varphi \, (ai + bj + ck)$, $a^2 + b^2 + c^2 = 1$, into the rotation of $\mathbb{R}^3$ around the oriented axis (a, b, c) through the angle 2φ; the coefficient 2 reflects the two-foldness of Rodrigues's mapping: it is actually a mapping $S^3/\mathbb{Z}_2 = \mathbb{R}P^3 \to SO(3)$, and as such is a diffeomorphism. However the verification that it is a diffeomorphism is complicated by a defect common to the two last constructions, namely the lack of homogeneity: in both cases a special verification of the smoothness near some singular set is needed (in the first case this is the image of the boundary sphere $S^2 \subset D^3$, and in the second case it is the image of the point $\varphi = 0$). The following third construction is free from this defect; the unit quaternion $x \in S^3$ is mapped to the conjugation $y \mapsto xyx^{-1}$ in the space $\mathbb{R}^3$ of purely imaginary quaternions. Quaternions x and $-x$ obviously define the same transformation of $\mathbb{R}^3$, and it is now quite easy to check that a diffeomorphism $S^3/\mathbb{Z}_2 = \mathbb{R}P^3 \ SO(3)$ is obtained. Note that the last mapping $S^3 \to SO(3)$ is also a group homomorphism of the multiplicative group of unit quaternions S^3 onto $SO(3)$.

The group $SO(4)$ is diffeomorphic to the product $S^3 \times SO(3)$ (note that though the sphere S^3 has a natural group structure – see above – the group $SO(4)$ is not isomorphic to the product of groups S^3 and $SO(3)$ – prove it!). Moreover, the fibration $SO(4) \to S^3$ considered above is trivial: it has a section which assigns to any point $x \in S^3$ the matrix with rows xi, xj, xk, x (here and above i, j, k are the quaternion units).

However, the group $SO(n)$ with $n \geqslant 5$ is not homeomorphic to any product of spheres; this may be deduced, for example, from the structure of the cohomology rings of the groups $SO(n)$ (see § 2.3.B).

B. The group $SU(2)$ is diffeomorphic to S^3; moreover, the group S^3 is isomorphic to $SU(2)$. This suggests the construction of a diffeomorphism $S^3 \to SU(2)$: it assigns to a point $x \in S^3$ the transformation of the space $\mathbb{H} = \mathbb{C}^2$ given by (left) multiplication by the quaternion x. More explicitly the diffeomorphism $S^3 \to SU(2)$ is given by the formula $(z_1, z_2) \mapsto \begin{pmatrix} \bar{z}_2 & -\bar{z}_1 \\ z_1 & z_2 \end{pmatrix}$ where $|z_1|^2 + |z_2|^2 = 1$.

Note that the group Spin(3) is isomorphic to $SU(2)$ and hence is also diffeomorphic to S^3. (The group Spin(4) is isomorphic and consequently diffeomorphic to $S^3 \times S^3$; note also the isomorphisms Spin(5) = $Sp(2)$ and Spin(6) = $SU(4)$.)

The groups $SU(n)$ with $n \geqslant 3$ are not homeomorphic to products of spheres, although they have the same cohomology rings as products of spheres. In particular, $SU(3)$ is not homeomorphic to $S^3 \times S^5$ (though $H^*(SU(3)) = H^*(S^3 \times S^5)$). This may be proved by comparing the homotopy groups $(\pi_4(SU(3)) = 0$ – see § 3. – while $\pi_4(S^3 \times S^5) = \mathbb{Z}_2)$ or considering the action of the Sq^2 operation (the mapping $Sq^2: H^3(SU(3); \mathbb{Z}_2) \to H^5(SU(3); \mathbb{Z}_2)$ is non-zero).

C. The group $Sp(1)$ is isomorphic to the groups $SU(2)$ and Spin(3) and hence is diffeomorphic to S^3. The groups $Sp(n)$ with $n \geqslant 2$ are not homeomorphic to products of spheres.

1.3. Homotopy Groups. A. Since $\pi_i(S^m) = 0$ for $i < m$, the homotopy sequences of the fiberings (2) imply that the homomorphisms

$$\pi_i(SO(n)) \to \pi_i(SO(n+1)) \quad \text{with } i < n - 1,$$

$$\pi_i(U(n)) \to \pi_i(U(n+1)) \quad \text{with } i < 2n,$$

$$\pi_i(Sp(n)) \to \pi_i(Sp(n+1)) \quad \text{with } i < 4n + 2,$$

induced by the canonical embeddings, are isomorphisms, and the same homomorphisms with $i = n - 1$, $i = 2n$, $i = 4n + 2$ are epimorphisms. Thus the sequences

$$\pi_i(SO(1)) \to \pi_i(SO(2)) \to \pi_i(SO(3)) \to \cdots$$

$$\pi_i(U(1)) \to \pi_i(U(2)) \to \pi_i(U(3)) \to \cdots$$

$$\pi_i(Sp(1)) \to \pi_i(Sp(2)) \to \pi_i(Sp(3)) \to \cdots$$

stabilize at certain terms (the terms $\pi_i(SO(i + 2)$, $\pi_i\left(U\left(\left[\frac{i+1}{2}\right]\right)\right)$, $\pi_i\left(Sp\left(\left[\frac{i+2}{4}\right]\right)\right)$ respectively). The stable groups are nothing else but the homotopy groups $\pi_i(SO)$, $\pi_i(U)$, $\pi_i(Sp)$ of the spaces SO, U, Sp.

B. The groups $\pi_i(SO)$, $\pi_i(U)$, $\pi_i(Sp)$ were calculated in 1957 by R. Bott.

Bott's Theorem. (1) *For $i \geqslant 1$*

$$\pi_i(SO) \cong \begin{cases} \mathbb{Z}_2 & \text{if } i \equiv 0, 1 \bmod 8, \\ \mathbb{Z} & \text{if } i \equiv 3, 7 \bmod 8, \\ 0 & \text{if } i \equiv 2, 4, 5, 6 \bmod 8. \end{cases}$$

(2) *For* $i \geqslant 1$

$$\pi_i(U) \cong \begin{cases} \mathbb{Z} & \text{if } i \equiv 1 \bmod 2, \\ 0 & \text{if } i \equiv 0 \bmod 2. \end{cases}$$

(3) *For* $i \geqslant 1$

$$\pi_i(Sp) \cong \pi_{i+4}(SO). \quad \square$$

It is seen from this theorem that (for $i \geqslant 1$) $\pi_i(SO) \cong \pi_{i+8}(SO)$, $\pi_i(U) \cong \pi_{i+2}(U)$, $\pi_i(Sp) \cong \pi_{i+8}(Sp)$. These three isomorphisms are known as the Bott periodicity isomorphisms. Note that Bott proved his periodicity in a more precise form. Namely, he showed that there exist homotopy equivalences $(\Omega^8 SO)_0 \sim SO$, $(\Omega^2 U)_0 \sim U$, $(\Omega^8 Sp)_0 \sim Sp$, where Ω denotes the loop space operator and $(\ )_0$ denotes taking the component of the base point. Bott also succeeded in determining the other spaces $(\Omega^i SO)_0$, $(\Omega^i U)_0$, $(\Omega^i Sp)_0$; here is his result.

Supplement to Bott's Theorem. (4) $(\Omega^1 SO)_0 \sim SO/U$, $(\Omega^2 SO)_0 \sim U/Sp$, $(\Omega^3 SO)_0 \sim BSp$, $(\Omega^4 SO)_0 \sim Sp$, $(\Omega^5 SO)_0 \sim Sp/U$, $(\Omega^6 SO)_0 \sim U/0$, $(\Omega^7 SO)_0 \sim BSO$; $(\Omega^1 U)_0 \sim BU$. $\square$

The original proof of Bott periodicity involved *Morse theory*; this proof is in the book Milnor (1963). With the rise of *K-theory* (stimulated a lot by Bott periodicity) other proofs of Bott's theorems appeared, which are simpler from the point of view of an algebraically oriented mathematician. These proofs may be found in Atiyah (1967), Atiyah (1966), Karoubi (1978).

C. The canonical mappings $U \rightleftarrows 0$, $U \rightleftarrows Sp$ (see § 1.1.B) induce the following mappings of the groups π_i:

$i \bmod 8$	$Sp \longrightarrow$	$U \longrightarrow$	$0 \longrightarrow$	$U \longrightarrow$	Sp
0	0	0	$\mathbb{Z}_2$	0	0
1	0	$\mathbb{Z} \xrightarrow{\text{epi}}$	$\mathbb{Z}_2$	$\mathbb{Z}$	0
2	0	0	0	0	0
3	$\mathbb{Z} \xrightarrow{2}$	$\mathbb{Z} \xrightarrow{1}$	$\mathbb{Z} \xrightarrow{2}$	$\mathbb{Z} \xrightarrow{1}$	$\mathbb{Z}$
4	$\mathbb{Z}_2$	0	0	0	$\mathbb{Z}_2$
5	$\mathbb{Z}_2$	$\mathbb{Z}$	0	$\mathbb{Z} \xrightarrow{\text{epi}}$	$\mathbb{Z}_2$
6	0	0	0	0	0
7	$\mathbb{Z} \xrightarrow{1}$	$\mathbb{Z} \xrightarrow{2}$	$\mathbb{Z} \xrightarrow{1}$	$\mathbb{Z} \xrightarrow{2}$	$\mathbb{Z}$

D. Many unstable homotopy groups of the manifolds $SO(n)$, $U(n)$, $Sp(n)$ are known as well. For instance, $\pi_{2n}(U(n)) \cong \mathbb{Z}_{n}$, $\pi_{4n+2}(Sp(n)) \cong \mathbb{Z}_{(2n+1)}$ for n odd, and $\pi_{4n+2}(Sp(n)) \cong \mathbb{Z}_{2((2n+1)}$ for n even. For more details see Fuks (1971).

§2. Lie Groups: Homological Information

In this section the symbols G_2, F_4, E_6, E_7, E_8 denote simply-connected compact Lie groups of the corresponding types.

2.1. Real Cohomology. With any compact Lie group we can associate a finite sequence of positive integers $d_1, \ldots, d_l$. Here l is in fact the rank of the group, and the d_i's are the degrees of the generators of the ring of polynomials on the Cartan subalgebra of the Lie algebra of our group which are invariant under the Weyl group. But all this is unnecessary for understanding the results listed below; we only need the numbers d_i. These numbers are given in Table 4.

Table 4

G	$d_1, \ldots, d_l$	G	$d_1, \ldots, d_l$
$SU(l+1)$	$1, \ldots, l$	G_2	$1, 5$
		F_4	$1, 5, 7, 11$
$SO(2l-1)$	$1, 3, \ldots, 2l-1$	E_6	$1, 4, 5, 7, 8, 11$
$Sp(2l)$	$1, 3, \ldots, 2l-1$	E_7	$1, 5, 7, 9, 11, 13, 17$
$SO(2l)$	$1, 3, \ldots, 2l-3, l-1$	E_8	$1, 7, 11, 13, 17, 19, 23, 29$

Theorem (E. Cartan, Pontryagin, Hopf, Samelson, Chevalley, Leray). $H^*(G; \mathbb{R})$ *is the exterior algebra on l generators of degrees $2d_1 - 1, \ldots, 2d_l - 1$. In other words, there is a multiplicative isomorphism*

$$H^*(G; \mathbb{R}) \cong H^*(S^{2d_1+1} \times \cdots \times S^{2d_l+1}; \mathbb{R}). \ \square$$

2.2. Cohomology Modulo "Good Primes". Integer Cohomology of $U(n)$ and $Sp(n)$. A. At present the integer cohomology rings of all compact Lie groups are known. It turns out, in particular, that for no prime p does the integer cohomology of a compact Lie group contain elements of order p^2. In view of this, we can restrict ourselves to cohomology rings with coefficients in $\mathbb{Z}_p$ for p prime. The results listed below are due to various authors (Miller, Borel, Baum, Browder, Araki, Shikata); there is a uniform exposition of these (and of some other results as well) in the preprint V.G. Kac "Torsion in the homology of compact Lie groups" (Berkeley, MSRI, 1984).

B. The following primes are regarded as "bad" for the following Lie groups. The prime 2 is bad for $SO(n)$, Spin(n) and all the exceptional groups; the prime 3 is bad for F_4, E_6, E_7, E_8; the prime 5 is bad for E_8. Other primes are regarded as "good".

C. Theorem. *If p is a good prime for the group G, then there is a multiplicative isomorphism*

$$H^*(G; \mathbb{Z}_p) \cong H^*(S^{2d_1-1} \times \cdots \times S^{2d_l-1}; \mathbb{Z}_p). \quad \square$$

D. The last theorem, together with the fact that all primes are good for the groups $U(n)$ and $Sp(n)$, implies the following result.

Theorem. *There are multiplicative isomorphisms*

$$H^*(U(n); \mathbb{Z}) \cong H^*(S^1 \times S^3 \times \cdots \times S^{2n-1}; \mathbb{Z});$$

$$H^*(Sp(n); \mathbb{Z}) \cong H^*(S^3 \times S^7 \times \cdots \times S^{4n-1}; \mathbb{Z}). \quad \square$$

2.3. Modulo 2 Cohomology of Orthogonal and Spinor Groups. A. Since the final result is rather cumbersome (see § 2.3.B below), we begin with the statement of an intermediate result which is easy to memorize (and relatively easy to prove – see Borel (1953)).

Theorem. *There is an* additive *isomorphism*

$$H^*(SO(n); \mathbb{Z}_2) \cong H^*(S^1 \times S^2 \times \cdots \times S^{n-1}; \mathbb{Z}_2);$$

Moreover, the ring $H^*(SO(n); \mathbb{Z}_2)$ has a system of generators $x_i \in Hi(SO(n); \mathbb{Z}_2)$, $i = 1, 2, \ldots, n-1$, such that the monomials of form

$$x_{i_1} x_{i_2} \ldots x_{i_q}, \qquad i_1 < i_2 < \cdots < i_q,$$

form an additive base in $H^*(SO(n); \mathbb{Z}_2)$. $\square$

B. Now we give the full statements. In these statements (and also in the statements of § 2.4), x_i, y_i denote i-dimensional cohomology classes, $A/(a, b, \ldots)$ denotes the quotient ring of the ring A over the ideal generated by $a, b, \ldots$, $A[x, y, \ldots]$ denotes the polynomial ring over A in $x, y, \ldots$, and $\Lambda_p(x, y, \ldots)$ denotes the exterior algebra over $\mathbb{Z}_p$ generated by $x, y, \ldots$

Theorem.

$$H^*(SO(n); \mathbb{Z}_2) = \mathbb{Z}_2[x_1, x_3, \ldots, x_m]/(x_1^{2^{k_1}}, \ldots, x_m^{2^{k_m}}),$$

where m is the largest odd number less than n, and the numbers k_s are defined by the formula $n \leqslant s \cdot 2^{k_s} < 2n$;

$$H^*(\text{Spin}(n); \mathbb{Z}_2) = \mathbb{Z}_2[x_3, \ldots, x_m]/(x_3^{2^{k_3}}, \ldots, x_m^{2^{k_3}}) \otimes \Lambda_2(y_{2^{k_1}-1}). \quad \square$$

2.4. Cohomology of the Exceptional Groups
Theorem

$$H^*(G_2; \mathbb{Z}_2) = \Lambda_2(x_5) \otimes \mathbb{Z}_2[x_3]/(x_3^4);$$

$$H^*(F_4; \mathbb{Z}_2) = \Lambda_2(x_5, x_{15}, x_{23}) \otimes \mathbb{Z}_2[x_3]/(x_3^4);$$

$$H^*(E_6; \mathbb{Z}_2) = \Lambda_2(x_5, x_9, x_{15}, x_{17}, x_{23}) \otimes \mathbb{Z}_2[x_3]/(x_3^4);$$

$$H^*(E_7; \mathbb{Z}_2) = \Lambda_2(x_{15}, x_{17}, x_{23}, x_{27}) \otimes \mathbb{Z}_2[x_3, x_5, x_9]/(x_3^4, x_5^4, x_9^4);$$

$$H^*(E_8; \mathbb{Z}_2) = \Lambda_2(x_{17}, x_{23}, x_{27}, x_{29}) \otimes \mathbb{Z}_2[x_3, x_5, x_9, x_{15}]/(x_3^{16}, x_5^8, x_9^4, x_{15}^4)$$

$$H^*(F_4; \mathbb{Z}_3) = \Lambda_3(x_3, x_7, x_{11}, x_{15}) \otimes \mathbb{Z}_3[x_8]/(x_8^3);$$

$$H^*(E_6; \mathbb{Z}_3) = \Lambda_3(x_3, x_7, x_9, x_{11}, x_{15}, x_{17}) \otimes \mathbb{Z}_3[x_8]/(x_8^3);$$

$$H^*(E_7; \mathbb{Z}_3) = \Lambda_3(x_3, x_7, x_{11}, x_{15}, x_{19}, x_{27}, x_{35}) \otimes \mathbb{Z}_3[x_8]/(x_8^3);$$

$$H^*(E_8; \mathbb{Z}_3) = \Lambda_3(x_3, x_7, x_{15}, x_{19}, x_{27}, x_{35}, x_{39}, x_{47}) \otimes \mathbb{Z}_3[x_8, x_{20}]/(x_8^3, x_{20}^3);$$

$$H^*(E_8; \mathbb{Z}_5) = \Lambda_5(x_3, x_{11}, x_{15}, x_{23}, x_{27}, x_{35}, x_{39}, x_{47}) \otimes \mathbb{Z}_5[x_{12}]/(x_{12}^5). \ \square$$

2.5. The K-functor. A compact Lie group G of rank l possesses l "fundamental representations" $\rho_i: G \to U(n_i)$ (see Bourbaki (1975) § 8.7.2); for example, in the case $G = U(n)$ these are the identity n-dimensional representation and its exterior powers. Each representation ρ_i determines a complex vector bundle over ΣG, and the stable equivalence class of this bundle is an element $[\rho_i]$ of the group $K^0(\Sigma G) = K^1(G)$, where K is the complex K-functor.

Theorem (L. Hodgkin). *The ring* $K^*(G) = K^0(G) \oplus K^1(G)$ *is isomorphic to*

$$\Lambda_\mathbb{Z}^*([\rho_i] \mid i = 1, \dots, l) = \Lambda_\mathbb{Z}^{\text{even}}([\rho_i] \oplus \Lambda_\mathbb{Z}^{\text{odd}}([\rho_i]). \ \square$$

For the details see Atiyah (1965).

§ 3. Stiefel Manifolds

3.1. Definitions. Geometrical and Homotopical Information. A. The (real) *Stiefel manifold* $\mathbb{R}V_{n,k}$ or $V_{n,k}$ is defined as the manifold of isometric embeddings $\mathbb{R}^k \to \mathbb{R}^n$, or equivalently, as the manifold of orthonormal k-forms in $\mathbb{R}^n$. Clearly

$$V_{n,k} = O(n)/O(n - k) = (\text{for } k < n) \, SO(n)/SO(n - k).$$

It is also evident that $V_{n,1} = S^{n-1}$, $V_{n,n-1} = SO(n)$, $V_{n,n} = O(n)$, and $V_{n,2}$ is the manifold of unit tangent vectors to S^{n-1}. One can also consider complex Stiefel manifolds

$$\mathbb{C}V_{n,k} = U(n)/U(n - k) = (\text{for } k < n)SU(n)/SU(n - k),$$

and quaternionic Stiefel manifolds

$$\mathbb{H}V_{n,k} = Sp(n)/Sp(n - k).$$

As in the real case, the manifolds $\mathbb{C}V_{n,k}$ and $\mathbb{H}V_{n,k}$ may be described as the manifolds of isometric $\mathbb{C}$-linear or $\mathbb{H}$-linear embeddings $\mathbb{C}^k \to \mathbb{C}^n$ or $\mathbb{H}^k \to \mathbb{H}^n$. Clearly $\mathbb{C}V_{n,1} = S^{2n-1}$, $\mathbb{C}V_{n,n-1} = SU(n)$, $\mathbb{C}V_{n,n} = U(n)$, $\mathbb{H}V_{n,1} = S^{4n-1}$, $\mathbb{H}V_{n,n} = Sp(n)$.

B. Theorem (Sutherland, 1964). *For $k \geqslant 2$ the manifolds $\mathbb{R}V_{n,k}$, $\mathbb{C}V_{n,k}$, $\mathbb{H}V_{n,k}$ are parallelizable.* $\square$

C. Information about the homotopy groups of Stiefel manifolds can be deduced, by means of homotopy sequences of fibrations, from information about the homotopy groups of Lie groups. The first non-trivial homotopy groups are of particular significance. (This stems from two facts. Firstly, the characteristic classes of Stiefel – Whitney and Chern of vector bundles are defined as the first obstructions to sections of the associated fiberings with Stiefel fibres, and these obstructions take values in the cohomology with coefficients in the first non-trivial homotopy group. Secondly, the classical universal bundles for the classical Lie groups are constructed as the limits of natural fibrations of Stiefel manifolds over the Grassmann manifolds, and to prove the universality of these fibrations it is necessary to know that Stiefel manifolds are aspherical up to high dimensions.)

Theorem.

$$\pi_i(\mathbb{R}V_{n,k}) = 0 \quad for \ i < n - k,$$

$$\pi_{n-k}(\mathbb{R}V_{n,k}) = \begin{cases} \mathbb{Z} & if \ k = 1 \ or \ n - k \ is \ odd, \\ \mathbb{Z}_2 & if \ k > 1 \ and \ n - k \ is \ even; \end{cases}$$

$$\pi_i(\mathbb{C}V_{n,k}) = \begin{cases} 0 & if \ i < 2(n - k) + 1, \\ \mathbb{Z} & if \ i = 2(n - k) + 1; \end{cases}$$

$$\pi_i(\mathbb{H}V_{n,k}) = \begin{cases} 0 & if \ i < 4(n - k) + 3, \\ \mathbb{Z} & if \ i = 4(n - k) + 3. \end{cases} \square$$

3.2. Cohomology. (See Borel (1953), §§ 10, 11 for proofs.)

A. **Theorem** (*compare § 2.2.D*). *There are multiplicative isomorphisms*

$$H^*(\mathbb{C}V_{n,k}; \mathbb{Z}) = H^*(S^{2n-(2k-1)} \times S^{2n-(2k-3)} \times \cdots \times S^{2n-1}; \mathbb{Z}),$$

$$H^*(\mathbb{H}V_{n,k}; \mathbb{Z}) = H^*(S^{4n-(4k-3)} \times S^{4n-(4k-7)} \times \cdots \times S^{4n-1}; \mathbb{Z}). \square$$

(Except for trivial cases, the spaces which have the same cohomology according to this theorem are not homomorphic – compare § 2.2.B–C.)

B. As to the cohomology of real Stiefel manifolds, we restrict ourselves to the following partial results.

Theorem (compare § 2.3). (1) *For* $\mathbb{K} = \mathbb{Q}$ *or* $\mathbb{Z}_p$ *with p an odd prime there are multiplicative isomorphisms*

$$H^*(V_{2m,2l}; \mathbb{K}) = H^*(\mathscr{S} \times S^{2m-1} \times S^{2l}; \mathbb{K},$$

$$H^*(V_{2m,2l+1}; \mathbb{K}) = H^*(\mathscr{S} \times S^{2m-1}; \mathbb{K}),$$

$$H^*(V_{2m-1,2l}; \mathbb{K}) = H^*(\mathscr{S} \times S^{2l}; \mathbb{K}),$$

$$H^*(V_{2m-1,2l+1}; \mathbb{K}) = H^*(\mathscr{S}; \mathbb{K}),$$

where $\mathscr{S} = S^{4l+3} \times S^{4l+7} \times \cdots \times S^{4m-5}$.

(2) *There is an additive isomorphism*

$$H^*(V_{n,k}; \mathbb{Z}_2) = H^*(S^{n-k} \times S^{n-(k-1)} \times \cdots \times S^{n-1}; \mathbb{Z}_2);$$

moreover, there exist elements $h_i \in H^i(V_{n,k}; \mathbb{Z}_2)$, $i = n - k, \ldots, n - 1$, *such that the products*

$$h_{i_1} \ldots h_{i_q}, \qquad n - k \leqslant i_1 < \cdots < i_q \leqslant n - 1,$$

form an additive base of $H^*(V_{n,k}; \mathbb{Z}_2)$. $\square$

Chapter 3
Grassmann Manifolds and Spaces

§1. Geometric Information

1.1. Definitions. In this chapter we deal with the following four series of manifolds.

A. Real *Grassmann manifolds* or simply Grassmann manifolds $\mathbb{R}G(m, n)$ or $G(m, n)^2$. By definition, $G(m, n)$ is the set of all n-dimensional subspaces of a $(m + n)$-dimensional real vector space. The topology and the differential structure are defined in $G(m, n)$ by means of the equality $G(m, n) = O(m + n)/O(m) \times O(n)$. The inclusion $\mathbb{R}^{m+n} \to \mathbb{R}^{m+n+1}$ induces the embedding $G(m, n) \to G(m + 1, n)$, and the isomorphism $\mathbb{R}^{m+n+1} = \mathbb{R}^{m+n} \times \mathbb{R}$ induces (by means of the formula $\pi \mapsto \pi \times \mathbb{R}$) an embedding $G(m, n) \to G(m, n + 1)$. These embeddings are converted into each other by the diffeomorphism $G(m, n) \approx G(n, m)$, acting by the formula $\pi \mapsto \pi^\perp$. We make identifications according to the above embeddings, and set $\bigcup_{m=1}^{\infty} G(m, n) = G(\infty, n)$, $\bigcup_{n=1}^{\infty} G(\infty, n) = G(\infty, \infty)$. The spaces $G(\infty, n)$ and $G(\infty, \infty)$ are called (real) *Grassmann spaces*. Note that under our identifications all real Grassmann manifolds and spaces are subspaces of the space $G(\infty, \infty)$.

B. Complex Grassmann manifolds and spaces are defined in exactly the same way as their real prototypes, except that the field $\mathbb{R}$ is replaced by the field $\mathbb{C}$. In this way we get the manifolds $\mathbb{C}G(m, n) = U(m + n)/U(m) \times U(n)$ and the spaces $\mathbb{C}G(\infty, n) = \bigcup_{m=1}^{\infty} \mathbb{C}G(m, n)$ and $\mathbb{C}G(\infty, \infty) = \bigcup_{n=1}^{\infty} \mathbb{C}G(\infty, n)$. What was said above for real Grassmannians remains valid in the complex case with appropriate changes of terms and notations.

C. Quaternionic Grassmann manifolds and spaces are defined in a similar way. However we must remember that the non-commutativity of quaternions forces

---

[2] Some authors use another system of notation in which our $G(m, n)$ is denoted by $G(m + n, n)$.

us to distinguish between left and right subspaces. These two possibilities give rise to two absolutely parallel theories, from which we have to choose one (it does not matter which). What was said above again remains valid.

D. There is no accepted name for the manifolds $G_+(m, n) = SO(m + n)/SO(m) \times SO(n)$. The term "oriented Grassmann manifolds" is evidently unsuitable because ordinary Grassmann manifolds are often orientable (see §1.2.C below) and may be oriented. The name "simply connected Grassmann manifolds" seems more appropriate, since $G_+(m, n)$ is simply connected except in the case $m = n = 1$, while $G(m, n)$ is simply connected only in the case when it reduces to a point (i.e. when $m = 0$ or $n = 0$). In this article we shall consider the manifolds $G_+(m, n)$ and the spaces $G_+(\infty, n)$, $G_+(\infty, \infty)$ without choosing any particular name for them. In other respects what said in §1.1.A remains true in the G_+ case with obvious small modifications.

E. Projective spaces are particular cases of Grassmann manifolds: $\mathbb{R}P^m = G(m, 1)$, $\mathbb{C}P^m = \mathbb{C}G(m, 1)$, $\mathbb{H}p^m = \mathbb{H}P(m, 1)$ $(0 \leqslant m \leqslant \infty)$. To these spaces should be added the Cayley projective plane $\mathbb{C}aP^2$ (see the details in Fuks, Rokhlin (1977)). In the G_+ case projective spaces are replaced by spheres: $G_+(m, 1) = S^m$.

1.2. General Information (see Fuks, Rokhlin (1977) for proofs). A. The dimension of $G(m, n)$ and $G_+(m, n)$ is equal to mn, the dimension of $\mathbb{C}G(m, n)$ is equal to $2mn$, the dimension of $\mathbb{H}G(m, n)$ is equal to $4mn$, the dimension of $\mathbb{C}aP^2$ is equal to 16.

B. The manifolds $\mathbb{C}G(m, n)$, $\mathbb{H}G(m, n)$, $\mathbb{C}aP^2$ and also the manifolds $G_+(m, n)$ with $(m, n) \neq (1, 1)$ are simply connected. The manifolds $G(m, n)$ with $m > 0$, $n > 0$ and $(m, n) \neq (1, 1)$ have fundamental group $\mathbb{Z}_2$ (for a non-contractible loop in $G(m, n)$ one can take any displacement of an n-subspace of R^{m+n} which returns it to the initial position with orientation reversed). The manifolds $G(1, 1)$ and $G_+(1, 1)$ are diffeomorphic to the circle and thus have fundamental group $\mathbb{Z}$.

As to higher homotopy groups of the Grassmannians, some information about them can be derived, from known facts about the homotopy groups of Lie groups (see Chapter 2) and the homotopy sequences of fiberings $O(m + n) \rightarrow G(m, n)$ (with fibre $O(m) \times O(n)$) etc. We point out the following useful statements. The space $\mathbb{R}P^\infty = G(\infty, 1)$ is $K(\mathbb{Z}_2, 1)$; the space $\mathbb{C}P^\infty = \mathbb{C}G(\infty, 1)$ is $K(\mathbb{Z}, 2)$; for any n, the spaces $G(\infty, n)$, $\mathbb{C}G(o\infty, n)$, $\mathbb{H}G(\infty, n)$, $G_+(\infty, n)$ are the classifying spaces for the groups $O(n)$, $U(n)$, $Sp(n)$, $SO(n)$, so these are isomorphisms $\pi_i(G(\infty, n)) = \pi_{i-1}(O(n))$, $\pi_i(\mathbb{C}G(\infty, n)) = \pi_{i-1}(U(n))$, $\pi_i(\mathbb{H}G(\infty, n)) = \pi_{i-1}(Sp(n))$, $\pi_i(G_+(\infty, n)) = \pi_{i-1}(SO(n))$.

C. The manifolds $G_+(m, n)$, $\mathbb{C}G(m, n)$, $\mathbb{H}G(m, n)$, $\mathbb{C}aP^2$ are orientable. The manifold $G(m, n)$ with $m > 0$, $n > 0$ is orientable if and only if $m + n$ is even.

D. The manifolds $G(m, n)$ are the bases of some important vector bundles. The total space of the "standard" n-dimensional vector bundle $\xi(m, n)$ over $G(m, n)$ (also called the "tautological" bundle) consists of pairs (π, x) with

$x \in \pi \in G(m, n)$, and projection $(\pi, x) \mapsto x$; one may say that the fibre of the tautological bundle over the point $\pi \in G(m, n)$ is π itself but considered as a vector space. If in the last definition the inclusion $x \in \pi$ is replaced by the inclusion $x \in \pi^{\perp}$, then we obtain the definition of the "dual standard" or "dual tautological" bundle $\xi^{\perp}(m, n)$ (which becomes $\xi(n, m)$ under the canonical diffeomorphism $G(m, n) \to G(n, m)$). Clearly $\xi(m, n) \oplus \xi^{\perp}(m, n) = m + n$ (which denotes the standard trivial $(m + n)$-dimensional bundle). Note also that $\xi(m, n)$ is the normal bundle of $G(m, n)$ in $G(m + 1, n)$, and $\xi^{\perp}(m, n)$ is the normal bundle of $G(m, n)$ in $G(m, n + 1)$.

The above definitions may be repeated word for word to obtain the definitions of oriented vector bundles $\xi_{+}(m, n)$ and $\xi_{+}^{\perp}(m, n)$ over $G_{+}(m, n)$, complex vector bundles $\mathbb{C}\xi(m, n)$ and $\mathbb{C}\xi^{\perp}(m, n)$ over $\mathbb{C}G(m, n)$, and quaternionic vector bundles $\mathbb{H}\xi(m, n)$ and $\mathbb{H}\xi^{\perp}(m, n)$ over $\mathbb{H}G(m, n)$. The properties of these bundles are similar to those of $\xi(m, n)$ and $\xi^{\perp}(m, n)$. The bundles $\xi(m, n)$, $\xi_{+}(m, n)$, $\mathbb{C}\xi(m, n)$, $\mathbb{H}\xi(m, n)$ admit the limit $m \to \infty$, which gives bundles $\xi(\infty, n)$, $\xi_{+}(\infty, n)$, $\mathbb{C}\xi(\infty, n)$, $\mathbb{H}\xi(\infty, n)$ over $G(\infty, n)$, $G_{+}(\infty, n)$, $\mathbb{C}G(\infty, n)$, $\mathbb{H}G(\infty, n)$. These four bundles are universal in their categories.

The tangent bundle to $G(m, n)$ is canonically isomorphic to $\text{Hom}(\xi(m, n), \xi^{\perp}(m, n))$ (a point of $G(m, n)$, sufficiently close to $\pi \in G(m, n)$, is the graph of a linear mapping $\pi \to \pi^{\perp}$, close to zero). Similar results are true for other Grassmann manifolds. The case $n = 1$, that is, the case of projective spaces, is of particular interest. In this case we denote by $\tau(m)$ the tangent bundle to $\mathbb{R}P^{m}$, put $\xi(m, 1) = \xi(m)$ and obtain

$$\tau(m) \oplus 1 = (m + 1)\xi(m)^{*},$$

where $*$ denotes conjugation. (*Proof*: $\tau(m) \oplus 1 = \text{Hom}(\xi(m), \xi^{\perp}(m)) \oplus \text{Hom}(\xi(m), \xi(m)) = \text{Hom}(\xi(m), \xi^{\perp}(m) \oplus \xi(m)) = \text{Hom}(\xi(m), m + 1) = (m + 1) \times \xi(m)^{*}$.) Exactly the same equality holds for other projective spaces. (In all cases there is a canonical isomorphism $\xi(m) = \xi(m)^{*}$ which allows the star to be removed in these equalities; however, in the complex case this isomorphism is not homorphic.)

1.3. Embeddings of the Manifolds $G(m, n)$, $\mathbb{C}G(m, n)$, $G_{+}(m, n)$ in Euclidean and Projective Spaces

A. The formula

$$(x_0 : x_1 : \ldots : x_n) \mapsto \{x_i x_j\} / \sqrt{\sum_{i \leqslant j} x_i^2 x_j^2}$$

defines an analytic embedding $\mathbb{R}P^{n} \to S^{2}\mathbb{R}^{n+1} = \mathbb{R}^{n(n+1)/2}$; this embedding is compatible with the canonical action of $O(n + 1)$ on $\mathbb{R}P^{n}$ and $S^{2}\mathbb{R}^{n+1}$. In a similar way (replacing the squares of the coordinates by the squares of the moduli of the coordinates under the root sign) real analytic embeddings $\mathbb{C}P^{n} \to \mathbb{C}^{n(n+1)/2}$ and $\mathbb{H}P^{n} \to \mathbb{H}^{n(n+1)/2}$ can be defined which are compatible with the actions of $U(n + 1)$ and $Sp(n + 1)$.

B. Let π be an n-dimensional subspace of $\mathbb{R}^{m+n}$ and $e_1, \ldots, e_n$ be a base of π. Consider the $n \times (m + n)$-matrix whose rows are the coordinates of $e_1, \ldots, e_n$ (in $\mathbb{R}^{m+n}$). The minors of order n of this matrix constitute a system of $\binom{m+n}{n}$ numbers (indexed by increasing sequences $\{i_1, \ldots, i_n\} \subset \{1, \ldots, m+n\}$). These numbers are called the *Plücker coordinates* of π (associated with the base $\{e_i\}$). The Plücker coordinates are not all zero and under a change of base are multiplied by the same non-zero number, namely by the determinant of the transformation from one base to another. Thus the Plücker coordinates define mappings

$$G(m, n) \to \mathbb{R}P^{\binom{m+n}{n}-1}, \qquad G_+(m, n) \to S^{\binom{m+n}{n}-1},$$

which are actually differential embeddings (see Fuks, Rokhlin (1977) 3.2.2.5). Moreover, these embedding are isometric with respect to the natural metrics (of group origin) in Grassmann manifolds, projective spaces and spheres.

The construction has an exact complex analogue which gives rise to a holomorphic isometric embedding

$$\mathbb{C}G(m, n) \to \mathbb{C}P^{\binom{m+n}{n}-1}.$$

C. The relations between the Plücker coordinates, that is the equations determining the images of the previous embeddings, are well known. The generating system of relations is

$$p_{i_1 \ldots i_n} p_{j_1 \ldots j_n} - \sum_{s=1}^{n} p_{i_1 \ldots i_{n-1} j_s} p_{j_1 \ldots j_{s-1} i_n j_{s+1} \ldots j_n} = 0,$$

where p stands for the Plücker coordinates (see the details in Hodge, Pedoe (1947)).

Note that the above implies that $\mathbb{C}G(m, n)$ is a projective algebraic manifold over $\mathbb{C}$; in particular, it is a Kähler manifold, and even a Hodge manifold.

D. The case $m = n = 2$ deserves to be considered separately. The manifolds $G(2, 2)$, $G_+(2, 2)$ and $\mathbb{C}G(2, 2)$ are defined, respectively, in $\mathbb{R}P^5$, S^5 or $\mathbb{C}P^5$ by a single equation

$$p_{12}p_{34} - p_{13}p_{24} + p_{14}p_{23} = 0 \tag{3}$$

($p_{12}, \ldots, p_{24}$ are the six coordinates in $\mathbb{R}P^5$, S^5 or $\mathbb{C}P^5$). Thus $\mathbb{C}G(2, 2)$ is a non-degenerate quadric in $\mathbb{C}P^5$, and $G(2, 2)$ is a non-degenerate quadric of signature $(3, 3)$ in $\mathbb{R}P^5$. As for the manifold $G_+(2, 2)$, it is defined in $\mathbb{R}^6$ by a system of two equations: the equation (3) and the equation $\sum p_{ij}^2 = 1$ of the sphere. This system is equivalent to the system

$$(p_{12} + p_{34})^2 + (p_{13} - p_{24})^2 + (p_{14} + p_{23})^2 = 1,$$
$$(p_{12} - p_{34})^2 + (p_{13} + p_{24})^2 + (p_{14} - p_{23})^2 = 1,$$

which shows that the manifold $G_+(2, 2)$ is diffeomorphic to the product $S^2 \times S^2$. (For another description of the diffeomorphism $G_+(2, 2) \approx S^2 \times S^2$ see Fuks, Rokhlin (1977) 3.2.3.4.)

E. An embedding of the manifold $G(m, n)$ in Euclidean space can be obtained by composing the embedding of §1.3.B with some embedding $\mathbb{R}P^{\binom{m+n}{n}-1} \to \mathbb{R}^N$ (for instance, with that of §1.3.A). These is a more economical embedding, namely the embedding $G(m, n) \to \text{Hom}(\mathbb{R}^{m+n}, \mathbb{R}^{m+n}) = \mathbb{R}^{(m+n)^2}$; it assigns to an n-dimensional subspace π of $\mathbb{R}^{m+n}$ the matrix of the operator of orthogonal projection of $\mathbb{R}^{m+n}$ onto π (regarded as a mapping $\mathbb{R}^{m+n} \to \mathbb{R}^{m+n}$). In exactly the same way, but using unitary projections, a real analytic embedding $\mathbb{C}G(m, n) \to \mathbb{C}^{(m+n)^2}$ can be defined.

§2. Homology Information

2.1. Cell Decomposition. A. Let $k_1, \ldots, k_q$ be an arbitrary (possibly empty) non-increasing sequence of positive integers, and let $m \geqslant q$, $n \geqslant k_1$. Denote by $e(k_1, \ldots, k_q)$ the subset of $G(m, n)$ composed of those π which satisfy the following conditions (we put $k_{q+1} = 0$):

$$\pi \supset \mathbb{R}^p \qquad \text{if } p \leqslant n - k_1,$$

$$\text{codim}_{\mathbb{R}^p}(\pi \cap \mathbb{R}^p) = s \quad \text{if } n - k_s + s \leqslant p < n - k_{s+1} + (s + 1),$$

$$\pi \subset \mathbb{R}^p \qquad \text{if } p \geqslant n + q + 1.$$

(We also give a more visual description of the set $e(k_1, \ldots, k_q)$. Consider the *Young diagram* of the sequence $k_1, \ldots, k_q$ (see Fig. 6) and arrange it as shown on Fig. 7. The heavy line in Fig. 7 is the graph of some non-decreasing function f,

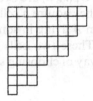

Fig. 6

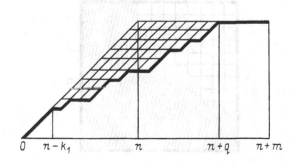

Fig. 7

and the set $e(k_1, \ldots, k_q)$ is defined by the condition $\dim(\pi \cap \mathbb{R}^p) = f(p)$. This description justifies the notation $e(\Delta)$ for the set $e(k_1, \ldots, k_q)$, where Δ denotes the Young diagram of the sequence $k_1, \ldots, k_q$.) It can be shown that the set $e(k_1, \ldots, k_q)$ is homeomorphic to $\mathbb{R}^k$ where $k = k_1 + \cdots + k$; moreover, these sets (with $q \leqslant m$ and $k_1 \leqslant n$) constitute a cell decomposition of $G(m, n)$ (see Fuks, Rokhlin (1977), 4.5.3.3 for the proof). This is called the *Schubert decomposition*, and its cells are called *Schubert cells*. The Schubert decomposition has the remarkable property that the canonical embeddings $G(m, n) \to G(m + 1, n)$, $G(m, n) \to G(m, n + 1)$ map the cell $e(k_1, \ldots, e_q)$ onto the cell with the same notation. Thus the cells $e(k_1, \ldots, k_q)$ corresponding to all possible sequences $k_1, \ldots, k_q$ constitute a cell decomposition of the space $G(\infty, \infty)$, and $G(m, n)$ is exactly the subspace of $G(\infty, \infty)$ composed of cells $e(k_1, \ldots, k_q)$ with $q \leqslant m$ and $k_1 \leqslant n$ (the case $m = \infty$ is not excluded). One can also say that $G(\infty, \infty)$ is decomposed into cells $e(\Delta)$ corresponding to all possible Young diagrams, and the dimension of the cell $e(\Delta)$ is equal to the number of squares of the diagram Δ; furthermore, the cell $e(\Delta)$ belongs to $G(m, n)$ if and only if the diagram is contained in the rectangle with horizontal side m and vertical side n (see Fig. 8; when $m = \infty$ the rectangle becomes a semi-infinite strip). The total number of cells of $G(m, n)$ is $\binom{m + n}{n}$.

   B. The Schubert cell decompositions of the spaces $\mathbb{C}G(m, n)$ and $\mathbb{H}G(m, n)$ are constructed in exactly the same way. The corresponding cells $\mathbb{C}e(k_1, \ldots, k_q)$ and $\mathbb{H}e(k_1, \ldots, k_q)$ have dimensions $2k$ and $4k$, where $k = k_1 + \cdots + k_q$. Otherwise what was said above for the real case remains true without change.

   C. The manifolds and spaces $G_+(m, n)$ are two-sheeted coverings of the manifolds and spaces $G(m, n)$. Thus they aquire a cell decomposition which contains two cells for any Young diagram Δ (lying in some rectangle or semi-strip if n is finite – compare with §2.1.A). These two cells may be denoted by $e_+(\Delta)$ and $e_-(\Delta)$; there is also a canonical way of choosing signs – see Fuks, Rokhlin (1977), 4.5.3.3.

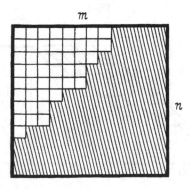

Fig. 8

When $n = 1$ the cell decomposition described above becomes the standard cell decomposition of projective spaces. Note that the Cayley projective plane possesses a similar cell decomposition; it has three cells of dimensions 0, 8, 16.

D. The cell $e(\Delta)$ is contained in the closure of the cell $e(\gamma)$ if and only if the Young diagram Δ is contained in the Young diagram γ. The same is true in the complex and quaternionic cases. It is also true that if $\Delta \subset \gamma$ then each of the cells $e_{\pm}(\Delta)$ is contained in the closure of each of the cells $e_{\pm}(\gamma)$, and if $\Delta \not\subset \gamma$, then none of these four inclusions takes place.

In the complex and quaternionic cases the dimensions of all cells are divisible by 2 or 4, so the problem of computing the incidence coefficients does not arise.

E. In the $G(m, n)$ case the incidence coefficient $[e(\gamma):e(\Delta)]$ may be non-zero only if γ is obtained from Δ by adjoining one square. In this case it is defined by the formula

$$[e(\gamma):e(\Delta)] = \begin{cases} 0 \text{ if } k + l \text{ is even,} \\ \pm 2 \text{ if } k + l \text{ is odd,} \end{cases}$$

where k, l are the coordinates of the square adjoined. The sign in the last formula is seldom needed for practical calculations but we give a way of determining it. An orientation of the cell $e(\Delta)$ is determined by the order of squares in the diagram Δ. We fix the order in which the squares in Δ are numbered in each row one after another from left to right, and the rows are numbered from top to bottom. Then the number $[e(\gamma):e(\Delta)]$, which is equal to ± 2, is equal to 2 if and only if the new square has an odd number in γ.

F. Finally, in $G_{+}(m, n)$ the number $[e_{\pm}(\gamma):e_{\pm}(\Delta)]$ is equal to ± 1 for any pair of diagrams γ, Δ in which γ is obtained from Δ by adjoining a single square. Here is the rule for the signs. Denote by N the number (in γ) of the new square and by k, l its coordinates. Then

$$[e_{\varepsilon_1}(\gamma):e_{\varepsilon_2}(\Delta)] = \begin{cases} (-1)^N \text{ if } k + l \text{ is even,} \\ (-1)^N \varepsilon_1 \varepsilon_2 \text{ if } k + l \text{ is odd.} \end{cases}$$

2.2. Homology and Cohomology: Cellular Calculations. A. It is quite easy to find the homology and cohomology of complex and quaternionic Grassmann manifolds and spaces. Since all the incidence coefficients are zero, the homology and cohomology groups are the same as the cellular chain and cochain groups. For the reader's convenience we give the statement of the result.

Theorem.

$$II_q(\mathbb{C}G(m, n); G) \cong H^q(\mathbb{C}G(m, n); G) \cong \underbrace{G \oplus \cdots \oplus G}_{d(q; m, n)}$$

where $d(q; m, n) = 0$ if q is odd, and $d(q; m, n)$ is the number of partitions of $q/2$ into the sum of no more than m parts less or equal to n (that is the number of Young diagrams of $q/2$ squares contained in the $m \times n$-rectangle) if q is even. $\square$

The quaternionic case is the same, except that the words "even", "odd" should be replaced by the words "divisible by 4", "not divisible by 4", and the fraction $q/2$ should be replaced by the fraction $q/4$.

The statement remains true if $m = \infty$ or $m = n = \infty$.

Note that all the isomorphisms of this theorem are canonical in virtue of what was said above. In particular, there are canonical bases in $H^*(\mathbb{C}G(m, n); \mathbb{Z})$ and $H^*(\mathbb{H}C(m, n); \mathbb{Z})$ whose elements are enumerated by Young diagrams. The element of this basis corresponding to the Young diagram $\varDelta$ will be denoted in the complex and quaternionic case by $c_\varDelta$ and $h_\varDelta$ respectively.

Note also that there is an easily memorizable formula for the Poincaré polynomial of complex and quaternionic Grassmann manifolds (the *Poincaré polynomial* – or the *Poincaré series* – of the space X is by definition $\sum b_q(X)t^q$, where b_q denotes the q-th Betti number). Namely, define the polynomial $\prod_k(\lambda)$ by the formula

$$\prod_k (\lambda) = (1 - \lambda^k)(1 - \lambda^{k-1})\ldots(1 - \lambda).$$

The Poincaré polynomials for $\mathbb{C}G(m, n)$ and $\mathbb{H}G(m, n)$ are equal to

$$\prod_{m+n} (t^2)/\prod_m (t^2) \prod_n (t^2),$$

$$\prod_{m+n} (t^4)/\prod_m (t^4) \prod_n (t^4),$$

respectively, In Chapter 4 we shall give generalizations of these formulae for flag manifolds (see Chapt.4, § 1.3.A).

B. In the real case the result is equally simple if we restrict ourselves to homology and cohomology with coefficients in $\mathbb{Z}_2$. Since all the incidence coefficients are even, these homology and cohomology groups are again the same as the corresponding cellular chain and cochain groups.

Theorem.

$$H_q(G(m, n); \mathbb{Z}_2) \cong H^q(G(m, n); \mathbb{Z}_2) \cong \underbrace{\mathbb{Z}_2 \oplus \cdots \oplus \mathbb{Z}_2,}_{d(q; m, n)}$$

where $d(q; m, n)$ is the number of partitions of q into sums of no more than m parts less than or equal to n (or of Young diagrams with q squares contained in the $m \times n$-rectangle). □

The isomorphisms of this theorem are also canonical. The element of $H^*(G(m, n); \mathbb{Z}_2)$ corresponding to the Young diagram $\varDelta$ is denoted by $w_\varDelta$.

For integer coefficients the situation is more complicated. A part of the integer cellular chain complex of the space $G(\infty, \infty)$ is shown on Fig. 9. We see that this complex decomposes into the sum of an infinite number of complexes, each one of which is the tensor product of some even number (possibly zero) of subcomplexes of the form $\cdots 0 \to \mathbb{Z} \xrightarrow{2} \mathbb{Z} \to 0 \cdots$ In order to enumerate all such subcomplexes we must extract their initial Young diagrams (i.e. the smallest

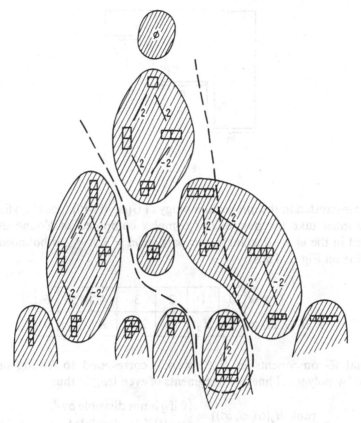

Fig. 9

Young diagram in each of the subcomplexes) and give a method of calculating the size of the complex for a given initial Young diagram. We shall do both of these: a Young diagram bounded by a polygonal line with segments of lengths $m_1, n_1, m_2, n_2, \ldots, m_r, n_r$ (see Fig. 10) is an initial diagram of a subcomplex of the form indicated if and only if the following mod 2 congruences hold:

$$m_1 + n_1 + \cdots + n_r \equiv 0, \quad n_1 \equiv m_2, \quad n_2 \equiv m_3, \ldots, n_{r-1} \equiv m_r;$$

this subcomplex is isomorphic to the tensor product of s copies of the complex $\cdots 0 \to \mathbb{Z} \overset{2}{\to} \mathbb{Z} \to 0 \cdots$, where s is the number of odd numbers in $m_1, n_1, \ldots, n_r$; the homology of this complex is of the form $\ldots, 0, \mathbb{Z}, 0, \ldots$ if $s = 0$ and is of the form $\ldots, \mathbb{Z}_2, (s-1)\mathbb{Z}_2, \binom{s-1}{2}\mathbb{Z}_2, \ldots, \mathbb{Z}_2, 0, \ldots$ if $s > 0$. In particular,

$q =$	0	1	2	3	4	5
$H_q(G(\infty, \infty)) \cong$	$\mathbb{Z}$	$\mathbb{Z}_2$	$\mathbb{Z}_2$	$2\mathbb{Z}_2$	$\mathbb{Z} \oplus 2\mathbb{Z}_2$	$5\mathbb{Z}_2$

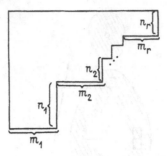

Fig. 10

If we are interested in the integer homology of $G(m, n)$ with (possibly) finite m, n then we must take the part of our complex composed of Young diagrams contained in the $m \times n$-rectangle; this part for $m = 3, n = 2$ is bounded by the dotted line on Fig. 9. Thus

$q =$	0	1	2	3	4	5	6
$H_q(G(3, 2)) \cong$	$\mathbb{Z}$	$\mathbb{Z}_2$	$\mathbb{Z}_2$	$2\mathbb{Z}_2$	$\mathbb{Z}$	$\mathbb{Z}_2$	0

Note that $\mathbb{Z}$-components in $H_*(G(\infty, \infty))$ correspond to Young diagrams bounded by polygonal lines with segments of even length; thus

$$\text{rank } H_q(G(\infty, \infty)) = \begin{cases} 0 \text{ if } q \text{ is not divisible by } 4, \\ p(q/4) \text{ if } q \text{ is divisible by } 4, \end{cases}$$

where p stands for the partition function. If at least one of the numbers m, n is even (or infinite), then $\mathbb{Z}$-components in $H_*(G(m, n))$ correspond to Young diagrams bounded by polygonal lines with the segments of even length contained in the $m \times n$-rectangle (or semi-strip); if both of m, n are odd, then the number of $\mathbb{Z}$-components is larger. Here is the exact statement (the notation $d(q; m, n)$ means the same as in § 2.1.B).

Theorem. *The numbers* rank $H_q(G(m, n))$ *are given by the following table*:

	m is even or infinite	m is odd
n is even or infinite	$d\left(\dfrac{q}{4}; \dfrac{m}{2}, \dfrac{n}{2}\right)$	$d\left(\dfrac{q}{4}; \dfrac{m-1}{2}, \dfrac{n}{2}\right)$
n is odd	$d\left(\dfrac{q}{4}; \dfrac{m}{2}, \dfrac{n-1}{2}\right)$	$d\left(\dfrac{q}{4}; \dfrac{m-1}{2}, \dfrac{n-1}{2}\right)$ $+ d\left(\dfrac{q-m-n+1}{4}; \dfrac{m-1}{2}, \dfrac{n-1}{2}\right)$

C. As to the spaces $G_+(m, n)$, we may add the following statement about their Betti numbers to what was said about them in § 2.1.F (see also § 2.3.F below).

Theorem. *The numbers* rank $H_q(G_+(m, n))$ *are given by the following table:*

	m is even or infinite	m is odd
n is even or infinite	$d\left(\dfrac{q}{4}; \dfrac{m}{2}, \dfrac{n}{2}\right)$ $+ d\left(\dfrac{q-m}{4}; \dfrac{m}{2}, \dfrac{n-2}{2}\right)$ $d\left(\dfrac{q-n}{4}; \dfrac{m-2}{2}, \dfrac{n}{2}\right)$	$d\left(\dfrac{q}{4}; \dfrac{m-1}{2}, \dfrac{n}{2}\right)$ $+ d\left(\dfrac{q-n}{4}; \dfrac{m-1}{2}, \dfrac{n}{2}\right)$
n is odd	$d\left(\dfrac{q}{4}; \dfrac{m}{2}, \dfrac{n-1}{2}\right)$ $+ d\left(\dfrac{q-m}{4}; \dfrac{m}{2}, \dfrac{n-1}{2}\right)$	$d\left(\dfrac{q}{4}; \dfrac{m-1}{2}, \dfrac{n-1}{2}\right)$ $+ d\left(\dfrac{q-m-n+1}{4}; \dfrac{m-1}{2}, \dfrac{n-1}{2}\right)$

2.3. The Cohomology Rings.

The geometrical information at our disposal is not sufficient for the calculation of cohomology rings of Grassmannians, but is sufficient for the formulation of the results.

A. Consider the mappings

$$(\mathbb{R}P^\infty)^n \to G(\infty, n), \qquad (\mathbb{C}P^\infty)^n \to \mathbb{C}G(\infty, n), \qquad (\mathbb{H}P^\infty)^n \to \mathbb{H}G(\infty, n) \quad (4)$$

that assign to a set of lines $l_1, \ldots, l_n$ in $\mathbb{R}^\infty$, $\mathbb{C}^\infty$ or $\mathbb{H}^\infty$, the subspace $l_1 \times \cdots \times l_n$ of the space $(\mathbb{R}^\infty)^n = \mathbb{R}^\infty$, $(\mathbb{C}^\infty)^n = \mathbb{C}^\infty$ or $(\mathbb{H}^\infty)^n = \mathbb{H}^\infty$. The mappings (4) can also be described as the mappings of classifying spaces corresponding to the group injections $O(1)^n \to O(n)$, $U(1)^n \to U(n)$, $Sp(1)^n \to Sp(n)$.

Recall that there are ring isomorphisms

$$H^*((\mathbb{R}P^\infty)^n; \mathbb{Z}_2) = \mathbb{Z}_2[x_1, \ldots, x_n],$$

$$H^*((\mathbb{C}P^\infty)^n; \mathbb{Z}) = \mathbb{Z}[y_1, \ldots, y_n],$$

$$H^*((\mathbb{H}P^\infty)^n; \mathbb{Z}) = \mathbb{Z}[z_1, \ldots, z_n],$$

where dim $x_i = 1$, dim $y_i = 2$, dim $z_i = 4$.

The Borel Theorem. *The mappings*

$$H^*(G(\infty, n); \mathbb{Z}_2) \to H^*((\mathbb{R}P^\infty)^n; \mathbb{Z}_2) = \mathbb{Z}_2[x_1, \ldots, x_n],$$

$$H^*(\mathbb{C}G(\infty, n); \mathbb{Z}) \to H^*((\mathbb{C}P^\infty)^n; \mathbb{Z}) = \mathbb{Z}[y_1, \ldots, y_n],$$

$$H^*(\mathbb{H}G(\infty, n); \mathbb{Z}) \to H^*((\mathbb{H}P^\infty)^n; \mathbb{Z}) = \mathbb{Z}[z_1, \ldots, z_n],$$

induced by the mappings (4) *are monomorphisms, and their images coincide with the spaces of symmetric polynomial in* x_i, y_i, z_i *respectively.* □

For the proof see Borel (1953).

B. Recall that in § 2.2 in the spaces $H^*(G(\infty, n); \mathbb{Z}_2)$, $H^*(\mathbb{C}G(\infty, n), \mathbb{Z})$, $H^*(\mathbb{H}G(\infty, n); \mathbb{Z})$, we described the additive bases $\{w_\Delta\}$, $\{c_\Delta\}$, $\{h_\Delta\}$, whose elements correspond to Young diagrams. Now we indicate the images of w_Δ, c_Δ, h_Δ under the identification of the rings $H^*(G(\infty, n); \mathbb{Z}_2)$, $H^*(\mathbb{C}G(\infty, n); \mathbb{Z})$, $H^*(\mathbb{H}G(\infty, n); \mathbb{Z})$ with the spaces of symmetric polynomials. In order to do this we need a definition from classical combinatorics. Let Δ be the Young diagram corresponding to the partition $k_1 \geqslant \cdots \geqslant k_r$ of the number q. The *Schur polynomial* $s_\Delta(t_1, \ldots, t_n)$ (where n is any number greater or equal to r) is defined by the formula

$$s_\Delta(t_1, \ldots, t_n) = \frac{\begin{vmatrix} t_1^{m_1} \ldots t_n^{m_1} \\ \cdots\cdots\cdots \\ t_1^{m_n} \ldots t_n^{m_n} \end{vmatrix}}{\prod_{i<j} (t_j - t_i)},$$

in which $(m_n, \ldots, m_1) = (n - 1 + k_1, \ldots, n - r + k_r, n - r - 1, \ldots, 1, 0)$. It is known that the Schur polynomials (corresponding to all partitions with $r \leqslant n$) constitute an additive basis in the space of symmetric polynomials in $t_1, \ldots, t_n$.

Theorem. *The elements* $w_\Delta, c_\Delta, h_\Delta$ *of the rings* $H^*(G(\infty, n); \mathbb{Z}_2)$, $H^*(\mathbb{C}G(\infty, n); \mathbb{Z})$, $H^*(\mathbb{H}G(\infty, n); \mathbb{Z})$ *correspond to the symmetric polynomials* $s_\Delta(x_1, \ldots, x_n)$, $s_\Delta(y_1, \ldots, y_n)$, $s_\Delta(z_1, \ldots, z_n)$. □

This theorem shows that in order to be able to represent the products of elements of the bases $\{w_\Delta\}$, $\{c_\Delta\}$, $\{h_\Delta\}$ in terms of these bases one must know the formulae expressing the products of Schur polynomials as linear combinations of Schur polynomials (e.g. $s_{\Delta_1} s_{\Delta_2} = \sum_\gamma a_\gamma s_\gamma \Leftrightarrow w_{\Delta_1} w_{\Delta_2} = \sum a_\gamma w_\gamma$). Such formulae are known: the classical Littlewood – Richardson rule yields a more or less effective way for calculating the coefficients a_γ (see details in Macdonald (1979)).

Note. Although the Littlewood – Richardson rule is very elegant, it is sometimes tedious to use it for explicit calculations. So we exhibit here two concrete facts concerning the cohomology rings of the Grassmann manifolds. We restrict ourselves to the complex case since the real case with coefficients in $\mathbb{Z}_2$ and the quaternionic case are quite similar.

1. Let the cells $\mathbb{C}e(\Delta)$ and $\Gamma e(\Delta')$ have complementary dimensions in $\mathbb{C}G(m, n)$. Then the product $c_\Delta c_{\Delta'}$ is non-zero if and only if the Young diagram Δ' is obtained from the complement of the Young diagram Δ in the $m \times n$-rectangle Ω by reflection in the centre of the rectangle (see Fig. 11). In this case the product is equal to c_Ω.

2. Let δ be the one-square Young diagram. Then

$$c_\delta^r = \sum_\Delta d_\Delta c_\Delta,$$

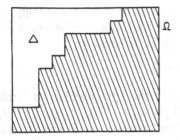 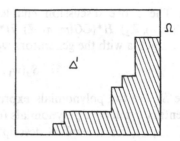

Fig. 11

where the summation is taken over all r-squared Young diagrams, and d_Δ is the dimension of the irreducible representation of the symmetric groups $S(r)$ corresponding to the Young diagram Δ. There are explicit formulae for d_Δ. For example, if Δ is the Young diagram of the partition $r_1, \ldots, r_s$ ($r_1 + \cdots + r_s = r, r_1 \geqslant \cdots \geqslant r_s$), then

$$d_\Delta = \frac{r! \prod_{i<j} (r_j - r_i)}{r_s!(r_{s-1} + 1)! \ldots (r_1 + s - 1)!}$$

(see also Knuth (1974), Chapt. 5, §1.4).

C. The spaces $G(\infty, n)$, $\mathbb{C}G(\infty, n)$, $\mathbb{H}G(\infty, n)$ are the bases of the n-dimensional "tautological bundles". These bundles have characteristic classes: the Stiefel – Whitney classes $w_1, \ldots, w_n \in H^*(G(\infty, n); \mathbb{Z}_2)$, the Chern classes $c_1, \ldots, c_n \in H^*(\mathbb{C}G(\infty, n); \mathbb{Z})$, and the Pontryagin classes $p_1, \ldots, p_n \in H^*(\mathbb{H}G(\infty, n); \mathbb{Z})$. The following fact is well known.

Theorem. *The classes w_i, c_i, p_i correspond to the elementary symmetric polynomials $e_i(x), e_i(y), e_i(z)$.* □

In virtue of the above results, we are now able to answer any question of the following form : what is the expression in terms of the basis $\{w_\Delta\}, \{c_\Delta\}$, or $\{h_\Delta\}$ (or what is the value on such and such a Schubert cell) of a given monomial in w_i, c_i, p_i? Namely, if

$$e_{i_1} \ldots e_{i_r} = \sum a_\Delta s_\Delta,$$

then the value of the monomial $w_{i_1} \ldots w_{i_r} \in H^{i_1 + \cdots + i_r}(G(\infty, n); \mathbb{Z}_2)$, $c_{i_1} \ldots c_{i_r} \in H^{2(i_1 + \cdots + i_r)}(\mathbb{C}G(\infty, n); \mathbb{Z}$, $p_{i_1} \ldots p_{i_r} \in H^{4(i_1 + \cdots + i_r)}(\mathbb{H}G(\infty, n); \mathbb{Z})$ on the Schubert cell $e(\Delta)$, $\mathbb{C}e(\Delta)$, $\mathbb{H}e(\Delta)$ is equal to a_Δ mod 2, a_Δ, a_Δ. For instance, $e_i = s_{\Delta_i}$, where Δ_i is the Young diagram consisting of a row of i squares. Thus

$$w_i = w_{\Delta_i}, \qquad c_i = c_{\Delta_i}, \qquad h_i = h_{\Delta_i},$$

i.e. the classes w_i, c_i, p_i take value 1 on the cells $e(\Delta_i), \mathbb{C}e(\Delta_i), \mathbb{H}e(\Delta_i)$ and take value 0 on all other cells. See Macdonald (1979) for a more or less explicit description of the coefficients a_Δ.

D. The above discussion enables us to describe the cohomology rings $H^*(G(m, n); \mathbb{Z}_2)$, $H^*(\mathbb{C}G(m, n); \mathbb{Z})$, $H^*(\mathbb{H}G(m, n); \mathbb{Z})$. Namely, $H^*(G(m, n); \mathbb{Z}_2)$ is the $\mathbb{Z}_2$-algebra with the generators $w_1, \ldots, w_n$ and the relations

$$S_{\Delta}(w_1, \ldots, w_n) = 0, \tag{4}$$

where S_{Δ} are the polynomials expressing the Schur polynomials in terms of elementary symmetric polynomials (that is $S_{\Delta}(e_1(t_1, \ldots, t_n), \ldots, e_n(t_1, \ldots, t_n) = s_{\Delta}(t_1, \ldots, t_n)$; for an explicit description of the polynomials S_{Δ} see Macdonald (1979) again), and Δ runs through Young diagrams with at most n columns which are not contained in the $m \times n$-rectangle. The system of relations (4) is certainly equivalent to some finite subsystem, but I do not know any reasonable way of extracting such a subsystem.

The rings $H^*(\mathbb{C}G(m, n); \mathbb{Z})$ and $H^*(\mathbb{H}G(m, n); \mathbb{Z})$ have the same description.

E. The structure of the ring $H^*(G(\infty, n); \mathbb{Z})$ and its interrelation with the Schubert cells can be reconstructed completely on the basis of the following two facts. (1) The mod 2 reduction $H^*(G(\infty, n); \mathbb{Z}) \to H^*(G(\infty, n); \mathbb{Z}_2)$ is a ring homomorphism. (2) For a Young diagram Δ, let Δ^+ denote the Young diagram obtained from Δ by subdividing each square into 4 smaller squares. The torsion-free part of the ring $H^*(G(\infty, n); \mathbb{Z})$ is isomorphic to the ring of symmetric polynomials of n variables of dimension 4. The isomorphism assigns to the Schur polynomial s_{Δ} the cohomology class of the cocycle corresponding to the cell $e(\Delta^+)$.

In addition, the values of the monomials of the Pontryagin classes for the tautological bundle on cells of the form $e(\Delta^+)$ are calculated using exactly the same formulae as those for the values of the Chern classes of the tautological bundle over $\mathbb{C}G(\infty, n)$ on the cells $\mathbb{C}e(\Delta)$; the values of the monomials of the Pontryagin classes of the tautological bundle on other cells are zero.

F. To compensate for the lack of information about the cohomology of $G_+(m, n)$ we state the following well known fact.

Theorem.

$$H^*(G_+(\infty, 2n + 1); \mathbb{R}) = \mathbb{R}[p_1, \ldots, p_n],$$

$$H^*(G_+(\infty, 2n); \mathbb{R}) = \mathbb{R}[p_1, \ldots, p_n, e].$$

Here p_i and e denote the Pontryagin classes and the Euler class of the tautological bundle over $G_+(\infty, 2n + 1)$, $G_+(\infty, 2n)$. □

2.4. The K-functor. In this subsection K denotes the complex K-functor.

Let τ_i, γ_i denote the elements of the group $K^0(\mathbb{C}G(m, n))$ represented by the ith exterior power of the tautological bundle and the ith exterior power of the dual tautological bundle respectively.

Theorem. $K^1(\mathbb{C}G(m, n)) = 0$; $K^0(\mathbb{C}G(m, n))$ *is the (unitary) ring with generators $\tau_1, \ldots, \tau_n; \gamma_1, \ldots, \gamma_n$ and the relations*

$$\tau_r + \sum_{i+j=r} \tau_i \gamma_j + \gamma_r = \binom{m+n}{r} \cdot 1 \qquad (1 \leqslant r \leqslant m+n). \ \square$$

For the proof of this theorem as well as some related results see Chapter 4 of the book Karoubi (1978).

Chapter 4
Some Other Important Homogeneous Spaces

§ 1. Flag Manifolds

1.1. Generalities. A. The *flag manifolds* dealt with in this section generalize the Grassmann manifolds as may already be seen from their notations. The *real flag manifold* $\mathbb{R}G(n_1, \ldots, n_r)$ is defined as the set of all chains (flags) of form $V_1 \subset \cdots \subset V_{r-1}$, where V_j are subspaces of $\mathbb{R}^n$ and $n = n_1 + \cdots + n_r$, dim $V_j = n_1 + \cdots + n_j$. The topology and the differential structure in $\mathbb{R}G(n_1, \ldots, n_r)$ are defined by means of the equality

$$\mathbb{R}G(n_1, \ldots, n_r) = O(n)/O(n_1) \times \cdots \times O(n_r).$$

In a similar way complex flag manifolds

$$\mathbb{C}G(n_1, \ldots, n_r) = U(n)/U(n_1) \times \cdots \times U(n_r)$$

and quaternionic flag manifolds

$$\mathbb{H}G(n_1, \ldots, n_r) = Sp(n)/Sp(n_1) \times \cdots \times Sp(n_r)$$

are defined. There is another definition of the manifolds $\mathbb{R}G(n_1, \ldots, n_r)$, $\mathbb{C}G(n_1, \ldots, n_r)$, $\mathbb{H}G(n_1, \ldots, n_r)$ as quotient spaces of the groups $GL(n, \mathbb{R})$ $GL(n, \mathbb{C})$, $GL(n, \mathbb{H})$ over appropriate parabolic subgroups.

B. The obvious fibrations

$$\mathbb{R}G(n_1, \ldots, n_r) \xrightarrow{G(n_i, n_{i+1})} \mathbb{R}G(n_1, \ldots, n_i + n_{i+1}, \ldots, n_r)$$

(the fibre is indicated above the arrow) and similar complex and quaternionic fibrations yield different representations of flag manifolds as "skew products" of Grassmannians. In particular they allow the calculation of the dimensions of flag manifolds (these can also be calculated using the formulae of § 1.1.A):

$$\dim \mathbb{R}G(n_1, \ldots, n_r) = \sum_{i \leqslant j} n_i n_j,$$

$$\dim \mathbb{C}G(n_1, \ldots, n_r) = 2 \sum_{i \leqslant j} n_i n_j,$$

$$\dim \mathbb{H}G(n_1, \ldots, n_r) = 4 \sum_{i \leqslant j} n_i n_j.$$

C. Complex and quaternionic flag manifolds are simply connected and hence orientable. The real flag manifold $\mathbb{R}G(n_1, \ldots, n_r)$ is orientable if and only if the numbers $n_1, \ldots, n_r$ are all even or all odd.

D. Without going into details, we can say that there exists an analogue of Plücker coordinates for flag manifolds. In particular, complex flag manifolds are complex analytic and even projective algebraic manifolds.

1.2. Cell Decompositions. A. Flag manifolds possess a natural cell decomposition which generalizes the Schubert cell decomposition of Grassman manifolds. The name Schubert is also applied to this decomposition and its cells. Schubert cells of a flag manifold are labelled with sets of dimensions d_{ij} of the intersections $V_i \cap L_j$, where L_j $(= \mathbb{R}^j, \mathbb{C}^j, \mathbb{H}^j)$ is the space of the first j coordinate axes. However, the numbers d_{ij} must satisfy a lot of cumbersome conditions, and we prefer the following, more reasonable, description of the Schubert cells.

The cells correspond to the sequences $k_1, \ldots, k_n$ of integers, precisely n_j of which are equal to j $(j = 1, \ldots, r)$. The cell associated with the sequence $k_1, \ldots, k_n$ is denoted in the real case by $\mathbb{R}e[k_1, \ldots, k_n]$, in the complex case by $\mathbb{C}e[k_1, \ldots, k_n]$, and in the quaternionic case by $\mathbb{H}e[k_1, \ldots, k_n]$. This cell consists of flags $V_1 \subset \cdots \subset V_{r-1}$ with

$$\dim \frac{V_i \cap L_j}{(V_{i-1} \cap L_j) + (V_i \cap L_{j-1})} = \delta_{ik_j}$$

(we take V_0 as zero and V_r as the whole space) or, equivalently,

$$\dim(V_i \cap L_j) = \mathrm{card}\{s \leqslant i \mid k_s \leqslant j\}.$$

The dimension of the cell $\mathbb{R}e[k_1, \ldots, k_n]$ is equal to the number of pairs (i, j) such that $i < j, k_i > k_j$; the dimensions of the cells $\mathbb{C}e[k_1, \ldots, k_n]$ and $\mathbb{H}e[k_1, \ldots, k_n]$ are respectively twice and four times larger.

B. In the Grassmann case the cell decomposition constructed is the same as the decomposition described in Chapter 3. In this case $r = 2$ and the sequence $k_1, \ldots, k_n$ contains the number 1 n_1 times and the number 2 n_2 times. Assign to this sequence an n-gonal line beginning at $(0, n_1)$ and ending at $(n_2, 0)$: all its segments are of length 1 and its ith segment is directed downward if $k_i = 1$ and is directed to the right if $k_i = 2$. This line bounds the Young diagram Δ and it is easy to see that $\mathbb{R}e[k_1, \ldots, k_n] = \mathbb{R}e(\Delta)$, $\mathbb{C}e[k_1, \ldots, k_n] = \mathbb{C}e(\Delta)$, $\mathbb{H}e[k_1, \ldots, k_n] = \mathbb{H}e(\Delta)$.

C. Alternatively the cells $\mathbb{R}e[k_1, \ldots, k_n]$, $\mathbb{C}e[k_1, \ldots, k_n]$, $\mathbb{H}e[k_1, \ldots, k_n]$ may be described as the orbits of the group of lower triangular matrices with units on the diagonal with respect to its natural action on the flag manifold. Namely, the cell associated with the sequence $k_1, \ldots, k_n$ is the orbit of the flag whose ith space is spanned by the coordinate vectors with the numbers s satisfying the condition $k_s \leqslant i$.

1.3. Homology and Cohomology. A. As in the Grassman case, cellular calculation of the homology and cohomology of complex and quaternionic flag
manifolds presents no difficulties (there are no cells of neighbouring dimensions,
so the homology and cohomology groups coincide with the chain and cochain
ones). Neither does the calculation in the real case if the coefficient domain is $\mathbb{Z}_2$
(all the incidence coefficients are even). It is convenient to formulate the result in
terms of the Poincaré polynomial.

Theorem. *The Poincaré polynomial of the manifold* $\mathbb{R}G(n_1, \ldots, n_r)$ *with coefficients in* $\mathbb{Z}_2$ *is equal to*

$$\frac{\Pi_n(t)}{\Pi_{n_1}(t) \ldots \Pi_{n_r}(t)}$$

where $\Pi_s(\lambda) = (1 - \lambda^s)(1 - \lambda^{s-1}) \ldots (1 - \lambda)$ *(compare Chapt. 3, §1.2.A). The
Poincaré polynomials of the manifolds* $\mathbb{C}G(n_1, \ldots, n_r)$ *and* $\mathbb{H}G(n_1, \ldots, n_r)$ *(with
coefficients in* $\mathbb{Z}$*) are the same with* t^2 *and* t^4 *instead of* t. $\square$

In order to calculate the integer homology of real flag manifolds, we have to
find the incidence coefficients between the Schubert cells (like those in the
Grassman case, they take on the values 0 and ± 2); if necessary, the reader can
perform the calculation following the model of Chapter 3, §2.2.

B. The following description of the multiplicative structure in the cohomology of flag manifolds is obtained by using spectral sequences (see Borel (1953)).
The relationship of this description with Schubert cells is not obvious; we give
details of this relationship only in the case of the manifold $\mathbb{C}G(1, \ldots, 1)$ of
(complex) complete flags (see §1.4.B).

Theorem. *The graded ring* $H^*(\mathbb{R}G(n_1, \ldots, n_r); \mathbb{Z}_2)$ *is isomorphic to the quotient
ring of the ring of polynomials over* $\mathbb{Z}_2$ *in* n *variables* $x_1, \ldots, x_n$ *of degree 1
invariant under permutations inside the groups* $\{x_{n_1 + \cdots + n_{s-1} + 1}, \ldots, x_{n_1 + \cdots + n_s}\}$
$(s = 1, \ldots, r)$ *over the ideal generated by polynomials without constant term
invariant under all permutations of* x*'s. The corresponding rings* $H^*(\mathbb{C}G(n_1, \ldots, n_r); \mathbb{Z})$ *and* $H^*(\mathbb{H}G(n_1, \ldots, n_r); \mathbb{Z})$ *are obtained by replacing* $\mathbb{Z}_2$ *by* $\mathbb{Z}$ *and the
variables* $x_1, \ldots, x_n$ *of degree 1 by the variables* $y_1, \ldots, y_n$ *of degree 2, and
variables* $z_1, \ldots, z_n$ *of degree 4, respectively.* $\square$

1.4. The Case of Complete Flag Manifolds. A. Let $\mathbb{R}F(n) = \mathbb{R}G(\underbrace{1, \ldots, 1}_{n})$,

$\mathbb{C}F(n) = \mathbb{C}G(\underbrace{1, \ldots, 1}_{n})$, $\mathbb{H}F(n) = \mathbb{H}G(\underbrace{1, \ldots, 1}_{n})$. These manifolds are called *com-*

plete flag manifolds. In this case the above theory acquires a very nice form. The
Schubert cells are parametrized with the usual permutations $(k_1, \ldots, k_n) \in S(n)$,
and the dimension of the cell is equal to the number of inversions in the
permutation, multiplied by 2 in the complex case and by 4 in the quaternionic

case. The cohomology ring with coefficients in $\mathbb{Z}_2$ in the real case and with coefficients in $\mathbb{Z}$ in the complex and quaternionic cases is isomorphic to the quotient ring of the ring of polynomials in n variables (which are of degree 1 in the real case, of degree 2 in the complex case, and of degree 4 in the quaternionic case) over the ideal generated by all symmetric polynomials without constant term.

B. Thus for a complete flag manifold with cohomology ring isomorphic to

$$\mathbb{K}[t_1, \ldots, t_n]/(e_1, \ldots, e_n),$$

where $\mathbb{K} = \mathbb{Z}$ or $\mathbb{Z}_2$ and $e_1, \ldots, e_n$ are the elementary symmetric polynomials, the Schubert cells define an additive base whose elements are labelled by permutations. We give a direct description of this base due to J.N. Bernstein, I.M. Gel'fand and S.I. Gel'fand (see Bernstein, Gel'fand I.M., Gel'fand S.I. (1973)).

To the permutation $w_0 = (n, n-1, \ldots, 1)$, which has the maximal possible number of inversions, assign the polynomial $P_{w_0} = \prod_{i \geqslant j}(t_i - t_j)$. If w is another permutation then decompose the permutation $w^{-1}w_0$ into the product of elementary transpositions (that is permutations with a single inversion):

$$w^{-1}w_0 = \gamma_1 \ldots \gamma_l$$

with minimal possible l. Then put $P_w = A\gamma_1 \ldots A\gamma_1 P_{w_0}$, where the operator A_γ is defined as follows. If $\gamma = (\ldots, i+1, i, \ldots)$, then

$$A_\gamma P(t_1, \ldots, t_n) = \frac{P(t_1, \ldots, t_n) - P(t_1, \ldots, t_{i-1}, t_{i+1}, t_i, t_{i+2}, \ldots, t_n)}{t_{i+1} - t_i}$$

1.5. Generalizations. In the subsection we use the simplest notions from the theory of Lie groups more freely than before.

A. The result of § 1.3.B admits a partial generalization to the case of the homogeneous space of a compact Lie group with a closed full rank subgroup as isotropy group.

Theorem. *Let G be a compact Lie group, and H be a closed subgroup whose rank n is equal to that of G. Let $x_1, \ldots, x_n$ be the coordinates in the common Cartan subalgebra of the Lie algebras of the groups G and H, and W_G, W_H be the Weyl groups of G, H. (i) If p is a prime such that there is no p-torsion in the cohomology of G and H, then*

$$H^*(G/H; \mathbb{Z}_p) = \mathbb{Z}_p[x_1, \ldots, x_n]^{W_H}/(\mathbb{Z}_p^+[x_1, \ldots, x_n]^{W_G}),$$

where $(\ldots)^W$ denotes the ring of W-invariants, $+$ stands for the subring of polynomials without constant term, the round brackets on the right-hand side denote the ideal generated by the set in the brackets and all x's have degree 2. (ii) If the cohomology of the groups G, H has no torsion at all, then

$$H^*(G/H; \mathbb{Z}) = \mathbb{Z}[x_1, \ldots, x_n]^{W_H}/(\mathbb{Z}^+[x_1, \ldots, x_n]^{W_G}). \quad \square$$

See Borel (1953) for the proof.

B. The results of 1.4 can be extended to the case of the quotient space of an arbitrary compact Lie group G over its maximal torus T. Without going into details we list the most important properties of this space. (i) The space G/T is decomposed into cells labelled with elements of the Weyl group W of the group G; the dimension of the cell is equal to twice the length of the corresponding element of the Weyl group (i.e. to twice the length of the shortest word representing this element as the product of simple reflections). (ii) $H^*(G/H; \mathbb{Q}) = \mathbb{Q}[x_1, \ldots, x_n]/(\mathbb{Q}^+[x_1, \ldots, x_n]^W)$, where $x_1, \ldots, x_n$ are coordinates in the Cartan subalgebra of the Lie algebra of G, and each variable x_i has degree 2. If the cohomology of G has no torsion (no p-torsion) then the same is true with coefficients in $\mathbb{Z}$ (in $\mathbb{Z}_p$). Note that (i) implies that the cohomology of G/T is always torsion-free. Note also that more detailed information on the cohomology of G/T may be found in the preprint of Kac which was referred to in Chapt. 2, §2.2.A, and that interrelations between the isomorphism (ii) and the cellular decomposition (i) are described in the article by Bernstein and the two Gel'fands which was referred to in §1.4.B.[3] (iii) The manifold G/T has a canonical complex structure. (This fact is of great importance for representation theory – see Bott (1957).)

§2. The Manifolds $U(n)/SO(n)$ and $U(n)/O(n)$

2.1. Generalities. A. The homogeneous space $U(n)/O(n)$ may be interpreted as the submanifold of the real Grassmannian $G(n, n)$ composed of real n-dimensional subspaces of the space $\mathbb{R}^{2n} = \mathbb{C}^n$ such that $V^\perp = \sqrt{-1}\, V$ (this is the $U(n)$-orbit in $G(n, n)$ of the real part $\mathbb{R}^n$ of $\mathbb{C}^n$; it is clear that the isotropy group is $O(n)$). In other words, $U(n)/O(n)$ is the manifold of *Lagrangian subspaces* of the standard symplectic space $\mathbb{R}^{2n}$, that is n-dimensional subspaces of $\mathbb{R}^{2n}$ to which the structure 2-form restricts trivially. Similarly, $U(n)/SO(n)$ is the manifold of oriented real n-dimensional subspaces V of $\mathbb{C}^n$ such that $V^\perp = \sqrt{-1}\, V$, or the manifold of oriented Lagrangian subspaces of $\mathbb{R}^{2n}$.

B. The manifolds $U(n)/O(n)$ and $U(n)/SO(n)$ are the bases of n-dimensional real *tautological vector bundles* of which the second is oriented. These bundles are the restrictions of the tautological bundles over $G(n, n)$ and $G_+(n, n)$; informally speaking, the fibre over a point $\pi \in U(n)/O(n)$ [$\in U(n)/SO(n)$] is π itself.

C. The significance of the spaces $U(n)/O(n)$ in symplectic geometry (better to say "in symplectic topology") originates mainly from the fact that if X is an [oriented] Lagrangian submanifold of the standard symplectic space $\mathbb{R}^{2n}$, then assigning to any point $x \in X$ the tangent space $T_x X$ translated to 0 determines

---

[3] This article also treats the case of the quotient space of G over an arbitrary parabolic subgroup, in particular the case of an arbitrary flag manifold. For example it contains a generalization of the results of 4B to arbitrary flag manifolds.

a mapping $X \to U(n)/O(n)$ $[\to U(n)/SO(n)]$. Hence the elements of the cohomology groups of $U(n)/O(n)$ [of $U(n)/SO(n)$] give rise to characteristic classes of [oriented] Lagrangian submanifolds of $\mathbb{R}^{2n}$.

Note also that our manifolds, or rather the Thom spaces of tautological bundles, play an essential role in V.I. Arnol'd's theory of Lagrangian and Legendrian cobordisms (see Chapter 6 of Arnol'd, Givental' (1985)).

D. In more traditional topological terms, the manifolds $U(n)/O(n)$ may be described as the classifying spaces for n-dimensional real vector bundles and n-dimensional oriented real vector bundles, whose complexifications are trivial and are furnished with a fixed homotopy class of trivializations.

2.2. Cellular Decompositions. A. The manifold $U(n)/O(n)$ possesses a convenient cellular decomposition which may be also called a Schubert decomposition, though it is due to V.I. Arnol'd. The cells of this decomposition are labelled by (possibly empty) sets $j_1, \ldots, j_r$ of integers such that $1 \leqslant j_1 < \cdots < j_r \leqslant n$. The cell $e\{j_1, \ldots, j_r\}$ is defined by the formula

$$e\{j_1, \ldots, j_r\} = \{\pi \in U(n)/O(n) | \dim[\text{Re}(\pi) \cap \mathbb{R}^j] = s \quad \text{for} \quad j_s \leqslant j < j_{s+1}\},$$

where Re: $\mathbb{C}^n \to \mathbb{R}^n$ is the "real part" operator, and we take $j_0 = 0, j_{r+1} = n + 1$. Alternatively, the cell $e\{j_1, \ldots, j_r\}$ may be described as the set of $\pi \in U(n)/O(n)$ such that the space $\text{Re}(\pi)$ has dimension r and belongs to the Schubert cell of the manifold $G(n - r, r)$ labelled with the Young diagram with lengths $j_r - r$, $j_{r-1} - (r - 1), \ldots, j_1 - 1$.

The dimension of the cell $e\{j_1, \ldots, j_r\}$ is equal to $j_1 + \cdots + j_r$. (The dimension of the underlying cell in $G(n - r, r)$ is equal to $j_1 + \cdots + j_r - \frac{1}{2}r(r + 1)$.)

B. The manifold $U(n)/SO(n)$ is a two-sheeted covering of $U(n)/O(n)$ and hence has a cellular decomposition $e_{\pm}\{j_1, \ldots, j_r\}$ whose cells are components of the inverse images of cells $e\{j_1, \ldots, j_r\}$ under the projection $U(n)/SO(n) \to U(n)/O(n)$. More precisely, the exact sequence

$$0 \to \text{Ker}(\text{Re}|\pi) \to \pi \to \text{Re}(\pi) \to 0$$

together with the canonical isomorphism $\text{Ker}(\text{Re}|\pi) = \text{Re}(\pi)^{\perp}$ (multiplying by $\sqrt{-1}$; $\perp$ is taken in $\mathbb{R}^n$) furnishes π with a canonical orientation (which depends continuously on π in the domain where $\dim \text{Re}(\pi)$ remains constant). We denote by $e_{+}\{j_1, \ldots, j_r\}$ [by $e_{-}\{j_1, \ldots, j_r\}$] the set of Lagrangian spaces π belonging to $e\{j_1, \ldots, j_r\}$ and furnished with the canonical orientation (with the opposite orientation).

C. The boundary of the cell $e\{j_1, \ldots, j_r\}$ includes the following cells with the coefficient ± 2 (we do not specify the sign):
(1) $e\{j_2, \ldots, j_r\}$ if $j_1 = 1$ and r is even;
(2) $e\{j_1, \ldots, j_{s-1}, j_s - 1, j_{s+1}, \ldots, j_r\}$ if $j_s - 1 > j_{s-1}$ and j_s is odd.
The boundary of the cell $e_{+}\{j_1, \ldots, j_r\}$ [the cell $e_{-}\{j_1, \ldots, j_r\}$] includes:
(1) the cells $e_{+}\{j, \ldots, j_r\}$ and $e_{-}\{j, \ldots, j_r\}$ with the coefficient ± 1 if $j_1 = 1$ (irrespective of the parity of r);

(2) the cell $e_+\{j_1, \ldots, j_{s-1}, j_s - 1, j_{s+1}, \ldots, j_r\}$ $[e_-\{j_1, \ldots, j_{s-1}, j_s - 1, j_{s+1}, \ldots, j_r\}]$ with the coefficient ± 2 if $j_s - 1 > j_{s-1}$ and j_s is odd.

2.3. Cellular Computation of Homology.

A. The computation of the homology of the manifolds $U(n)/O(n)$ and $U(n)/SO(n)$ is similar to that of the Grassman manifolds. The easiest job is, of course, to find the homology of $U(n)/O(n)$ with coefficients in $\mathbb{Z}_2$: since all the incidence coefficients are even, this homology coincides with the corresponding chain groups. (Compare with assertion 2° of Theorem of §2.3.C and with Theorem of §2.4.A)

Some results of the calculation of the integer homologies are presented in Tables 5, 6.

Table 5

k \ n	1	2	3	4	5	≥6
0	Z	Z	Z	Z	Z	Z
1	Z	Z	Z	Z	Z	Z
2	0	$\mathbb{Z}_2$	$\mathbb{Z}_2$	$\mathbb{Z}_2$	$\mathbb{Z}_2$	$\mathbb{Z}_2$
3	0	0	$\mathbb{Z}_2$	$\mathbb{Z}_2$	$\mathbb{Z}_2$	$\mathbb{Z}_2$
4	0	0	0	$\mathbb{Z}_2$	$\mathbb{Z}_2$	$\mathbb{Z}_2$
5	0	0	Z	Z	$\mathbb{Z} \oplus \mathbb{Z}_2$	$\mathbb{Z} \oplus \mathbb{Z}_2$
6	0	0	Z	$\mathbb{Z} \oplus \mathbb{Z}_2$	$\mathbb{Z} \oplus \mathbb{Z}_2$	$\mathbb{Z} \oplus \mathbb{Z}_2 \oplus \mathbb{Z}_2$

Groups $H_k(U(n)/O(n))$

Table 6

k \ n	1	2	3	4	5	≥6
0	Z	Z	Z	Z	Z	Z
1	Z	Z	Z	Z	Z	Z
2	0	Z	$\mathbb{Z}_2$	$\mathbb{Z}_2$	$\mathbb{Z}_2$	$\mathbb{Z}_2$
3	0	Z	$\mathbb{Z}_2$	$\mathbb{Z}_2$	$\mathbb{Z}_2$	$\mathbb{Z}_2$
4	0	0	0	Z	$\mathbb{Z}_2$	$\mathbb{Z}_2$
5	0	0	Z	Z	$\mathbb{Z} \oplus \mathbb{Z}_2$	$\mathbb{Z} \oplus \mathbb{Z}_2$
6	0	0	Z	$\mathbb{Z} \oplus \mathbb{Z}_2$	$\mathbb{Z} \oplus \mathbb{Z}_2$	$\mathbb{Z} \oplus \mathbb{Z}_2 \oplus \mathbb{Z}_2$

Groups $H_k(U(n)/SO(n))$

B. $\mathbb{Z}$-components in the integral homology of $U(n)/O(n)$ correspond to cells $e\{j_1, \ldots, j_r\}$ with the following property: if j_s is even then $s < r$ and $j_{s+1} = j_s + 1$. In particular, the qth Betti number $b_q(n)$ of $U(n)/O(n)$ is equal to the number of partitions of the number q into the sum of different numbers of form $4m + 1$ less than $2n$.

Further, the Betti numbers $b_q^+(n)$ of $U(n)/SO(n)$ are given by the formula

$$b_q^+(n) = \begin{cases} b_q(n) & \text{if } n \text{ is odd,} \\ b_q(n) + \dfrac{b_{n(n+1)}}{2} - q(n) & \text{if } n \text{ is even.} \end{cases}$$

C. More elaborate analysis of the cellular complex gives the following result.

Theorem. 1°. *If the incidence coefficients* $[e\{1, j_2, \ldots, j_r\} : e\{j_2, \ldots, j_r\}]$ *with even r in the cellular complex of $U(n)/O(n)$ are put equal to zero, then a complex is obtained whose homology is $H_*(U(n)/SO(n))$.*
2°. $H_*(U(n)/O(n); \mathbb{Z}_2) \cong H_*(U(n)/SO(n); \mathbb{Z}_2)$ *for any n.*
3°. $H_*(U(n)/O(n)) \cong H_*(U(n)/SO(n))$ *for n odd.* □

Note that the isomorphisms of 2° and 3° are not induced by the projection $U(n)/O(n) \to U(n)/SO(n)$.

2.4. The Cohomology Rings
A. **Theorem.** *There is a multiplicative isomorphism*

$$H^*(U(n)/O(n); \mathbb{Z}_2) = H^*(S^1 \times S^2 \times \cdots \times S^n; \mathbb{Z}_2).\ \square$$

We make two additional remarks about this statement. Firstly, the multiplicative generators of the algebra $H^*(U(n)/O(n); \mathbb{Z}_2)$ arising from this theorem are precisely the Stiefel – Whitney classes $w_1, \ldots, w_n$ of the tautological bundle. Thus

$$H^*(U(n)/O(n); \mathbb{Z}_2) = \mathbb{Z}_2[w_1, \ldots, w_n]/(w_1^2, \ldots, w_n^2).$$

Secondly, the preceeding statement interacts nicely with the cellular structure of $U(n)/O(n)$ given in § 2.2.A. Namely, if $1 \leqslant j_1 < \cdots < j_r \leqslant n, 1 \leqslant i_1 < \cdots < i_s \leqslant n$, then

$$\langle w_{i_1} \cdots w_{i_s}, e\{j_1, \ldots, j_r\}\rangle = \begin{cases} 1 & \text{if } \{i_1, \ldots, i_s\} = \{j_1, \ldots, j_r\}, \\ 0 & \text{otherwise.} \end{cases}$$

I do not know the multiplicative structure of the cohomology of $U(n)/SO(n)$ (its additive structure is given by the assertion 2° of the theorem of § 2.3.C).

B. **Theorem.** *If $\mathbb{K} = \mathbb{Q}$ or $\mathbb{K} = \mathbb{Z}_p$ for an odd prime p, then there are multiplicative isomorphisms*

$$H^*(U(2m-1)/O(2m-1); \mathbb{K}) = H^*(U(2m)/O(2m); \mathbb{K})$$

$$= H^*(U(2m-1)/SO(2m-1); \mathbb{K}) = H^*(\mathscr{S}; \mathbb{K});$$

$$H^*(U(2m)/SO(2m); \mathbb{K}) = H^*(\mathscr{S} \times S^{2m}; \mathbb{K});$$

where $\mathscr{S} = S^1 \times S^5 \times \cdots \times S^{4m-3}$. □

Some other details, such as the connection with the characteristic classes of the tautological bundles, are given in Fuks (1968).

§ 3. The Manifolds $SO(2n)/U(n)$ and $U(2n)/Sp(n)$

We restrict ourselves here to some statements concerning the cohomology of these manifolds; see the proofs in Borel (1953), § 31.

A. **Theorem.** *There is an additive isomorphism*

$$H^*(SO(2n)/U(n); \mathbb{Z}) = H^*(S^2 \times S^4 \times \cdots \times S^{2n-2}; \mathbb{Z}).$$

If p is an odd prime, then there is a multiplicative isomorphism

$$H^*(SO(2n)/U(n); \mathbb{Z}_p) = S_p(x_1, \ldots, x_n)/(e_1(x_1^2, \ldots, x_n^2), \ldots, e_{n-1}(x_1^2, \ldots, x_n^2), x_1 \ldots x_n),$$

where $S_p(x_1, \ldots, x_n)$ is the ring of symmetric polynomials in the variables $x_1, \ldots, x_n$ (of degree 2), and e_i is the i-the elementary symmetric polynomial.
    The mapping

$$H^*(SO(2n)/U(n); \mathbb{Z}_2) \to H^*(SO(2n); \mathbb{Z}_2)$$

induced by the projection $SO(2n) \to SO(2n)/U(n)$ is monomorphic. Its image is generated (in the notation of Theorem 2.3.B) by the squares $x_1^2, x_3^2, \ldots, x_l^2$, where l is the greatest odd integer less than n. □

Comparing this result with Theorem of §2.3.B we obtain the following unexpected

Corollary. *There is a multiplicative isomorphism*

$$H^*(SO(n); \mathbb{Z}_2) = H^*(SO(2n)/U(n); \mathbb{Z}_2),$$

doubling the grading (that is, taking $H^i(SO(n); \mathbb{Z}_2)$ into $H^{2i}(SO(2n)/U(n); \mathbb{Z}_2)$). □

B. **Theorem.** *There is a multiplicative isomorphism*

$$H^*(U(2n)/Sp(n); \mathbb{Z}) = H^*(S^1 \times S^5 \times \cdots \times S^{4n-3}; \mathbb{Z}).$$ □

Remark. The projection $U(2n) \to U(2n)/Sp(n)$ induces a monomorphism

$$H^*(U(2n)/Sp(n); \mathbb{Z}) \to H^*(U(2n); \mathbb{Z}),$$

but I do not know any reasonable description of the image of this monomorphism.

Chapter 5

Some Manifolds of Low Dimension

§ 1. Closed Surfaces

In spite of the seeming simplicity of the subject, the theory of closed 2-manifolds is an extensive area of classical topology with many striking and deep results. It is not easy to select material for a short section dealing with this subject, but it seems to me that any list of classical manifolds not containing two-dimensional spheres with handles would be incomplete. Those who wish to learn the theory in detail may read the books Zieschang, Vogt, Coldewey (1980), Zieschang (1981).

1.1. The Standard Surfaces. A. Theorem. *Any closed smooth (PL, topological) 2-manifold is diffeomorphic (PL-equivalent, homeomorphic) to one of the following standard surfaces (see Fig. 12):*
  (1) *the 2-sphere with g handles (g = 0, 1, ...);*
  (2) *the projective plane with g handles (g = 0, 1, ...);*
  (3) *the Klein bottle with g handles (g = 0, 1, ...).*
The surfaces listed are not homeomorphic to each other. □

(For the proof of the differential version of this theorem see Fuks, Rokhlin (1977),
  It may be shown that the connected sum (see Chapt. 1, § 2.1.A) of two projective planes is diffeomorphic to the Klein bottle, and the connected sum of three

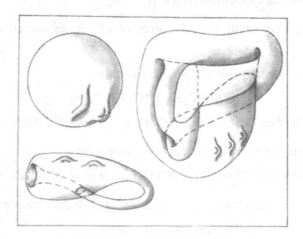

Fig. 12

projective planes is diffeomorphic to the projective plane with one handle. Therefore all the surfaces of the types (2) and (3) may be characterized as connected sums of projective planes. Furthermore, it is clear that forming a connected sum with the projective plane is equivalent to attaching a crosscap (i.e. cutting out a small disc and pasting in a *Möbius band*). We thus obtain a convenient statement of the theorem; any closed surface is diffeomorphic (*PL*-equivalent, homeomorphic) to either the 2-sphere with g handles ($g = 0, 1, \ldots$) or the 2-sphere with h crosscaps ($h = 1, 2, \ldots$) (notations: P_g, Q_h). Note that the surface P_g is orientable, while the surface Q_h is not. The surface P_g is also called a *surface of genus g*. In addition any compact surface (possibly with boundary) is diffeomorphic (*PL* equivalent, homeomorphic) to either a sphere with handles and holes or a sphere with crosscaps and holes.

B. Recall the classical *PL* realization of P_g and Q_h. The surface P_g ($g \geqslant 1$) may be obtained from a regular $4g$-gon by pasting together the $(4i + 1)$st side with the $(4i + 3)$rd side and the $(4i + 2)$nd side with the $(4i + 4)$th side, with reversed orientation ($i = 0, \ldots, g - 1$). The surface Q_h ($h \geqslant 2$) may be obtained from a regular $2h$-gon by pasting together the $(2i + 1)$st side with the $(2i + 2)$nd side, preserving the orientation ($i = 0, \ldots, h - 1$). (In both cases the sides are numbered in cyclic order and are oriented in the positive direction.) This realization furnishes the standard surfaces with a standard cellular decomposition: the surface P_g has one two-dimensional cell, $2g$ one-dimensional cells, and the zero-dimensional cell, the surface Q_h has one two-dimensional cell, h one-dimensional cells, and one zero-dimensional cell. (The surfaces P_0 and Q_1 also have such cellular decompositions.)

1.2. Homotopy Properties. A. $\pi_1(P_g)$ is the group with $2g$ generators $a_1, b_1,$ $\ldots, a_g, b_g$ (see Fig. 13) and one relation $a_1 b_1 a_1^{-1} b_1^{-1} \ldots a_g b_g a_g^{-1} b_g^{-1} = 1$; $\pi_1(Q_h)$ is the group with h generators $c_1, \ldots, c_h$ (see Fig. 14) and one relation $c_1^2 \ldots c_h^2 = 1$. (This follows immediately from the cellular decompositions of §1.1.B.)

B. $\pi_i(P_g) = 0$ for $i \geqslant 2$, $g \geqslant 1$; $\pi_i(Q_h) = 0$ for $i \geqslant 2$, $h \geqslant 2$. (This follows from the fact that the universal covering of P_g with $g \geqslant 1$ and Q_h with $h \geqslant 2$ is

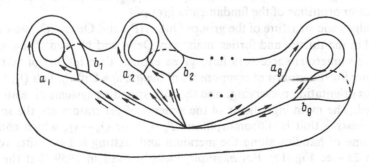

Fig. 13

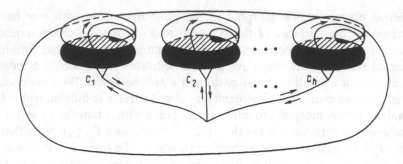

homeomorphic to the Euclidean plane; for a more visual proof see Bourbaki [1975], 5.3.4.)

C. $H_1(P_g) = \mathbb{Z} \oplus \cdots \oplus \mathbb{Z}$; the intersection numbers define a symplectic struc-

$$\underbrace{\phantom{\mathbb{Z} \oplus \cdots \oplus \mathbb{Z}}}_{2g}$$

ture in $H_1(P_g)$. $H_1(Q_h) = \underbrace{\mathbb{Z} \oplus \cdots \oplus \mathbb{Z}}_{h-1} \oplus \mathbb{Z}_2$. The Euler characteristic of P_g is

equal to $2 - 2g$, the Euler characteristic of Q_h is equal to $2 - h$.

1.3. Automorphisms. (See the details in Rieschang (1981) or Stillwell (1980).) For any smooth two-dimensional manifold the embeddings of the group of self-diffeomorphisms into the group of *PL* self-equivalences and into the group of self-homeomorphisms are homotopy equivalences. Hence the results of this subsection, which are formulated for diffeomorphisms, are also valid for *PL* equivalences and homeomorphisms.

A. The embeddings $O(3) \to \mathrm{Diff}\, S^2$, $SO(3) \to \mathrm{Diff}\, \mathbb{R}P^2$ are homotopy equivalences.

B. For $g \geqslant 1$ ($h \geqslant 2$) the identity component of the group $\mathrm{Diff}\, P_g$ (the group $\mathrm{Diff}\, Q_h$) is contractible. The group of components of the group $\mathrm{Diff}\, P_g$ (of the group $\mathrm{Diff}\, Q_h$) coincides with the group $\mathrm{Out}\,\pi_1(P_g)$ (the group $\mathrm{Out}\,\pi_1(Q_h)$) of outer automorphisms of the fundamental group.

The algebraic structure of the groups $\mathrm{Out}\,\pi_1(P_g)$ and $\mathrm{Out}\,\pi_1(Q_h)$ was studied in detail in the thirties and forties, mainly by Dehn and Nielsen. These authors and their numerous successors have given explicit descriptions of systems of generators for the groups of components of the groups $\mathrm{Diff}\, P_g$, $\mathrm{Diff}\, Q_h$, $\mathrm{Diff}_+ P_g$ (+ means orientation preserving), and the corresponding systems of relations are presented. The main ingredients of the systems of generators are the so-called *Dehn's twists* – that is diffeomorphisms $P_g \to P_g$ or $Q_h \to Q_h$ which consist in cutting one of handles along the meridian and pasting it back after rotation through 2π (see Fig. 15). For example, Dehn proved in 1939 that the group $\mathrm{Comp}(\mathrm{Diff}\, P_g)$ for $g \geqslant 2$ is generated by 3 generators, one of which has in-

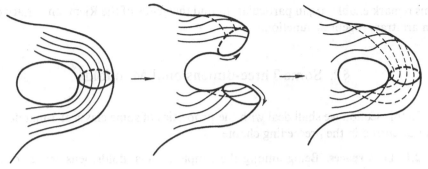

Fig. 15

finite order (Dehn's twist) while two others are of finite order (the group Comp (Diff P_1) = $SL(2, \mathbb{Z})$ is generated by two elements of finite order). In 1964 Lickorisch proved that for $g \geqslant 1$ the group Comp(Diff$_+$ P_g) is generated by $3g - 1$ Dehn's twists.

1.4. Complex Structures. The surface P_g has complex structures. For $P_0 = S^2$ this structure is unique up to equivalence (the Riemann sphere), the torus P_1 has a one-parameter family of (classes of) complex structures, for $g \geqslant 2$ (the class of) a complex structure on P_g depends on $3g - 3$ parameters (moduli). Each of these structures makes P_g into a projective algebraic variety. For any g the set of these structures has itself a natural complex projective algebraic manifold structure.

Some of the complex structures on closed surfaces have simple explicit realizations. For instance, a non-singular curve of degree m in $\mathbb{C}P^2$ has genus $\dfrac{(m - 1)(m - 2)}{2}$. More generally, a non-singular complete intersection in $\mathbb{C}P^{r+1}$ of multidegree $m_1, \ldots, m_r$ has genus

$$\frac{m_1 \ldots m_r(m_1 + \cdots + m_r - (r + 2)) + 2}{2}. \tag{5}$$

Note that not every integer is representable in the form (5), so some spheres with handles are not homeomorphic to any projective complete intersection; this is the case, for example, for the surface of genus 2. (But any sphere with handles is homeomorphic to an algebraic submanifold of $\mathbb{C}P^1 \times \mathbb{C}P^1$: a non-singular curve of degree m_1 in the coordinates of the first factor and of degree m_2 in the coordinates of the second factor has genus $(m_1 - 1)(m_2 - 1)$.)

In addition, if the complex curve X is an m-sheeted covering of the Riemann sphere with r branch points, and if these points have $m_1, \ldots, m_r$ pre-images respectively, then the genus of X is equal to

$$\frac{(r - 2)m - (m_1 + \cdots + m_r) + 2}{2};$$

this remark enables us, in particular, to find the genus of the Riemann surface of an arbitrary algebraic function.

§2. Some Three-dimensional Manifolds

In this section we shall deal with the properties of some classical 3-manifolds not discussed in the preceeding chapters.

2.1. Lens spaces. Being among the simplest 3-manifolds, lens spaces often appear in various constructions of geometry and analysis (e.g. in the theory of singularities). Their popularity in topology stems from the fact they yield the simplest situation when the specific character of the topology of non-simply connected manifolds becomes apparent.

A. Fix relatively prime positive integers p, q with $p \geqslant 3$, $1 \leqslant q < p$. The (three-dimensional) *lens space* $L(p, q)$ is defined as the quotient space of the sphere $S^3 = \{(z_1, z_2) \in \mathbb{C}^2 | |z_1|^2 + |z_2|^2 = 1\}$ over the group $\mathbb{Z}_p$ whose generator F acts on S^3 according to the formula

$$F(z_1, z_2) = (z_1 e^{2\pi i/p}, z_2 e^{2\pi i q/p}). \tag{6}$$

Exercises. (1) Show that the manifold obtained from two solid tori by pasting along some homeomorphism of the boundary tori is homeomorphic to either the sphere S^2 or the projective space $\mathbb{R}P^3$ or the product space $S^1 \times S^2$ or a lens space, and that all lens spaces may be constructed in this way. (2) Show that the manifold of unit tangent vectors to $\mathbb{R}P^2$ is homeomorphic to the lens space $L(4, 1)$.

(In a similar way lens spaces of any odd dimension and of infinite dimension can be defined; the latter are interesting as examples of $K(\mathbb{Z}_p, 1)$ spaces.)

B. It is easy to construct a cellular decomposition of the sphere S^3 (not depending on q) invariant under the action (6). This decomposition contains p 0-cells $\sigma_k^0 = (0, e^{2\pi i k/p})$, p 1-cells $\sigma_k^1 = \left\{(0, e^{2\pi i \theta}) \left| \dfrac{k}{p} < \theta < \dfrac{k+1}{p}\right.\right\}$, p 2-cells $\sigma_k^2 = \{(\rho e^{2\pi i k/p}, z) | 0 < \rho \leqslant 1, |z| = \sqrt{1 - \rho^2}\}$, and p 3-cells $\sigma_k^3 = \left\{(\rho e^{2\pi i \theta}, z) | 0 < \rho \leqslant 1, \dfrac{k}{p} < \theta < \dfrac{k+1}{p}|z| = \sqrt{1 - \rho^2}\right\}$ (in all cases $k = 0, 1, \ldots, p - 1$). Evidently,

$$\partial \sigma_k^1 = \sigma_{k+1}^0 - \sigma_k^0, \partial \sigma_k^2 = \sigma_0^1 + \cdots + \sigma_{p-1}^1, \partial \sigma_k^3 = \sigma_{k+1}^2 - \sigma_k^2;$$

$$F(\sigma_k^0) = \sigma_{k+q}^0, F(\sigma_k^1) = \sigma_{k+q}^1, F(\sigma_k^2) = \sigma_{k+1}^2, F(\sigma_k^3) = \sigma_{k+1}^3 \tag{7}$$

(in all these formulae the subscripts are elements of the group $\mathbb{Z}_p$). Factoring by $\mathbb{Z}_p$ converts this decomposition into a cellular decomposition of the lens space containing one cell in each dimension: σ^0, σ^1, σ^2, σ^3; it satisfies the relations $\partial \sigma^1 = 0$, $\partial \sigma^2 = p\sigma^1$, $\partial \sigma^3 = 0$. These formulae enable us to find the fundamental groups of lens spaces and their homology with arbitrary coefficients (formulae (7) allow us to calculate homology with arbitrary local coefficients). In particular,

$$\pi_1(L(p, q)) = \mathbb{Z}_p,$$

$i =$	0	1	2	3
$H_i(L(p, q)) =$	$\mathbb{Z}$	$\mathbb{Z}_p$	0	$\mathbb{Z}$
$H_i(L(p, q); \mathbb{Z}_p) =$	$\mathbb{Z}_p$	$\mathbb{Z}_p$	$\mathbb{Z}_p$	$\mathbb{Z}_p$

C. The homotopy invariants of lens spaces calculated in 2.1.B do not depend on q, but nevertheless the spaces $L(p, q)$ and $L(p, q'')$ may be homotopically different. The homotopy invariant distinguishing between them is the "self-linking number" of the generator α of the group $H_1(L(p, q); \mathbb{Z}_p)$. (In more familiar terms this is the residue $\langle \alpha \beta(\alpha) \rangle, [L(p, q)]_p \rangle \in \mathbb{Z}_p$ where β is the Bockstein homomorphism and $[\ ]_p$ denotes the fundamental class mod p.) This residue depends on the choice of the generator α of $H_1(L(p, q); \mathbb{Z}_p)$ and of the fundamental class; hence it is defined up to sign change and to multiplication by the square of an arbitrary invertible element of the ring $\mathbb{Z}_p$. An easy calculation (using the above cellular decomposition) shows that up to the indicated factors the invariant is equal to q. Moreover, it turns out that the necessary condition of homotopy equivalence of lens spaces which arises from this invariant is, in fact, also sufficient. We obtain

Theorem. *The lens spaces $L(p, q')$ and $L(p, q'')$ are homotopy equivalent if and only if*

$$q' \equiv \pm m^2 q'' \bmod p$$

for some $m \in \mathbb{Z}$. □

For example, the lens spaces $L(5, 1)$ and $L(5, 2)$ are not homotopy equivalent.

D. It also turns out that homotopy equivalent lens spaces may be non-diffeomorphic (and even non-homeomorphic, which, however, is the same thing in the three-dimensional case).

Theorem. *The lens spaces $L(p, q')$ and $L(p, q'')$ are diffeomorphic if and only if either*

$$q' \equiv \pm q'' \bmod p$$

or

$$q'q'' \equiv \pm 1 \bmod p.$$ □

For example, the lens spaces $L(7, 1)$ and $L(7, 2)$ are homotopy-equivalent but are not diffeomorphic to each other.

In order to prove this theorem (or, rather, its *PL* version) a special invariant was invented in early thirties, *the Reidemeister–Franz–de Rham torsion*. Later this invariant proved to be very important for topology and some other mathematical disciplines. The details may be found in the original works of Reidemeister (very well written, by the way) and also in the survey Rham, Maumary, Kervaire (1967) or in the textbook Dubrovin, Novikov, Fomenko (1984) (see Exercises 5–8 in §11).

E. Note in addition that the sign $\pm$ in the last two theorems may be removed if only orientation-preserving homotopy equivalences and diffeomorphisms are considered (we mean the orientation of lens spaces induced by the orientation of the sphere S^3). This remark is particularly interesting in the case of self-mappings of lens spaces. Namely, we have

Theorem. *A degree* -1 *mapping* $L(p, q) \to L(p, q)$ *exists if and only if* -1 *is a quadratic residue modulo p. A degree* -1 *diffeomorphism (or, equivalently, homeomorphism)* $L(p, q) \to L(p, q)$ *exists if and only if* $q^2 \equiv -1$ mod p. $\square$

For example, there exists no degree -1 mapping of $L(3, 1)$ into itself; a purely geometrical proof of this fact was found by Kneser in the twenties (long before the appearance of the theory outlined above).

2.2. The Poincaré Sphere. Let I be the group of orientation-preserving linear transformations of the icosahedron and let I^* be the inverse image of I under the canonical projection of $S^3 = \mathrm{Spin}(3)$ into $SO(3)$. The group I^* is of order 120; as an abstract group, it can be described by the system of two generators a, b and two relations $a^5 = (ab)^5 = b^3$. (*Exercise*: find two quaternions a, b of unit length which generate in S^3 a subgroup isomorphic to I^*. Hint: we must have $a^5 = (ab)^2 = b^3 = -1$.) The coset space S^3/I^* is called the *Poincaré sphere* or the *icosahedron space*, or the *dodecahedron space*; we denote it by $\mathscr{P}$.

B. Evidently $\mathscr{P}$ is a closed oriented three-dimensional manifold with fundamental group I^*. Since $[I^*, I^*] = I^*$ (if the relation $ab = ba$ is added, then the group I^* becomes trivial), $H_1(\mathscr{P}) = 0$. Thus $\mathscr{P}$ has the same integer homology as the three-dimensional sphere (i.e. $\mathscr{P}$ is a "three-dimensional homology sphere").

C. Poincaré invented the manifold $\mathscr{P}$ as a counter-example to his own conjecture which stated that any three-dimensional homology sphere was homeomorphic to S^3. This counter-example, firstly, served him as motivation for the definition of the fundamental group, and, secondly, impelled him to improve his conjecture, restricting it to three-dimensional homotopy spheres. In its final form the conjecture asserts that any closed simply connected three-dimensional manifold is homeomorphic (or, equivalently, diffeomorphic) to S^3. Today this conjecture is the most famous problem in topology.

There are many different constructions of the Poincaré sphere. Eight of them are described in the vividly written article Kirby, Scharlemann (1977). One of them was presented in §2.2.4. Another is implicitly contained in the theorem of

Chapt. 1, §2.2.B: the system of equations given there defines the Poincaré sphere in the case $n = 1$. One more construction is described below. (For the remaining five constructions see the article referred to above.)

Consider the simplest knot in S^3 – the trefoil (see Fig. 4 on page 205) and fix its framing in such a way that the curve traced by the end of the first vector of the framing has linking number with the knot equal to -1. (The trefoil knot of Fig. 4 is regarded as situated in $\mathbb{R}^3$ with the standard orientation.) This framing defines (up to isotopy) a diffeomorphism between the tubular neighbourhood of the trefoil knot and the product $S^1 \times D^2$. Attach the handle $D^2 \times D^2$ to the ball D^4 (with boundary S^3) along this homeomorphism. The boundary of the resulting 4-manifold is again $\mathscr{P}$.

§3. Some Four-dimensional Manifolds

There is a conjecture that any smooth closed simply connected four-dimensional manifold is diffeomorphic to the connected sum of manifolds from the following short list: (1) $S^2 \times S^2$; (2) $\mathbb{C}P^2$; (3) a non-singular surface of degree 4 in $\mathbb{C}P^3$. It is not easy to say if this conjecture is reliable because, for instance, it includes the four-dimensional differential Poincaré conjecture, which has been neither proved nor disproved up to now. One can only assert that no counter-example to this conjecture is known and that all smooth closed simply connected 4-manifolds which are known today are homeomorphic to connected sums of the form indicated. However, it seems reasonable to devote a few lines of our survey to these three manifolds. At the end of this section we shall briefly discuss a wider class of 4-manifolds, namely the class of non-singular complex surfaces in $\mathbb{C}P^3$.

A. The manifolds $S^2 \times S^2$ and $\mathbb{C}P^2$ belong to classes of manifolds considered in previous chapters. Therefore we restrict ourselves to one important property of these manifolds which is characteristic of the topology of four-dimensional manifolds. Some of their other properties may be found in the Exercises below.

Let α be a two-dimensional integer homology class of a smooth manifold X. Any arbitrary oriented submanifold of X diffeomorphic to the 2-sphere with g handles, whose fundamental class is α, is called *a genus g realization* of α.

The Rokhlin Theorem (see Rokhlin (1971)). (1) *If there exists a genus g realization of the class* $m \in H_2(\mathbb{C}P^2) = \mathbb{Z}$, *then*

$$g \geq \begin{cases} \dfrac{1}{4}m^2 - 1 & \text{if } m \text{ is even,} \\[2mm] \dfrac{q^2 - 1}{4q^2}m^2 - 1 & \text{if } q \text{ is a prime power and } q \text{ divides } m. \end{cases}$$

(2) *If there exists a genus g realization of the class* $(m_1, m_2) \in H_2(S^2 \times S^2) = \mathbb{Z} \oplus \mathbb{Z}$ *and* q *is a power of an odd prime which divides both* m_1 *and* m_2, *then*

$$g \geqslant \frac{q^2 - 1}{2q^2} |m_1 m_2| - 1. \quad \square$$

In addition, it follows from an earlier theorem of Rokhlin that the class $(m_1, m_2) \in H_2(S^2 \times S^2)$ has no genus 0 realization if both numbers m_1, m_2 are congruent to 2 modulo 4. No other lower bounds for the genus of realizations of two-dimensional homology classes of $S^2 \times S^2$ and $\mathbb{C}P^2$ are known. On the other hand, algebraic curves have minimal genus among known realizations (see § 1.4). It can be seen that the problem of computing the minimal genus of realizations is still far from its final solution.

Exercises. (1) Show that the quotient space

$$\mathbb{C}P^2/(z_0:z_1:z_2) \sim (\bar{z}_0:\bar{z}_1:\bar{z}_2)$$

is homeomorphic to S^4. (2) Show that the quotient space

$$S^2 \times S^2/(x_1, x_2) \sim (Rx_1, Rx_2),$$

where R is the reflection in the equatorial plane, is homeomorphic to S^4. (3) Show that the quotient space

$$S^2 \times S^2/(x_1, x_2) \sim (x_2, x_1)$$

is homeomorphic to $\mathbb{C}P^2$.

B. Non-singular surfaces of degree 4 (quartics) in $\mathbb{C}P^3$ are of exceptional interest in complex analysis, because they are the most accessible representatives of the famous class of non-singular "$K3$-surfaces". We restrict ourselves to the description of their cohomology ring.

The Betti numbers of a non-singular quartic Q are equal, respectively, to 1, 0, 22, 0, 1. The cup-product $H^2(Q; \mathbb{Z}) \times H^2(Q; \mathbb{Z}) \to H^4(Q; \mathbb{Z}) = \mathbb{Z}$ is an unimodular integral quadratic form of rank 22, which is isomorphic to the direct sum of five irreducible forms: two forms E_8 (where E_8 is the positive rank 8 quadratic form whose matrix is represented by the Dynkin diagram E_8) and three rank 2 forms U whose matrix is $\begin{pmatrix} 0 & 1 \\ 1 & 0 \end{pmatrix}$.

C. The Betti numbers of a nonsingular surface V_m of degree m in $\mathbb{C}P^3$ are equal to 1, 0, $m^3 - 4m^2 + 6m - 2$, 0, 1. The cup-product $H^2(V_m; \mathbb{Z}) \times H^2(V_m; \mathbb{Z}) \to H^4(V_m; \mathbb{Z}) = \mathbb{Z}$ is an unimodular integral quadratic form of signature $m(4 - m^2)/3$ which is equivalent (over $\mathbb{Z}$) to a diagonal form if m is odd and to the direct sum of $m(m^2 - 4)/24$ forms E_8 and $(m^3 - 6m^2 + 11m - 3)/3$ forms U if m is even.

Exercises. (1) The manifold V_2 is diffeomorphic to $S^2 \times S^2$. (2) The manifold V_3 is diffeomorphic to the connected sum of 7 copies of $\mathbb{C}P^2$, one of which is equipped with the standard orientation while the rest are equipped with the opposite orientation (we write $V_3 \approx \mathbb{C}P^2 \# 6(-\mathbb{C}P^2)$).

Note in addition that the classical Pontryagin – Whitehead Theorem asserts that closed simply connected 4-manifolds with isomorphic integer cohomology rings are homotopy equivalent, and the recent Freedman Theorem implies that smooth closed simply connected 4-manifolds with isomorphic integer cohomology rings are homeomorphic. Hence V_5 is homeomorphic to $9\mathbb{C}P^2$ $\#(-44\mathbb{C}P^2)$, V_6 is homeomorphic to $4Q \# 9(S^2 \times S^2)$ etc. No obstruction for these homeomorphisms to be diffeomorphisms is known, but neither is any non-trivial decomposition into a connected sum known for any non-singular surface in $\mathbb{C}P^3$ of degree $\geqslant 5$.

References*

Adams, J.F.: Lectures on Lie Groups. New York–Amsterdam: Benjamin, 1969, Zbl.206,316

Arnol'd, V.I., Givental', A.B.: Symplectic geometry. In: Itogi Nauki Tekh., Ser. Sovrem. Probl. Mat., Fundam. Napravleniya 4, 1985, 7–139, Zbl.592.58030. English transl. in: Encycl. Math. Sc. 4, 1–136. Berlin, Heidelberg, New York: Springer-Verlag, 1990

Atiyah, M.F.: K-theory. Notes by D.W. Anderson. New York–Amsterdam: Benjamin, 1967, Zbl.159,533

Atiyah, M.F.: K-theory and reality. Q. J. Math. Oxf., II. Ser. *17* (1966) 367–386, Zbl.146,191

Atiyah, M.F.: On the K-theory of compact Lie groups. Topology *4* (1965) 95–99, Zbl.136,210

Bernstein, J.N., Gel'fand, I.M., Gel'fand, S.I.: Schubert cells and cohomology of spaces G/P. Usp. Mat. Nauk *28*, No. 3 (1973) 3–26. English transl.: Russ. Math. Surv. *28*, No. 3 (1973) 1–26, Zbl.286.57025

Borel, A.: Sur la cohomologie des espaces fibrés principaux et des espaces homogènes de groupes de Lie compacts. Ann. Math. II. Ser., *57* (1953) 115–207, Zbl.52,400

Bott, R.: Homogeneous vector bundles. Ann. Math., II. Ser. *66*, No. 2 (1957) 203–248, Zbl.94,357

Bourbaki, N.: Groupes et algèbres de Lie. Fasc. 38. Paris: Hermann, 1975, Zbl.329.17002

Browder, W.: Surgery on Simply Connected Manifolds. Berlin, Heidelberg, New York: Springer-Verlag, 1972, Zbl.239.57016

Douady, A.: Noeuds et structures de contact en dimension 3 (d'après Daniel Bennequin). Semin. Bourbaki, Astérisque (1983) 105–106, 129–148, Zbl.522.53034

Dubrovin, B.A., Novikov, S.P., Fomenko, A.T.: Modern Geometry. Methods of Homology Theory. Moscow: Nauka, 1984, Zbl.582.55001. English transl.: Grad. Texts Math. 124. New York, Berlin, Heidelberg: Springer-Verlag, 1990, 416 pp.

Fuks, D.B., Fomenko, A.T., Gutenmacher, V.L.: Homotopic Topology. Moscow: Moscow Univ. Press, 1969, Zbl.189,540

Fuks, D.B.: On the Maslov–Arnold characteristic classes. Dokl. Akad. Nauk SSSR *178*, No. 2 (1968) 303–306. English transl. Sov. Math., Dokl. *9* (1968) 96 99, Zbl.175,203

Fuks, D.B.: Homotopic Topology. Algebra. Geometry. Topology. Moscow: VINITI, 1971, 71–122

Fuks, D.B.: Foliations. Itogi Nauki Tekh., Ser. Algebra Topol. Geom. *18* (1981) 151–213. English transl.: J. Sov. Math. *18* (1982) 255–291, Zbl.479.57014

Fuks, D.B., Rokhlin, V.A.: Beginner's Course in Topology. Moscow: Nauka, 1977. English transl.: Berlin, Heidelberg, New York: Springer-Verlag. 1984, Zbl.417.55002

Hodge, W.V.D., Pedoe, D.: Methods of algebraic Geometry. Vol. 1, Cambridge: Cambridge University Press 1947, Zbl.48,145

Karoubi, M.: K-Theory. An Introduction. Berlin, Heidelberg; New York: Springer-Verlag, 1978, Zbl.382.55002

---

*For the convenience of the reader, references to reviews in Zentralblatt für Mathematik (Zbl.), complied using the MATH database, have, as far as possible, been included in this bibliography.

Kirby, R., Scharleman, M.G.: Eight Faces of the Poincaré Homological 3-sphere. Geometric Topology, Proc. Conf., Athens/Ga. 1977. New York–San Francisco–London: Academic Press, 1979, 113–146, Zbl.469.57006

Knuth, D.E.: The Art of Computer Programming. Vol. 3. Sorting and Searching. Reading, Mass.: Addison–Wesley, 1974, Zbl.302.68010

Macdonald, I.G.: Symmetric Functions and Hall Polynomials. Oxford: Clarendon Press, 1979, Zbl.487.20007

Milnor, J.: Morse Theory. Princeton: Princeton Univ. Press, 1963, Zbl.108,104

Milnor, J., Kervaire, M.: Groups of homotopy spheres. I. Ann. Math., II. Ser. 77, No. 3 (1963) 504–537, Zbl.115,405

Mosher, R.E., Tangora, M.C.: Cohomology Operations and Applications in Homotopy Theory. New York – Evanston – London: Harper and Row, 1968, Zbl.153,533

Pontryagin, L.S.: Smooth Manifolds and their Applications in Homotopy Theory. 2nd ed. Moscow.: Nauka, 1976, 1st ed. 1955, Zbl.64,174

Postnikov, M.M.: Lectures in Algebraic Topology. Homotopy of CW Complexes. Moscow: Nauka, 1985, Zbl.578.55001

Rham, G. de, Maumary, S., Kervaire, M.A.: Torsion et type simple d'homotopie. Lect. Notes Math. 48, 1967, Zbl.153,539

Rokhlin, V.A.: Two-dimensional submanifolds of four-dimensional manifolds. Funkts. Anal. Prilozh. 5, No. 1 (1991) 48–60. English transl.: Funct. Anal. Appl. 5, No. 1 (1971) 39–48, Zbl.268.57019

Schwartz, J.T.: Differential Geometry and Topology. New York: Gordon & Breach, 1968, Zbl. 187,450

Spanier, E.H.: Algebraic Topology. New York: McGraw Hill Book Company, 1966, Zbl.145,453

Steenrod, N.: Topology of Fibre Bundles. Princeton, 1951, Zbl.54,71

Stillwell, J.: Classical Topology and Combinatorial Group Theory. Berlin, Heidelberg, New York: Springer-Verlag, 1980, Zbl.453.57001

Toda, H.: Composition methods in the homotopy groups of spheres. Ann. Math. Stud. 49 (1962) Zbl.101,407

Whitehead, G.W.: Recent Advances in Homotopy Theory. Providence, R.I.: Am. Math. Soc., 1970, Zbl.212,486

Zieschang, H., Vogt, E., Coldewey, H.D.: Surfaces and planar discontinuous groups. Lect. Notes Math. 835, 1980, Zbl.438.57001

Zieschang, H.: Finite groups of mapping classes of surfaces. Lect. Notes Math. 875, 1981, Zbl.472.57006

Index